AF578227

Current Topics in Microbiology 232 and Immunology

Springer
Berlin
Heidelberg
New York
Barcelona
Budapest
Hong Kong
London
Milan
Paris
Santa Clara
Singapore
Tokyo

Antigen Presentation

Edited by J. Lindsay Whitton

With 11 Figures

J. Lindsay Whitton, M.D., Ph. D.
Associate Professor
Department of Neuropharmacology, CVN-9
The Scripps Research Institute
10550 N. Torrey Pines Road
La Jolla, CA 92037
USA

Cover Illustration: A diagrammatic summary of the antigen presentation pathways. This cartoon represents a professional APC, in which both class I and class II pathways (yellow and green respectively) are active. In the class I pathway, an endogenously synthesized protein (purple-white-green) is tagged by ubiquitin (Ub), and thereby is targeted to the proteasome for degradation. The resulting peptides enter the ER, where the highest-affinity peptide (green, in this example) binds to class I MHC, and is transported to the cell surface to induce $CD8^+$ (usually cytotoxic) T cells. The upper part of the cartoon shows a soluble protein (light blue-dark blue-yellow) entering the cell by endocytosis; it is degraded within the endosome, and one of the resulting peptides (in this case, dark blue) is bound by class II MHC, and transported to the cell surface to induce $CD4^+$ (usually helper) T cells.

Cover Design: Design & Production GmbH, Heidelberg

ISSN 0070-217X
ISBN 3-540-63813-X Springer-Verlag Berlin Heidelberg New York

Library of Congress Catalog Card Number 15-12910
Printed in Germany

SPIN: 10498166 27/3136 – 5 4 3 2 1 0 – Printed on acid-free paper

Preface

Antigen presentation is a subject most often considered at the molecular level, but its results are, of course, also relevant to the whole organism. This compilation of chapters from experts is aimed at reviewing the broad spectrum from molecular, biochemical and pharmacological considerations through to the effects on host immunity and the diseases which result from defects in these pathways. We begin with the class I MHC pathway. Drs. Nandi, Marusina and Monaco review how the endogenous peptides are generated within the cell and how they are transported into the endoplasmic reticulum. There, the story is taken up by Drs. Fourie and Yang, who relate how the complexes are assembled and transported to the cell membrane. Drs. Hudrisier and Gairin then describe the structure of the mature class I complex and reflect upon the pharmacology and possible therapeutic use of peptide antagonists. The following two chapters illustrate the consequences of low class I MHC expression. The defects in immune responsiveness which result from globally defective class I presentation are reviewed by Drs. Frelinger and Serody, who describe the resultant defects in the ability to control challenge by different microbial pathogens, while Dr. Rall argues that the restricted class I expression, seen in certain tissues of normal animals, is most probably beneficial to the host, preventing immunopathologically mediated cell death. Then, the interface between the host and several viral pathogens is addressed in Chaps. 7 and 8; Drs. Sparer and Gooding describe how adenoviruses interfere with antigen presentation and, thus, with the host's ability to respond appropriately to the virus challenge. Drs. Johnson and Hill show how, even in a single virus family (the herpes viruses), many different approaches to immune evasion have been taken. In Chap. 9, Drs. Nordeng, Gorvel and Bakke provide an excellent review of the MHC class II pathway, demonstrating that the invariant chain has many and varied functions. In the final chapter, Drs. Eibl and Wolff discuss the consequences of a defective MHC class II pathway in transgenic mouse models and in humans.

I am grateful to all of the authors for the time and effort expended in providing what we hope is a timely and broad-ranging introduction to this important topic.

La Jolla, January 1998 J. LINDSAY WHITTON

Contents

List of Contributors

(Their addresses can be found at the beginning of their respective chapters.)

An Overview of Antigen Presentation and Its Central Role in the Immune Response*

J. LINDSAY WHITTON

1 Introduction

In order to more fully appreciate the roles of the antigen presentation pathways, it is important to understand how they interact with other facets of the host immune response and how they relate to – and have evolved along with – the microbial universe in which we bathe. Therefore, this overview chapter is intended less as a summary of antigen presentation, and more as a frame of biological relevance through which antigen presentation can be viewed.

---

*This is publication number 10946-NP from the Scripps Research Institute

Department of Neuropharmacology, CVN-9, The Scripps Research Institute, 10550 N. Torrey Pines Rd., La Jolla, CA 92037, USA

Our environment is rich in microbes. With some of these we live in symbiosis; for example, we provide food and shelter for many enteric bacteria, some of which perform vital functions on our behalf and whose eradication results in disease. Other organisms, when encountered, are almost invariably pathogenic; for example, human immunodeficiency virus (HIV) infection usually leads to disease unless appropriate treatment is taken. Between these extremes, there are many microbes whose ability to cause disease depends on a variety of factors, including the microbial genotype and the host's ability to counter the primary challenge. The importance of the immune system in combating microbial challenge has long been understood, and the molecular mechanisms underlying these host-microbe interactions are becoming increasingly clear.

The immune system, including the antigen presentation pathways, has evolved along with infectious agents, each constantly trying to outwit the other. In many cases, host immunity triumphs, and the molecular events leading to this are described in several of the following chapters. In other cases, the microbe is victorious, and two chapters are devoted to describing how certain viruses attempt to circumvent the antigen presentation pathways and thereby to escape the overall thrust of host immunity. In the ongoing interaction between infectious agents and the vertebrate immune system, each of the opponents is sufficiently malleable to be shaped by the other. For example, immunological pressures on microbes result in selection of those agents best able to evade the immune response. Microbial evolution is therefore driven, at least in part, by the efforts of the immune system to eradicate infection. In similar ways, the immune system, including the antigen presentation pathways, have been shaped by the nature of the microbes encountered. For example, introduction of a virus into a previously unexposed population can result in devastating illness and the death of many infected hosts. Survivors are commonly those who are genetically best suited to resist the particular infection, and this genetic fitness will be passed on to their progeny. In this way, the genotype of the host population is altered by the infectious agent. This scenario has happened often in the past and will doubtless recur. For instance, the introduction, by the conquistadores, of smallpox and measles to the New World resulted in devastating disease and mortality in the native peoples, while subsequent generations were less susceptible to the microbes. Another outstanding example was the deliberate introduction of myxoma virus to wild Australian rabbits in an attempt to control the exponentially increasing rabbit population in Australia. Although initially successful in that many of the infected rabbits died, the survivors bred hardy offspring which were more resistant to the agent, and before long the rabbit population had rebounded. Plans are afoot to repeat this attempt at biological control of this mammalian "pest" using a different virus; time will tell whether this second endeavor will be any more successful than the first or whether the resilience and adaptability of mammalian immunity, coupled with the legendary fecundity of rabbits, will thwart our best efforts. These and many other examples clearly show that microbes and host immunity exert selective pressures on each other and that both can evolve as a result.

2 Immune Response

The immune system can be subdivided in various ways. For example, it is common to categorize immunity as "humoral" or "cellular". The former term often is considered synonymous with antibody-mediated immunity, but more correctly it refers to fluid-borne factors in the immune response (e.g., antibodies, cytokines); and cellular immunity, often thought to consist only of T lymphocytes, in fact includes natural killer (NK) cells, polymorphs, macrophages etc. A second method of categorizing the immune system is by specificity, i.e., nonspecific immunity and antigen-specific immunity. The relationships between humoral and cellular immunity and antigen specificity are diagrammed in Fig. 1. Nonspecific immunity includes such things as the skin barrier, gastric acid, mucus, and, in some tissues, ciliary defense mechanisms, cells such as NK, macrophages, and polymorphs, and soluble factors such as certain cytokines (e.g., interferon-γ), none of whose activities require specific antigen recognition. Antigen-specific responses rely on lymphocytes, the progenitors of which are hematopoietic stem cells.

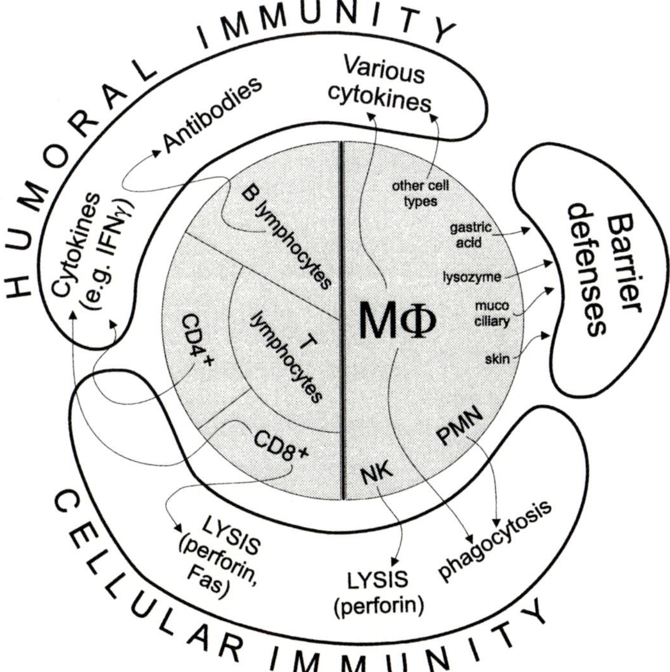

Fig. 1. Immune response. The *central gray circle* represents the immune response, divided into antigen-specific immunity (*left*) and nonspecific immunity (*right*). The effector pathways are shown, as are their designations as humoral, cellular, or barrier immunity. *NK*, natural killer; *PMN*, polymorphonuclear leukocytes; *MΦ*, macrophages; *IFN*, interferon

There are two classes of lymphocyte: T (thymus-derived) lymphocytes (which pass through the thymus during their maturation) and B lymphocytes (which give rise to antibody-producing plasma cells). Lymphocytes entering the thymus undergo two types of selection: positive selection of those cells able to recognize the appropriate molecules on cell surfaces, and negative selection of that pool to eradicate those lymphocytes that may recognize self cells too well, with possible autoimmune results. Only approximately 2% of the input lymphocytes successfully exit the thymus as mature T cells, with the remaining 98% being deleted by the selection processes. B lymphocytes recognize antigen through their encoded antibodies, and T lymphocytes recognize antigen presented by class I or class II major histocompatibility complex (MHC) proteins on the cell surface, as described in detail below and in other chapters in this volume. When one attempts to pigeonhole biological processes, overlaps frequently result, as can be seen in Fig. 1. For example, while interferon-γ is nonspecific in its actions, it may be secreted locally by antigen-specific T lymphocytes to counter a virus infection, and under these circumstances it might be argued that this cytokine comes under the umbrella of an effector molecule in the antigen-specific immune response. Similarly, macrophages can be considered part of the cellular response, since they engulf microbes, but they are also important to the humoral response, since they secrete many cytokines.

3 T Lymphocytes and Their Interactions with MHC

All T lymphocytes carry T cell receptors on their cell surfaces; these receptors are similar in structure to the Fab portion of an antibody and confer upon the T cells their ability to recognize specific antigens. There are two general classes of T lymphocyte, which differ in their functions and in the antigens they recognize; the classes are most easily distinguished by the presence of a surface protein, either CD8 or CD4. $CD8^+$ T lymphocytes recognize peptide fragments (epitopes) presented on the cell surface by MHC class I molecules; MHC class I is present on almost all nucleated cells, and therefore most somatic cells can be surveyed by $CD8^+$ T cells. These lymphocytes synthesize perforin, a protein which, when inoculated into the membrane of a target cell, undergoes self-assembly into a pore, resulting in the lysis of the target cell. For this reason, $CD8^+$ T lymphocytes often are known as cytolytic T lymphocytes (CTL), although they also can control infection by noncytolytic means, by cytokine production (Ramsay et al. 1993). $CD4^+$ T lymphocytes recognize peptide fragments presented on the cell surface by MHC class II molecules; MHC class II expression is restricted to specialized antigen-presenting cells (APC) such as dendritic cells, macrophages, and B lymphocytes. $CD4^+$ lymphocytes therefore cannot survey the same large cell population as can $CD8^+$ cells; instead, their main role is to recognize signals from APC and to provide help (in the way of cytokines) to B cells for the development of fully functional antibody responses, and to $CD8^+$ T cells to facilitate the production and maintenance of memory cells (von Herrath et al. 1996).

4 Relative Contributions of Antibodies and T Cells in Combating Microbial Infections

This book is devoted to antigen presentation; I shall therefore focus on antigen-specific immune responses, both arms of which play important roles in combating infection. In general terms, antibodies act to decrease antigen load and to diminish or abolish microbial infectivity, while $CD8^+$ T cells recognize and eradicate intracellular infections. Therefore, during a virus infection, antibodies and T cells act in complementary fashions. Antibodies limit virus dissemination and therefore limit the number of cells infected, while $CD8^+$ T cells eradicate infected cells, thus diminishing the production of infectious particles.

The relative importance of antibodies and T cells in controlling microbial infection depends on the nature of the invading microorganism. In humans, the contribution of each arm of the antigen-specific immune response can be approximately evaluated by studying "experiments of nature," i.e., human genetic diseases in which the immune responses are impaired. Children born with genetic deficiencies in antibody production, and who have no detectable immunoglobulins, nevertheless appear well able to control many primary or subsequent virus infections. Both the incidence of viral disease and its severity are similar to those in normal children (GOOD and ZAK 1956; GOOD 1991). However, agammaglobulinemic children show a marked increase in susceptibility to bacterial infection. Furthermore, genetic defects in components of the complement cascade (an important effector mechanism in many antibody-mediated responses) result in a reduced capacity to cope with bacterial, particularly meningococcal, infection, although resistance to virus infection and disease appears normal (ORREN et al. 1987; MCBRIDE et al. 1991).

Taken together, these observations suggest that antibodies are essential for the control of bacterial infections, but are not essential for control of many viral infections. That antibodies are not essential for control of many viral infections does not, of course, mean that they are without value. Indeed, the benefits of antiviral antibodies in almost all viral infections are abundantly clear. High levels of neutralizing antibody can protect animals from infection and disease, and in certain instances (e.g., rabies, hepatitis B virus, and Junin virus infections) postexposure antibody therapy is often recommended and efficacious. While antibody responses contribute greatly to the control of both bacterial and viral infections, $CD8^+$ T cells usually contribute only to control of the latter. Human experiments of nature are not quite so compelling in this instance, since it is difficult to find syndromes complementary to the agammaglobulinemias (i.e., syndromes in which antibodies are unaffected, while T cell responses are reduced). Nevertheless, human disease does provide a clue. In humans with impaired T cell responses, e.g., in patients with Di George's syndrome (congenital thymic aplasia), acquired immunodeficiency syndrome (AIDS), or leukemia or in patients receiving immunosuppressive drugs, both the severity and frequency of virus infections are greatly increased (NAHMIAS et al. 1967; SIEGEL et al. 1977; GRAY et al. 1987; KERNAHAN et al. 1987). However, in these instances the antibody responses, too, are often compromised. The availability of transgenic animal models has greatly enhanced our ability to dissect the relative contributions of both arms of the antigen-specific immune system. These topics are addressed in the chapters by Frelinger and Serody (class I/CD8$^+$ T cells) and Eibl and Wolf (class II/CD4$^+$ T cells) in this volume.

Thus, in broad terms, antibodies are important in the control of virus and bacterial infections, while CD8$^+$ T cells are important in the control of viral, but not bacterial infections. Intriguingly, virus infections induce in the host both antibody and T cell responses, while bacterial infections induce antibodies, but most often fail to induce CD8$^+$ T cells. Thus the immune response mounted by the host is that which is best suited to eradicate the particular type of microorganism being faced. How does the infected host "know" what type of immune response to mount?

5 Class I and Class II MHC Antigen Presentation Pathways

Since these pathways are described in great detail in the forthcoming chapters, I will provide only a brief summary here. Class I and class II MHC molecules present peptide antigens at the surface of cells. The overall structures of the MHC-peptide complexes are similar for both class I and class II; both molecules contain a groove in which a peptide is carried and thereby "presented" to circulating T cells (BJORKMAN et al. 1987; BROWN et al. 1993). Furthermore, for both MHC classes there is polymorphism around the groove, ensuring that different MHC alleles will bind to, and present, different peptides from the microbial universe. Why has evolution demanded that two such apparently similar molecules be produced? The answer is relatively straightforward. Peptides presented by the class I system are generated from proteins which have been synthesized within the cell and then broken down by a cytosolic organelle, the proteasome; the peptides are transferred into the endoplasmic reticulum (see the chapter by NANDI, MARUSINA, and MONACO), where they associate with class I molecules and are transported to the cell surface (see the chapter by FOURIE and YANG). In contrast, peptides associated with class II molecules are generated from proteins which have been taken up from the fluid phase by specialized APC and degraded inside endosomes (see the chapter by NORDENG, GORVEL, and BAKKE). Thus, when the different classes of T cell distinguish between a class I complex and a class II complex (CD8$^+$/class I and CD4$^+$/class II), they are discriminating between intracellular antigen and soluble (extracellular) antigen. Therefore, CD8$^+$ T cells, which interact with class I complexes, are activated by intracellular organisms, such as viruses. In contrast, CD4$^+$ T cells, which interact with class II complexes, are activated by peptide sequences derived from proteins taken from the fluid phase.

In the above lies the answer, in molecular terms, to the question of how the host "knows" what type of immune response to mount. Viruses infect cells, and their proteins are synthesized intracellularly; therefore, viral proteins efficiently enter the class I MHC pathway, thereby inducing CD8$^+$ T cell responses. Furthermore, virus proteins may be secreted from cells or released from cells following cytolysis, to interact with B lymphocytes, thus inducing antibody production. In addition, they may be endocytosed by specialized APC, where their epitope peptides will be presented by class II MHC, therefore inducing CD4$^+$ T cell responses to provide help to the B lymphocyte and CD8$^+$ T cell responses. Virus infection thus induces both B lymphocyte and T lymphocyte (CD4$^+$ and CD8$^+$) responses. In contrast, bacterial infections are only rarely intracellular. Most bacteria exist and replicate extracellularly and therefore their antigens do not efficiently enter the class I MHC pathway. For this reason, most bacterial infections fail to induce effective CD8$^+$ T cell responses. However, antibody responses are efficiently induced, as are CD4$^+$ T cell responses, which are activated, as for viral infection, by the uptake of bacterial antigen by APC. Furthermore, phagocytosis of bacteria, or bacterial debris, by macrophages also may contribute to the presentation of antigen by the class II pathway.

The host-microbe interactions and the central role of antigen presentation are represented diagrammatically in Fig. 2. A review of this figure will permit the experiments of nature to be understood at a more molecular level. Children lacking antibody responses will be unable to control extracellular microbes, since antibodies are the main bulwark against these agents. In contrast, such children remain relatively healthy when faced with infection by an intracellular microbe, since CD8$^+$ CTL are available to combat this type of infection. One might also predict, from Fig. 2, that ablation of the CD8$^+$ T cell pathway alone would have minimal effect on the control of bacterial infections, where it plays no role, but may have some effect on the control of virus infections; since the antibody and CD4$^+$ T cell pathways would remain largely intact, one might also anticipate that some level of control of virus infections would remain. As stated above, there are no "clean" clinical syndromes in which the

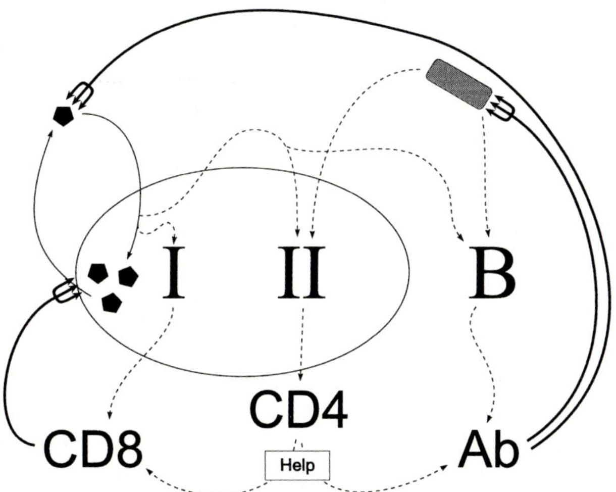

Fig. 2. Central role played by the antigen presentation pathways in determining the nature of the immune response mounted to the invading microbe. Intracellular and extracellular microbes are represented by a virus (*black polygon*) and bacterium (*gray rectangle*), respectively. *Interrupted lines* show entry into the antigen presentation pathways (I and II) or activation of B lymphocytes and the responses which result. The effector mechanisms are shown as *thicker lines* terminating in *tridents*

class I MHC presentation pathway, or the $CD8^+$ T cell response, is defective. However, these expectations are borne out by studies of transgenic knockout mice, as described by FRELINGER and SERODY (this volume). The experimental results are quite complex; in summary, ablation of the $CD8^+$ CTL response has a marked effect on the host's ability to clear intracellular (usually virus) infection, but the immune responses which remain are often able to control the infection, if sometimes only partially. The class II pathway also has been disrupted (class II knockout mice and CD4 knockout mice), and certain approximately parallel human disease states have been described; the results are detailed by EIBL and WOLF (this volume).

Of course, as with most generalizations, there are a number of exceptions. For example, not all bacteria are extracellular; some, e.g., Listeria monocytogenes, have a largely intracellular lifestyle, and $CD8^+$ cells are important to the control of these bacterial infections (HARTY and BEVAN 1992; KAGI et al. 1994). In addition, there are exceptions to the "class I/endogenous, class II/exogenous" rule; it appears that some specialized cells (perhaps macrophages) can take up exogenous protein and introduce its epitopes into the class I pathway (ROCK 1996).

6 Some Problems Associated with Antigen Presentation

Antigen presentation by the MHC controls T cell activation and allows the surveillance of intracellular proteins. The advantages to the host, particularly when facing infestation by an intracellular microbe, are obvious. However, the system has some limitations.

6.1 Immune Responses Fail if Host MHC Alleles Cannot Present Any Microbial Peptides

The polymorphic nature of the MHC alleles ensures that different MHC molecules will present different microbial epitopes. At first sight, the advantage this offers seems clear, since the presence of multiple MHC alleles would permit the infected host to present a greater variety of microbial peptides, thus perhaps increasing the chance that an effective immune response would be induced. However, the haploid mouse and human genomes contain only a small number of class I MHC alleles (usually three). This limits the ability of the individual host to present microbial fragments and may even eliminate an individual's ability to respond to a vaccine or to a pathogen. For example, the susceptibility of a naive population to a new microbe and the selective advantage of hosts able to respond to the agent were described above; many of the host genes which determine susceptibility or resistance are concerned with antigen presentation, and the genes encoding the MHC molecules are frequently involved. Furthermore, we have shown that vaccine failure may result from the host's inability to present CTL epitopes, which itself results from the lack of a single MHC allele (KLAVINSKIS et al. 1989).

Why has evolution risked such failures by limiting the number of MHC alleles to be carried by an individual, when the presence of multiple alleles would enhance our ability to present microbial peptides to our circulating T cells? The most probable explanation involves the maturation of T cells in the thymus. As stated above, T cells undergo negative selection, in which cells able to interact too strongly with self peptide-MHC complexes are deleted. If an individual expressed a large number of MHC alleles, then T cell deletion would be greatly increased, resulting in a marked diminution in the number of mature T cells emerging from the thymus. Thus the number of MHC alleles available to us most probably represents an evolutionary balance. Too many MHC alleles would result in deletion of too many T cells, resulting in a severely restricted peripheral T cell repertoire. In contrast, too few MHC alleles would too greatly limit our ability to present microbial antigen. At both extremes, albeit for different reasons, our ability to mount an effective immune response would be reduced, and we would be more susceptible to microbial infection and consequent disease. Of course, this argument presupposes that given MHC alleles will have certain affinities for peptides and therefore will be able to present a certain number of peptides. If MHC molecules were more sequence specific in their binding (and therefore could present fewer peptide sequences, perhaps binding them with higher affinity), then the number of evolutionarily permitted alleles per genome might alter to reflect this change in order to retain the same overall antigen presentation capacity of the host. In this light, HUDRISIER and GAIRIN point out in this volume that there is indeed a limit in the affinity of MHC-peptide interaction, and they suggest that one underlying reason might be to ensure that MHC alleles remain capable of binding and presenting a relatively large number of peptide sequences.

6.2 MHC Expression Is Not Invariably Beneficial and Is Restricted in Certain Tissues

Modification of antigen presentation pathways in different cell types also plays an important role in the host-microbe interaction. For example, as described by RALL (this volume), expression of certain genes involved in class I antigen presentation, including the MHC class I genes themselves, is reduced in neurons. This may benefit the host, by preventing lysis of virus-infected neurons by $CD8^+$ CTL. However, this immune privilege may also benefit the virus, providing it with a haven in which it is "invisible" to the immune system. As RALL points out, this may explain why so many viruses are neurotropic; they may have evolved to take advantage of the host's unwillingness to destroy these important and irreplaceable cells. The inappropriate presence of MHC on certain cells, in the absence of secondary signal molecules, might also cause "anergy" of peripheral T cells, crippling the host's ability to mount an effective response.

6.3 Antigen Presentation Provides a Target for Viral Interference

Viruses have developed many methods to evade host immune responses. Indeed, from the evasion techniques employed, one can infer the responses most able to paralyze the microbe. For example, influenza virus uses antigenic drift and shift to escape antibody responses, implicating these immune effectors in virus eradication. Consistent with the importance of $CD8^+$ T cells in the control of virus infections, many viruses have found methods of avoiding contact with these cells, and many of these do so by avoiding, or interrupting, the class I MHC antigen presentation pathway. They may do so by restricting virus protein expression, by turning off host protein expression, by inhibiting transport of peptides into the endoplasmic reticulum, by preventing transport of the peptide-MHC complex, and by a variety of other elegant intercessions. It appears that, if there is an opening, a virus has exploited it. The methods used by adenoviruses are reviewed in this book by SPARER and GOODING, and herpesvirus trickery is detailed by JOHNSON and HILL.

6.4 Antigen Presentation May Be Associated with Immunopathological Phenomena

The antigen presentation pathway and MHC molecules do not themselves distinguish between microbial and self proteins and peptides. Both class I and class II MHC will present peptide sequences from host proteins. This may be advantageous (e.g., allowing a tumor cell antigen to be displayed to the immune system) and should at worst be neutral; T cells capable of detrimental interactions with self epitopes should be deleted in the thymus. In practice, however, T cells may fail to be deleted, or normally harmless T cells may be activated by microbial infection and then may cross-react with self epitopes to the detriment of the host ("molecular mimicry"). There are many examples of diseases linked to class I and class II MHC loci. For example, ankylosing spondylitis is tightly linked to the class I HLA-B27 allele, while multiple sclerosis and juvenile-onset diabetes are associated with certain class II alleles. Thus we pay a price for the ability to display our cellular contents to our own potentially lethal immune effectors.

6.5 Antigen Presentation and Histoincompatibility

Presentation of self epitopes by MHC is a primary feature underlying graft rejection. Indeed, this phenomenon gave the MHC its name. Our need to protect ourselves against intracellular organisms probably drove the evolution of the class I MHC pathway, which in turn has greatly complicated our recent desire to transplant organs.

7 Ways of Exploiting the Antigen Presentation Pathways

As our knowledge of the pathways increases, so should our ability to manipulate them to our advantage.

7.1 Enhancing Cytolytic T Lymphocyte Induction by Improving Antigen Entry into the MHC Class I Pathway

7.1.1 Synthetic Peptides

As would be expected from the foregoing, soluble proteins are poor inducers of CTL. However it was anticipated that short peptides containing precise CTL epitopes might be able to bind to appropriate cell surface class I alleles, thus inducing good $CD8^+$ T cell responses. We now understand that efficient entry into the class I pathway requires intracellular peptide, and regular synthetic peptides usually are rather ineffective unless administered along with large doses of adjuvant. However, addition of a lipid tail renders peptides more immunogenic (DERES et al. 1989); although the underlying mechanism has not been determined, it is plausible that interaction with, and perhaps passage through, the cell membrane is important.

7.1.2 DNA Immunization

The endogenous synthesis of encoded protein is one major advantage of this new approach to vaccination (HASSETT and WHITTON 1996; ULMER et al. 1996). We have succeeded in improving the technique by the stable attachment of ubiquitin to the viral protein. The resulting protein is so rapidly degraded by the proteasome that it is almost undetectable by biochemical means; class I MHC presentation is enhanced, as are CTL induction and antiviral protection (RODRIGUEZ et al. 1997). However, a cautionary note must be sounded; the ubiquitinated protein does not induce antibodies, perhaps because the rapid degradation prevents intact soluble material from inducing $CD4^+$ T cell and B cell responses.

7.2 Enhancing Virus Clearance by Improving Antigen Presentation

As described by RALL, neurons are a frequent target for persistent virus infection, and the lack of class I MHC molecules (JOLY et al. 1991), and perhaps of other proteins important in antigen presentation (JOLY and OLDSTONE 1992), may render the virus invisible to the host. If class I MHC levels are increased virus clearance may result; RALL reports that, intriguingly, such clearance may not be overtly harmful to the host.

7.3 Preventing Immunopathology by Interrupting Antigen Presentation

T cell-mediated immunopathology requires an appropriate peptide-MHC target. In theory, formation of such a complex might be prevented by expression of a competitor peptide, able to bind to the MHC allele with high affinity (thus excluding all other peptides); the resulting complex would obviously have to be nonimmunogenic. Although this may appear a tall order, nature has already lit the way; hepatitis B virus can express peptides which are nonimmunogenic, but which can compete with the immunogenic virus peptides, thus diminishing their expression and preventing the host from responding (Bertoletti et al. 1994). A peptide has been developed which binds with high affinity to one murine allele (Gairin and Oldstone 1992); the potential therapeutic use of such peptides is discussed in this volume by Hudrisier and Gairin.

8 Coda

Although the following chapters demonstrate a detailed knowledge of the classical pathways, much remains to be done. In what other ways can we exploit these pathways? Can we control Th1 and Th2 skewing of the immune response? What do γδ T cells see and do? What are the functions of the various nonclassical MHC molecules? How are non-protein antigens processed and presented? In the field of antigen presentation, the harvest has just begun.

Acknowledgements. I am grateful to Terry Calhoun for skilled secretarial assistance. My studies on antigen presentation and vaccine development are supported by NIH grants AI37186 and AI27028.

References

Bertoletti A, Sette A, Chisari FV, Penna A, Levrero M, De Carli M, Fiaccadori F, Ferrari C (1994) Natural variants of cytotoxic epitopes are T-cell receptor antagonists for antiviral cytotoxic T cells. Nature 369:407–410

Bjorkman PJ, Saper MA, Samraoui B, Bennett WS, Strominger JL, Wiley DC (1987) The foreign antigen binding site and T cell recognition regions of class I histocompatibility antigens. Nature 329:512–518

Brown JH, Jardetzky TS, Gorga JC, Stern LJ, Urban RG, Strominger JL, Wiley DC (1993) Three-dimensional structure of the human class II histocompatibility antigen HLA-DR1. Nature 364:33–39

Deres K, Schild H, Wiesmuller KH, Jung G, Rammensee HG (1989) In vivo priming of virus-specific cytotoxic T lymphocytes with synthetic lipopeptide vaccine. Nature 342:561–564

Gairin JE, Oldstone MB (1992) Design of high-affinity major histocompatibility complex-specific antagonist peptides that inhibit cytotoxic T-lymphocyte activity: implications for control of viral disease. J Virol 66:6755–6762

Good RA (1991) Experiments of nature in the development of modern immunology. Immunol Today 12:283–286

Good RA, Zak SJ (1956) Disturbance in gamma-globulin synthesis as "experiments of nature". Pediatrics 18:109–149

Gray MM, Hann IM, Glass S, Eden OB, Jones PM, Stevens RF (1987) Mortality and morbidity caused by measles in children with malignant disease attending four major treatment centres: a retrospective review. Br Med J (Clin Res) 295:19–22

Harty JT, Bevan MJ (1992) CD8+ T cells specific for a single nonamer epitope of Listeria monocytogenes are protective in vivo. J Exp Med 175:1531–1538

Hassett DE, Whitton JL (1996) DNA Immunization. Trends Microbiol 4:307–312

Joly E, Oldstone MBA (1992) Neuronal cells are deficient in loading peptides onto MHC class I molecules. Neuron 8:1185–1190

Joly E, Mucke L, Oldstone MBA (1991) Viral persistence in neurons explained by lack of major histocompatibility class I expression. Science 253:1283–1285

Kagi D, Ledermann B, Burki K, Hengartner H, Zinkernagel RM (1994) CD8+ T cell-mediated protection against an intracellular bacterium by perforin-dependent cytotoxicity. Eur J Immunol 24:3068–3072

Kernahan J, McQuillin J, Craft AW (1987) Measles in children who have malignant disease. Br Med J (Clin Res) 295:15–18

Klavinskis LS, Whitton JL, Oldstone MBA (1989) Molecularly engineered vaccine which expresses an immunodominant T-cell epitope induces cytotoxic T lymphocytes that confer protection from lethal virus infection. J Virol 63:4311–4316

McBride SJ, McCluskey DR, Jackson PT (1991) Selective C7 complement deficiency causing recurrent meningococcal infection. J Infect 22:273–276

Nahmias AJ, Griffith D, Salsbury C, Yoshida K (1967) Thymic aplasia with lymphopenia, plasma cells, and normal immunoglobulins. Relation to measles virus infection. JAMA 201:729–734

Orren A, Potter PC, Cooper RC, du Toit E (1987) Deficiency of the sixth component of complement and susceptibility to Neisseria meningitidis infections: studies in 10 families and five isolated cases. Immunology 62:249–253

Ramsay AJ, Ruby J, Ramshaw IA (1993) A case for cytokines as effector molecules in the resolution of virus infection. Immunol Today 14:155–157

Rock KL (1996) A new foreign policy: MHC class I molecules monitor the outside world. Immunol Today 17:131–137

Rodriguez F, Zhang J, Whitton JL (1997) DNA immunization: ubiquitination of a viral protein enhances CTL induction, and antiviral protection, but abrogates antibody induction. J Virol 71:8497–8503

Siegel MM, Walter TK, Ablin AR (1977) Measles pneumonia in childhood leukemia. Pediatrics 60:38–40

Ulmer JB, Sadoff JC, Liu MA (1996) DNA vaccines. Curr Opin Immunol 8:531–536

von Herrath MG, Yokoyama M, Dockter J, Oldstone MBA, Whitton JL (1996) CD4-deficient mice have reduced levels of memory cytotoxic T lymphocytes after immunization and show diminished resistance to subsequent virus challenge. J Virol 70:1072–1079

How Do Endogenous Proteins Become Peptides and Reach the Endoplasmic Reticulum

DIPANKAR NANDI,[2,3] KATE MARUSINA,[2,4] and JOHN J. MONACO[1,2]

1 Introduction

T lymphocytes, via specific T cell receptors (TCR), recognize antigenic peptides bound to major histocompatibility complex (MHC)-encoded molecules. The recent crystallization of TCR-MHC complexes has enriched our understanding of this recognition process (GARCIA et al. 1996a; GARBOCZI et al. 1996). Broadly speaking, there are two types of MHC molecules, i.e., classical and nonclassical molecules, and

[1]Howard Hughes Medical Institute, University of Cincinnati, 231 Bethesda Avenue, Cincinnati, OH 45267–0524, USA

[2]Department of Molecular Genetics, University of Cincinnati, 231 Bethesda Avenue, Cincinnati, OH 45267–0524, USA

[3]*Present address:* Department of Biochemistry, Indian Institute of Science, Bangalore 560012, India

[4]*Present address:* Howard Hughes Medical Institute, University of California at San Francisco, San Francisco, CA 94143–0724, USA

the majority of TCR recognize classical MHC molecules. Classical MHC molecules can be further divided into two types, i.e., class I and class II. $CD8^+$ T cells, or cytotoxic T lymphocytes (CTL), recognize MHC class I molecules bound to peptides derived from endogenous proteins (i.e., proteins that are synthesized within the cell or artificially introduced directly into the cytoplasm or nucleus) that are primarily degraded in the cytoplasm. The CD8 accessory molecule present on most CTL appears to augment the formation of TCR-MHC/peptide complexes (GARCIA et al. 1996b; GAO et al. 1997). In addition to presenting peptides to T cells, expression of MHC class I molecules protects cells from lysis by natural killer (NK) cells (LANIER 1997). On the other hand, $CD4^+$ T lymphocytes or helper T lymphocytes recognize MHC class II molecules that bind peptides derived from exogenous or membrane proteins which enter the cell via endocytosis or phagocytosis and are primarily degraded by lysosomal proteases. However, this division between the two pathways is not always straightforward, and multiple mechanisms may operate in specialized antigen-presenting cells, e.g., macrophages and dendritic cells, to allow peptides derived from exogenous proteins access to MHC class I molecules (BEVAN 1995; JONDAL et al. 1996; ROCK 1996). These mechanisms allow for a wide variety of peptides generated from cytoplasmic and lysosomal compartments to be presented to T lymphocytes. Usually, peptides are derived from self cellular proteins for presentation to T lymphocytes that are tolerized to self peptides as a result of thymic education. However, in the event of an infection or oncogenic transformation, foreign or abnormal peptides are presented on MHC molecules, resulting in a T cell-mediated immune response. Therefore, studying the mechanisms by which peptides are generated within cells and form complexes with MHC molecules for presentation to T lymphocytes has important ramifications in our understanding of the basic cellular immune response, the generation of peptide vaccines, transplantation, and autoimmunity. This review will attempt to summarize our current knowledge of MHC class I antigen processing. The major molecules and events involved in this pathway are depicted in Fig. 1.

MHC class I molecules consist of three components: heavy chain, β_2-microglobulin, and a peptide. Formation of this trimeric complex is essential for efficient cell surface expression, and deficiency or lack of any one of these constituents results in reduced class I cell surface expression (SHIMIZU and DEMARS 1989; ZIJLSTRA et al. 1990; VAN KAER et al. 1992). In the absence of peptide, "empty" dimers consisting of MHC class I heavy chain and β_2-microglobulin are formed, and these dimers are unstable at 37°C. The resulting defect in MHC class I expression can be corrected by incubating cells at lower temperatures, which stabilizes cell surface expression of empty MHC class I molecules, or by supplying synthetic peptides exogenously that can bind to empty MHC class I molecules at the cell surface (LJUNGGREN et al. 1990; SCHUMACHER et al. 1990; ORTIZ-NAVARETTE and HÄMMERLING 1991; ROCK et al. 1991). Therefore, peptides play a key role in stabilizing cell surface expression of MHC class I molecules. However, in vivo, native proteins need to be introduced into the cytoplasm to generate peptides capable of binding MHC class I molecules (BENNINK et al. 1982; GOODING and O'CONNELL 1983; TOWNSEND et al. 1986; MOORE et al. 1988; YEWDELL et al. 1988). MHC class I-binding peptides are generated in the cytosol and are therefore insensitive to agents that block the generation of MHC class

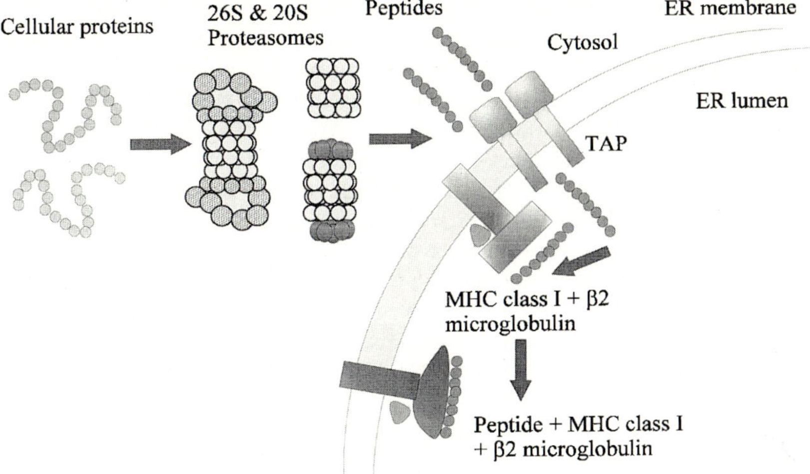

Fig. 1. General model of MHC class I antigen processing. *ER*, endoplasmic reticulum; *TAP*, transporter associated with antigen processing; *MHC*, major histocompatibility complex. See text for details

II-binding peptides, such as the lysosomotropic agents chloroquine or ammonium chloride (MORRISON et al. 1986). This suggests that nonlysosomal mechanisms are utilized to generate peptides that bind MHC class I molecules. The general model of MHC class I antigen processing involves at least three steps: First, endogenous proteins in the cytosol, including membrane proteins (SILICIANO and SOLOSKI 1995), are proteolytically cleaved by enzymes, presumably proteasomes or proteasome-associated complexes, to produce peptides. Second, these peptides are transported into the endoplasmic reticulum (ER) by the transporter associated with antigen processing (TAP). In the ER, peptides bind newly synthesized dimers of MHC class I heavy chain and β_2-microglobulin; this trimeric complex traffics subsequently to the cell surface for perusal by $CD8^+$ T cells (reviewed by HEEMELS and PLOEGH 1995; YORK and ROCK 1996). We shall consider the key players involved in this pathway, i.e., MHC class I molecules, proteasomes, and TAP.

2 MHC Class I Molecules

Classical MHC class I, or class Ia, molecules consist of a heavy chain (45 kDa), encoded by genes present on chromosome 6 in humans (HLA-A, HLA-B, HLA-C) or chromosome 17 in mice (H-2K, H-2D, H-2L), which associates noncovalently with a 12-kDa protein, β_2-microglobulin, which is encoded on chromosome 15 in humans and chromosome 2 in mice, and a short peptide (usually eight to ten amino acids long) derived from cellular proteins or foreign antigen (RAMMENSEE et al. 1995;

RAMMENSEE 1996). MHC-encoded heavy chain molecules are highly polymorphic (PARHAM and OHTA 1996), and most of the polymorphic residues are located in the two extracellular domains, α_1 and α_2, which form the peptide-binding groove, giving different allelic forms of these MHC molecules the ability to bind distinct arrays of peptides (FALK et al. 1991). Peptide binding to MHC class I molecules is also determined by "anchor" residues, i.e., residues of the peptide whose side chains are buried in pockets within the MHC peptide-binding groove. For most MHC class I molecules, the peptides have two primary anchor residues, one at the carboxyl terminus and one at an internal residue whose position varies depending on the identity of the class I allele (ENGELHARD 1994; RAMMENSEE et al. 1995). In general, MHC class I molecules optimally bind peptides that are nine amino acids in length; however, peptides of up to 33 amino acids have been found to be associated with some MHC class I molecules (JOYCE et al. 1994; URBAN et al. 1994).

3 Proteasomes

Proteasomes are multisubunit enzymes that are responsible for degrading unfolded proteins and peptides in an ATP-independent fashion; they are abundant in eukaryotic cells, often constituting 0.5%–1% of total cellular protein (COUX et al. 1996). They are essential components of eukaryotic cells, performing multiple proteolytic functions, and are also responsible for the generation of most of the peptides that bind MHC class I molecules (CERUNDOLO et al. 1997; CRAIU et al. 1997). Proteasome particles are cylindrical in shape with dimensions of approximately 150 Å in height and 110 Å in diameter; they have a molecular weight of approximately 700 kDa and a sedimentation coefficient of 20S. They are composed of four rings with seven subunits in each ring; the outer rings are made up of α-subunits, whereas the inner rings are made up of β-subunits. The α-subunits are important in assembly (ZWICKL et al. 1994) and are the elements to which regulators bind (PETERS et al. 1993; YOSHIMURA et al. 1993; GRAY et al. 1994), whereas the β-subunits are important for catalytic activity (FENTEANY et al. 1995; SEEMÜLLER et al. 1995; CRAIU et al. 1997). Unlike α-subunits, most β-subunits are synthesized as precursors that subsequently undergo processing, and recent studies have shown that the processing of precursor β subunits is autocatalytic (CHEN and HOCHSTRASSER 1996; SCHMIDTKE et al. 1996; SEEMÜLLER et al. 1996). Proteasomes in their simplest form, such as those from the archaebacterium *Thermoplasma acidophilum,* contain only one type of α- and one type of β-subunit (LÖWE et al. 1995). Recently, genes encoding two different α- and two different β-subunits have been discovered in *Rhodococcus*; the organization of these subunits in proteasomes is not yet known (TAMURA et al. 1995), but they can assemble with random stoichiometry in vitro. Yeasts contain seven different α- and seven different β-proteasome genes, with the product of each gene occupying a specific and nonredundant position within each seven-membered ring. Interestingly, mammals (and probably most vertebrates) contain seven different α- and ten different β-proteasome genes; the three "extra" proteasome β-subunits in mammals are inducible with interferon (IFN)-γ (MONACO and NANDI 1995) and, unlike the other 14 proteasome subunit genes, are constitutively expressed only in certain tissues. Remarkably, despite the diversity in the numbers of different subunits, the quaternary structure of proteasomes is strongly conserved from *Thermoplasma* to humans (PÜHLER et al. 1994).

There are at least two features that distinguish proteasomes from most other conventional enzymes. First, proteasomes are multicatalytic, i.e., a single proteasome molecule contains independent active sites and may possess as many as five to nine different activities (ORLOWSKI et al. 1993; STEIN et al. 1996). Second, they apparently degrade substrates in a processive manner, i.e., they degrade one substrate molecule to completion (but not to single amino acids) before attacking another substrate molecule (AKOPIAN et al. 1997). 20S proteasomes belong to a superfamily of N-terminal nucleophile hydrolases (BRANNIGAN et al. 1995), and catalytic β-subunits possess an amino terminal threonine that resides in the catalytic site, is critical for proteolytic activity (SEEMÜLLER et al. 1995), and is covalently modified by the proteasome-specific inhibitor lactacystin (FENTEANY et al. 1995; CRAIU et al. 1997). Two other residues in β-subunits are also important for proteasome catalytic activity: a lysine residue and a negatively charged residue (SEEMÜLLER et al. 1995), although the exact role of these two residues is not known. X-ray crystallographic studies on *Thermoplasma* (LÖWE et al. 1995) and yeast (GROLL et al. 1997) proteasomes have localized the active sites to a large central chamber, resulting in the isolation of the active sites from intracellular contents, thereby preventing nonspecific proteolysis. Interestingly, access to this proteolytic chamber may differ in *Thermoplasma* and yeast proteasomes: denatured proteins or peptides may enter via narrow pores (13Å) at the ends of *Thermoplasma* proteasomes (LÖWE et al. 1995); however, these pores at the ends of yeast proteasomes are blocked, and access to the central proteolytic chamber may occur via small openings located in the interface between the α- and β-rings (GROLL et al. 1997).

IFN-γ is produced by Th1, Tc1, and NK cells in response to immune activation, and the IFN-γ-mediated upregulation of the components involved in MHC class I antigen processing may play an important role in the amplification of the immune response. In fact, the key components in the MHC class I pathway, namely MHC class I heavy chain, β_2-microglobulin, TAP, three proteasome subunits – LMP2 (or PSMB9), LMP7 (or PSMB8), and MECL1 (or PSMB10) – and the proteasome regulators PA28α and PA28β, are all inducible with IFN-γ (BOEHM et al. 1997). Nevertheless, other cytokines, e.g., tumor necrosis factor (TNF)-α and IFN-α/β, may also upregulate TAP and the proteasome subunits LMP2 and LMP7 (EPPERSON et al. 1992; AKI et al. 1994). However, it should be stressed that considerable constitutive expression of the inducible proteasome subunits occurs in normal tissues, particularly in hematopoietic cells, and there is no significant difference in the expression of the three IFN-γ-inducible proteasome subunits in mice lacking IFN-γ as compared to their wild-type littermates (M. Cruz and J.J. Monaco, unpublished observations). The presence of LMP2 and LMP7 subunits, together with MHC class I molecules, in jawed vertebrates but their absence in jawless vertebrates suggests that these IFN-γ-inducible proteasome subunits have evolved with the adaptive immune system (KANDIL et al. 1996).

The IFN-γ-inducible subunits compete with their constitutive counterparts for incorporation during the assembly of newly synthesized proteasomes (AKI et al. 1994), and LMP2, LMP7, and MECL1 replace the constitutive subunits delta (PSMB6), X/MB1 (PSMB5), and Z (PSMB7), respectively, in different subsets of mammalian proteasomes (GROETTRUP et al. 1996b; HISAMATSU et al. 1996; NANDI et al. 1996a). The high expression of the IFN-γ-inducible subunits in lymphoid tissues or upregulation in the presence of IFN-γ (GROETTRUP et al. 1996b; HISAMATSU et al. 1996; NANDI et al. 1996a; STOHWASSER et al. 1997; M. Cruz and J.J. Monaco, unpublished observations) appears to result in preferential incorporation of these subunits into proteasomes. Moreover, two of the IFN-γ-inducible proteasome subunits, LMP2 (PSMB9) and MECL1 (PSMB10), are incorporated earlier than their constitutive counterparts during proteasome assembly, which may also facilitate the formation of proteasomes containing primarily IFN-γ-inducible subunits (NANDI et al. 1997). In addition, the presence of one IFN-γ-inducible subunit may influence the incorporation of other IFN-γ-inducible subunits into catalytically active 20S proteasomes; in the absence of LMP2, MECL1 is not efficiently incorporated into 20S proteasomes and, in the absence of LMP7, both LMP2 and MECL1 are not efficiently incorporated into 20S proteasomes (T. Griffin, D. Nandi, J.J. Monaco, and R.A. Colbert, unpublished observations).

Interestingly, only the three pairs of constitutive and IFN-γ-inducible subunits possess the amino-terminal threonine that is critical for proteolytic activity (SEEMÜLLER et al. 1995). The two members of each pair are considerably more closely related in sequence than any other pairwise comparisons among proteasome subunit sequences (HISAMATSU et al. 1996) and probably occupy a single position in the proteasome structure in a mutually exclusive fashion (KOPP et al. 1993, 1995, 1997; PETERS et al. 1993). Although differences in enzymatic activities are observed in proteasomes containing the constitutive or IFN-γ-inducible subunits, the nature and magnitude of these changes are still controversial (DRISCOLL et al. 1993; GACZYNSKA et al. 1993, 1994; AKI et al. 1994; BOES et al. 1994; VAN KAER et al. 1994; GROETTRUP et al. 1995; KUCKELKORN et al. 1995; USTRELL et al. 1995a,b; EHRING et al. 1996; STOHWASSER et al. 1996), most likely due to several factors, including the fact that the third inducible subunit (MECL1) had not yet been characterized (and hence was not controlled for) at the time these studies were done, nor was the preferential assembly of subunits described above, or the potential contribution of the IFN-γ-inducible 11S proteasome regulator (see below), appreciated at the time. Moreover, most of these studies were performed using fluorogenic peptide substrates consisting of three or four amino acids, and the significance of peptidase activity against these substrates versus the in vivo activity against natural protein substrates is not well established. Further work is needed to understand the rules that govern the generation of peptides by proteasome subsets containing different β-subunits.

The ability of an anti-MHC alloantiserum to immunoprecipitate proteasomes and the genetic localization of two polymorphic proteasome subunit genes, *Lmp2* and *Lmp7*, together with the transporter subunit genes *Tap1* and *Tap2*, in the MHC complex strongly suggested that proteasomes were involved in the generation of MHC class I-binding peptides (for a review, see MONACO 1992). Two of the IFN-γ-inducible proteasome subunits, LMP2 and LMP7, are important in the processing of some antigens, and proteasomes isolated from LMP2- or LMP7-deficient mice demonstrate altered peptidase profiles (FEHLING et al. 1994; VAN KAER et al. 1994; STOHWASSER et al. 1996). Mice lacking LMP2 exhibit reduced numbers of $CD8^+$ T cells and a lower ability to mount a CTL response to influenza virus (VAN KAER et al. 1994). On the other hand, mice lacking LMP7 exhibit reduced MHC class I cell surface expression and do not present the male-specific antigen, H-Y, efficiently (FEHLING et al. 1994). Although incorporation of LMP2 and LMP7 subunits changes the peptidase activity against fluorogenic substrates (see discussion above), no change was found in the cleavage profiles of two proteins digested with proteasomes containing constitutive subunits versus proteasomes containing LMP2 and LMP7 (KUCKELKORN et al. 1995; EHRING et al. 1996). However, incorporation of LMP2 and LMP7 into proteasomes affects the digestion pattern of other proteins (BOES et al. 1994; CERUNDOLO et al. 1995), consistent with a role of these subunits in the processing of some, but not all antigens (ARNOLD et al. 1992). Furthermore, the role of the third IFN-γ-inducible subunit, MECL1, in the processing of antigens has not been studied. Together, the data are consistent with the possibility that the IFN-γ-inducible proteasome subunits have evolved in mammals to broaden the repertoire of peptides produced from individual proteins, or to optimize the generation of peptides capable of binding MHC class I molecules and eliciting a CTL response, or both.

Apart from the role of proteasomes containing LMP2 and LMP7 in antigen processing, there is additional evidence for the role of 20S proteasomes in this process. Proteasomes preferentially generate peptides that are in the size range of MHC-binding peptides (NIEDERMANN et al. 1996; YANG et al. 1996). Moreover, digestion of two model proteins with purified proteasomes produced measurable amounts of the exact epitopes that bind MHC class I molecules (DICK et al. 1994). In addition, the cleavage preferences of proteasomes appear to correlate with generation of CTL epitopes (NIEDERMANN et al. 1995; OSSENDORP et al. 1996). Finally, studies with proteasome inhibitors also implicate proteasomes in antigen processing.

The use of peptide aldehydes and other proteasome inhibitors has contributed significantly to our understanding of the roles of proteasomes in vivo. The peptide aldehydes readily enter mammalian cells and inhibit the activities of 20S and 26S proteasomes; consequently, they also reversibly block the degradation of most abnormal, short-lived, and long-lived proteins (ROCK et al. 1994), suggesting that proteasomes are responsible for most nonlysosomal protein degradation. These inhibitors also inhibit assembly of MHC class I molecules as well as presentation of antigens introduced in the cytoplasm (ROCK et al. 1994; HARDING et al. 1995; HUGHES et al. 1996: SUH et al. 1996). Unfortunately, these peptide inhibitors are not proteasome specific and inhibit other proteases, such as calpain and cathepsin B (ROCK et al. 1994). It is also of concern that a peptide aldehyde inhibitor inhibits signal peptidase, an enzyme unrelated to proteasomes (HUGHES et al 1996); it is therefore possible that the inhibition of MHC class I antigen processing using these compounds is a result of inhibition of hitherto unidentified enzymes and not proteasomes.

However, the discovery of a proteasome-specific inhibitor, lactacystin, a fungal metabolite (FENTEANY et al. 1995), has confirmed these earlier results on the role of proteasomes in protein turnover and the generation of peptides for MHC class I antigen processing (CERUNDULO et al. 1997; CRAIU et al. 1997). Lactacystin is rapidly converted to its active form, clastolactacystin-β-lactone (DICK et al. 1997), which covalently modifies the terminal threonines of all mammalian processed β-subunits containing this catalytically important residue and greatly reduces proteasomal activity (CRAIU et al. 1997). Together these different lines of evidence point to an important role of proteasomes as the key, if not the only enzyme involved in the generation of peptides that bind MHC class I molecules.

Proteasomes may bind to at least two types of regulators, i.e., 19S and 11S. 26S proteasome complexes (approximately 2000 kDa) are composed of 19S regulatory components attached to both ends of a catalytically active 20S proteasome core (PETERS et al. 1993; YOSHIMURA et al. 1993). The 19S regulatory components consist of approximately 21 different subunits, some (approximately six) of which are ATPases belonging to the AAA family (ATPases associated with a variety of cellular activities). The other 15 subunits are non-ATPase subunits that are unrelated to each other (COUX et al. 1996; TANAKA and TSURUMI 1997). The role of some of the 19S regulator subunits is to bind ubiqutinated proteins and present unfolded proteins to the 20S proteasome core for degradation. 26S proteasome complexes are responsible for the majority of nonlysosomal and ATP-dependent degradation of proteins involved in cell cycle regulation, metabolic adaptation, removal of abnormal proteins, processing of inactive transcription factor precursors, and in the degradation of some membrane proteins (COUX et al. 1996). The 26S proteasome/ubiquitin pathway is also responsible for degradation by the N-end rule, so-named because the amino-terminal residue of a protein correlates well with its in vivo half-life (VARSHAVSKY 1996). Usually, 26S proteasomes degrade poly-ubiquitin-conjugated proteins, although this complex is known to degrade certain nonubiquitinated proteins, e.g., ornithine decarboxylase, in an ATP-dependent fashion (MURAKAMI et al. 1992).

The evidence for a role of the ubiquitin-ATP-26S proteasome-dependent pathway in antigen processing is widely accepted, but still somewhat controversial. In the late 1980s, TOWNSEND et al. (1988) showed that ubiquitination of the target protein increased the efficiency of antigenic peptide generation. Using cell lines that possess a temperature-sensitive defect in one of the enzymes required for ubiquitination, a model antigen, ovalbumin, was presented less efficiently at nonpermissive temperatures (MICHALEK et al. 1993). However, these experiments were not reproduced by another group, in whose hands the mutant cells used by the earlier investigators were able to ubiquitinate proteins even at nonpermissive temperatures (COX et al. 1995). Using a different strategy, GRANT et al. (1995) confirmed the critical role of the amino terminus in determining the rate of degradation and demonstrated that antigen processing and presentation could be prevented by blocking potential ubiquitination sites on the protein. This approach needs to be studied on additional proteins before the role of ubiquitination and 26S proteasomes in MHC class I processing can be generalized.

In addition to binding 19S regulator subunits, 20S proteasomes also bind to 11S regulators, which are composed of two homologous subunits, PA28α and PA28β (DUBIEL et al. 1992; MA et al. 1992; REALINI et al. 1994; AHN et al. 1995). Although the individual molecular weights of these subunits are approximately 30 kDa, these subunits form hexameric rings with a molecular weight of approximately 200 kDa which bind the ends of 20S proteasomes (GRAY et al.1994). The expression of the two PA28 subunits is increased with IFN-γ (AHN et al. 1995; JIANG and MONACO 1997), and the genes encoding these subunits are linked (H. Jiang and J.J. Monaco, unpublished observation). However, the tissue expression pattern of the PA28 subunits differs from that observed for the IFN-γ-inducible proteasome β-subunits: unlike the preferential expression of the IFN-γ-inducible proteasome β-subunits in lymphoid tissues (STOHWASSER et al. 1997; M. Cruz and J.J. Monaco, unpublished observation), both the PA28 subunits are expressed in all tissues (JIANG et al. 1997). Interestingly, the transcription of Ki, a nuclear protein homologous to the PA28 subunits, is also increased by IFN-γ (AHN et al. 1995; JIANG and MONACO 1997). However, the induction by IFN-γ is transient and is tenfold less than that observed for the PA28 subunits in a hepatoma cell line (JIANG and MONACO 1997); moreover, protein levels do not necessarily correlate with mRNA levels in these cells. Thus the significance of Ki in the activation of proteasomes and its role in MHC class I antigen processing requires additional work. The association of PA28 with proteasomes greatly stimulates proteasome-mediated activity against certain fluorogenic peptide substrates (GROETTRUP et al. 1995; USTRELL et al. 1995b). Importantly, incubation of purified proteasomes with PA28 enhances "double" cleavage events, resulting in an enhanced yield of peptides optimal for TAP and MHC class I loading (DICK et al. 1996). Thus both PA28 and the incorporation of the IFN-γ-inducible subunits into proteasomes may skew the repertoire of peptides generated by proteasomes to favor MHC class I-binding (GROETRRUP et al. 1995b). Indeed, expression of the PA28α subunit in a fibroblast cell line improves the processing and presentation of two viral epitopes, suggesting an important in vivo role for this regulator in the MHC class I antigen-processing pathway (GROETTRUP et al. 1996a). Recombinant PA28α expressed in *Escherichia coli* is functionally indistinguishable from that of purified native PA28 activator (REALINI et al. 1994). As discussed above, transfection of PA28α alone enhances MHC class I antigen processing; therefore, the role and function of PA28β is not well understood.

4 Transporter Associated with Antigen Processing

The delivery of peptides from the cytoplasm into the ER is facilitated by TAP, which is composed of two homologous integral membrane proteins, TAP1 and TAP2, belonging to the family of ATP binding cassette (ABC) transporters (DEVERSON et al. 1990; MONACO et al. 1990; SPIES et al. 1990; TROWSDALE et al. 1990; KELLY et al. 1992). Both TAP1 and TAP2 have amino-terminal transmembrane domains, containing either six or eight transmembrane-spanning segments, and a C-terminal domain containing a pair of motifs indicative of nucleotide binding. The localization of TAP in the ER membrane and *cis*-Golgi complex (KLEIJMEER et al. 1992) is consistent with its functional role. Recombinant TAP1 expressed in *E. coli* spans the membrane eight times with several large loops exposed in the lumen of the ER and both the amino and carboxy termini, including theABC, residing in the cytoplasm (GILEADI and HIGGINS 1997). Defects in either or both TAP subunits result in greatly reduced surface expression (usually less than 5% of wild-type levels) of conformationally intact MHC class I molecules and impaired recognition by CTL (LJUNGGREN et al. 1990; ROCK et al. 1991). Disruption of TAP1 by homologous recombination in TAP1 knockout mice leads to decreased numbers of $CD8^+$ T cells as well as functional NK cells, whereas $CD4^+$ T cells are not affected (VAN KAER et al. 1992; LJUNGGREN et al. 1994). In humans, an inherited TAP deficiency, resulting in low numbers of $CD8^+$ T cells, was identified in two siblings from Morocco (DE LA SALLE et al. 1994). Transfection of *TAP* into mutant cells with defective *TAP* genes restores cell surface MHC class I expression and the ability to be recognized by CTL (POWIS et al. 1991a; SPIES and DEMARS 1991; ARNOLD et al. 1992; ATTAYA et al. 1992; SPIES et al. 1992). Additionally, TAP is required for optimal cell surface expression of some nonclassical MHC class I molecules (ATTAYA et al. 1992; ALDRICH et al. 1994a; TABACZEWSKI and STROYNOWSKI 1994; LEE et al. 1995) but not others (HANAU et al. 1994; BRUTKIEWICZ et al. 1995; HOLCOMBE et al. 1995). Although presentation of most MHC class I-binding peptides is dependent on TAP, some peptides may be delivered to the ER by one or more TAP-independent pathways. A subset of HLA-A2 molecules bind ER translocation signal peptides, which are already present in the ER, thereby circumventing the requirement for TAP (HENDERSON et al. 1992; WEI and CRESSWELL 1992). CTL can also be generated to some viral proteins containing ER insertion sequences (ANDERSON et al. 1991; HAMMOND et al. 1993; ZHOU et al. 1993a; BACIK et al. 1994; KHANNA et al. 1994; LEE et al. 1996), although this does not necessarily indicate TAP independence. Finally, $CD8^+$ T cells isolated from $TAP^{-/-}$ mice are still present in low but significant numbers and possess epitope-specific cytotoxic activity (ALDRICH et al. 1994b; VAN SANTEN et al. 1995; SANDBERG et al. 1996).

Peptide affinity for TAP ranges from 410 to 1600 n*M*, with approximately 10^5–10^6 binding sites per cell (VAN ENDERT et al. 1994a; UEBEL et al. 1995; YANG and BRACIALE 1995; KOOPMAN et al. 1996). Overexpression of TAP1 and TAP2 in insect cells has shown that these two proteins form a heterodimer capable of transporting peptides in the presence of ATP without any requirement for accessory proteins (MEYER et al. 1994; UEBEL et al. 1994; VAN ENDERT et al. 1994a). Normal peptide binding and translocation occurs only in the presence of both TAP1 and TAP2 (ANDROLEWICZ et al. 1994a,b; VAN ENDERT et al. 1994a; NIJENHUIS et al. 1996; NIJENHUIS and HÄMMERLING 1996; MARUSINA et al. 1997a). The region in TAP1 and TAP2 that contribute to peptide binding corresponds to the transmembrane domain close to the cytosolic ABC (NIJENHUIS et al. 1996; NIJENHUIS and HÄMMERLING 1996). The ABC domains of TAP expressed in *E. coli* and *D. melanogaster* bind ATP (MULLER et al. 1994; WANG et al. 1994), and it is clear that ATP hydrolysis is required for TAP1/2 complexes to translocate peptides into the ER, because nonhydrolyzable ATP analogues fail to support TAP-dependent transport (ANDROLEWICZ et al. 1993; NEEFJES et al. 1993; SHEPHERD et al. 1993). However, ATP is not required for peptide binding, only for transport (ANDROLEWICZ et al. 1994b; VAN ENDERT et al. 1994a; WANG et al. 1996a,b; MARUSINA et al. 1997a).

During MHC class I assembly, TAP associates with preformed MHC class I heavy chain-β_2-microglobulin dimers lacking peptide; on binding peptides, the trimeric complexes consisting of MHC class I heavy chain, β_2-microglobulin, and peptide dissociate from TAP and egress to the cell surface (ORTMANN et al. 1994; SUH et al. 1994; CARRENO et al. 1995; SADASIVAN et al. 1996; WANG et al. 1996b). This TAP-MHC binding may be responsible for efficient loading of peptides on MHC class I molecules (GRANDEA et al. 1995; PEACE BREWER et al. 1996). Other molecules are also involved in TAP-MHC binding: calreticulin, an ER luminal protein, and TAP-associated glycoprotein (tapasin), which may serve to bridge complexes containing TAP, MHC class I molecules, and calreticulin (SADASIVAN et al. 1996; SOLHEIM et al. 1997). Although binding to TAP is not essential for peptide loading of class I molecules, cells with defects in the tapasin gene are defective in TAP-class I interaction and have partial defects in the class I antigen-processing pathway. Surprisingly, TAP1 alone, but not TAP2, can bind MHC class I molecules (ANDROLEWICZ et al. 1994a; SUH et al. 1994), suggesting that the MHC class I-binding site is on TAP1. However, it is possible that TAP1 may form homodimers that can bind MHC class I molecules and transport some peptides (GABATHULER et al. 1994). Free MHC class I heavy chain, in the absence of β_2-microglobulin, does not bind TAP (SADASIVAN et al. 1996); however, free β_2-microglobulin, in the absence of MHC class I heavy chain, can bind TAP (CARRENO et al. 1995; SOLHEIM et al. 1997). TAP binds MHC class I molecules with different efficiencies: most HLA-A and HLA-C alleles interact efficiently with TAP, whereas several HLA-B alleles associate very inefficiently (NEISIG et al. 1996). The site of MHC class I heavy chain binding is controversial; residues in the α_2 domain of the class I heavy chain are thought to be involved in TAP interaction because HLA alleles interact with TAP with different affinities (NEISIG et al. 1996). Comparison of the sequences of different HLA alleles suggested that polymorphisms in the α_1- and α_2-domains or the peptide-binding groove are responsible for the differential ability of HLA molecules to associate with TAP (NEISIG et al. 1996). In addition, mutations in the α_2-domain of HLA-A2 ablated TAP association (LEWIS et al. 1996; PEACE-BREWER et al. 1996). These findings are in contrast with other studies (CARRENO et al. 1995; SUH et al. 1996) which identified residues critical for interaction with TAP in the α_3 domain of MHC class I molecules. This discrepancy may stem from the fact that universal rules for TAP-MHC class I interactions may not exist. For example, both soluble and membrane-bound forms of L^d are associated with TAP (CARRENO et al. 1995), whereas only membrane-bound but not soluble forms of HLA-G co-immunoprecipitated with TAP (LEE et al. 1995). Alternatively, mutations in different domains of MHC class I molecules may affect its conformation or affect TAP association with accessory molecules, e.g., tapasin, thereby indirectly affecting TAP-MHC class I association.

5 Transcriptional Activation of Interferon-γ-Inducible Proteasome Subunits and TAP

Lmp2 (Psmb9) and *Lmp7 (Psmb8)* are tightly linked to *Tap1* and *Tap2*; this cluster of genes, important in antigen processing, is localized in the class II region of the MHC complex (reviewed in TROWSDALE 1996). However, the gene encoding the third

IFN-γ-inducible subunit, *MECL1 (Psmb10)*, is located on chromosome 16 in humans (LARSEN et al. 1993) and on chromosome 8 in mice (M. Cruz and J.J. Monaco, in press). Interestingly, *Lmp2* and *Tap1* are transcribed divergently from a single bidirectional promoter (KISHI et al. 1993; WRIGHT et al. 1995), which "works" stronger in the direction of *Tap1* expression (MIN et al. 1996). The transcription factor IRF-1 has been shown to bind an IFN-γ-stimulated response element (ISRE) present in this region (WHITE et al. 1996). Mice lacking IRF1 have low numbers of $CD8^+$ T cells, and low expression of TAP1 and LMP2 may be responsible, at least in part, for this phenotype (WHITE et al. 1996). Thus not only are *Tap1* and *Lmp2* closely linked, but they are also coordinately expressed, suggesting a strong functional relationship between proteasomes and TAP in the MHC class I antigen-processing pathway. The *Lmp7 (Psmb8)* promoter contains a single ISRE block, whereas the *MECL1 (Psmb10)* promoter contains two blocks of ISRE (L. Elenich and J. Monaco, unpublished observation).

The mouse *Tap1* transcriptional start site has been mapped 256 bp upstream from the ATG codon (WRIGHT et al. 1995), whereas the *Tap2* transcriptional start site is undetermined. The *Tap1* promoter contains five overlapping consensus transcription factor binding sites, three of which bind the IFN-γ-inducible factors Stat1 and IRF1, and two of which bind transcriptional activators NFκB and Sp1 (MIN et al. 1996). Interestingly, activation of transcription of *Tap1* occurs more rapidly than for MHC class I (EPPERSON et al. 1992; GASKINS et al. 1992), and the rapid activation of Stat1 following IFN-γ treatment may account for the quicker induction of *Tap1* expression (MIN et al. 1996). The quicker induction of TAP may be designed to ensure that an adequate supply of peptides is present in the ER prior to synthesis of high levels of MHC class I molecules.

6 Role of Polymorphisms in LMP2 (PSMB9), LMP7 (PSMB8), and TAP

Susceptibility to many autoimmune diseases is genetically linked to the MHC locus, with particular MHC class I and class II alleles being elevated in patient populations as compared to controls (SINHA et al. 1990). The presence of genes encoding LMP2, LMP7, TAP1, and TAP2 in the MHC class II region and their role in the generation and transport of MHC class I-binding peptides has engendered considerable interest in the possibility that polymorphisms in these genes may be causally associated with this genetically determined disease susceptibility. The existence of structural differences between different alleles of LMP2, LMP7, TAP1, and TAP2 raises the question of whether functional differences between such alleles exists and what the potential net effect of such differences are on the repertoire of MHC class I-presented peptides. However, caution must be exercised in interpreting any such associations between these genes and autoimmune diseases due to the possibility of linkage disequilibrium with alleles of other MHC genes.

The IFN-γ-inducible proteasome subunits LMP2 and LMP7 are polymorphic, and three alleles of each of these molecules are found in inbred mice (MONACO and MCDEVITT 1986; ZHOU et al. 1993b; ZANELLI et al. 1995; NANDI et al. 1996b). Polymorphisms in mouse LMP7 (PSMB8) are located in the carboxyl terminus of the protein. A polymorphism (glycine to arginine) at position 272 affects not only the pI of the molecule but, surprisingly, also affects its apparent molecular weight (NANDI et al. 1996b). Out of the five polymorphic residues in human LMP7, four are located in intron 6 and are therefore not present in the coding sequence (BECK et al. 1992; FRÜH et al. 1992; KIM et al. 1996); the fifth polymorphism results in a coding change (lysine to glutamine) at position 49 of human LMP7 (MAKSYMOWYCH et al. 1995). However, the functional significance of this change is uncertain, since it occurs in the protein presequence that is removed by post-translational processing and hence does not result in a difference in the mature, catalytically active molecule. Interestingly, one polymorphism (histidine to arginine) at position 60 of LMP2 (PSMB9) is conserved, both in position and in the identity of the two alternative amino acid residues, between humans and mice (ZHOU et al. 1993b; MAKSYMOWYCH et al. 1994; NANDI et al. 1996b). This is the only known polymorphism in human LMP2 and is correlated with the development of acute anterior uveitis (AAU) and peripheral arthritis in HLA-B27 individuals suffering from ankylosing spondylitis (AS) (MAKSYMOWYCH et al. 1994), AAU in unselected patients with AS (MAKSYMOWYCH and RUSSELL 1995), and HLA-B27-associated juvenile rheumatoid arthritis (PRYHUBER et al. 1996). However, no correlation was found between LMP2 and LMP7 polymorphisms and the development of adult rheumatoid arthritis (MAKSYMOWYCH et al. 1995). A polymorphism in intron 6 of LMP7 is strongly associated with insulin-dependent diabetes mellitus (IDDM), and the polymorphism in LMP2 at position 60 is associated with the development of IDDM in patients containing a particular MHC class II allele (DENG et al. 1995). Therefore, it is possible that polymorphism in PSMB8 (LMP7) and PSMB9 (LMP2) may affect proteasomal activity, resulting in the generation and presentation of different epitopes of the same antigen to T cells in different individuals that may affect the immune response.

Both TAP1 and TAP2 in mice, rats, and humans are polymorphic (COLONNA et al. 1992; POWIS et al. 1992a–c, 1993; JOLY et al. 1994; MARUSINA et al. 1997b). Rats possess two main forms of TAP2, A or B, differing by 25 amino acids, but only four of these (residues 217, 218, 374, and 380) apparently contribute to specificity of peptide transport (POWIS et al. 1992a,b; JOLY et al 1994; ARMANDOLA et al. 1996; MOMBURG et al. 1996). As judged from the hydropathy analysis of the TAP2 protein, these segments are located on the two cytoplasmic loops closest to the ABC. Interestingly, a point mutation at position 374 in human TAP2 alters transport specificity, suggesting that this residue is important in determining peptide specificity (ARMANDOLA et al. 1996). In vitro peptide translocation assays have shown that rat TAP2A does not have a strong preference for the C-terminal amino acid of the peptides to be transported, similar to human TAP; on the other hand, rat TAP2B prefers to transport peptides ending in hydrophobic amino acids, similar to mouse TAP. Thus different sets of peptides are translocated in rats containing TAP2A or TAP2B (HEEMELS et al. 1993; MOMBURG et al. 1994a; ARMANDOLA et al. 1996; MOMBURG et al. 1996). The rat MHC class I molecule RT1.A^{a} prefers to bind peptides containing an arginine residue at their carboxyl termini and is efficiently loaded with peptides in cells containing TAP2A (Powis et al. 1996). Not surprisingly, trafficking of RT1.A^{a} is slower in case of TAP2B-containing cells, presumably due to a lack of efficient transport of appropriate peptides (Powis et al. 1991b). Similar functional polymorphism has been observed for TAP in Syrian hamsters (Lobigs et al. 1995). Polymorphism in class I molecules of rats and hamsters is severely limited as compared to mice and humans. It is possible that these species have taken different evolutionary routes for diversifying their CD8 T cell repertoires, and polymorphism in TAP may compensate in some sense for limited polymorphism in class I molecules.

Mouse TAP are also polymorphic (Pearce et al. 1993; Marusina et al. 1997b), and a comprehensive sequence comparison of TAP1 and TAP2 from five major *Mus musculus* haplotypes and from one *Mus castaneus* subtype was published recently (Marusina et al. 1997). Each haplotype has a unique TAP1 and TAP2 heterodimer, and the extent of polymorphism is somewhat intermediate between the highly polymorphic rat and conserved human TAP. In contrast to rat TAP alleles, no significant functional differences have been detected between mouse TAP alleles in vitro (Schumacher et al. 1994a,b). Thus, in mice, it is improbable that polymorphism in TAP grossly influences the set of peptides that are transported into the lumen of the ER. Nevertheless, subtle differences in transport specificity between mouse TAP alleles cannot be ruled out. The potential input of minor sequence changes on specificity and/or molecular topology of a transporter is emphasized by the observation that a single amino acid change in the primary structure of TAP is known to affect its function (Armandola et al. 1996).

Structural polymorphism has also been described for human TAP transporters, but allelic variation is more limited, with an average of two to five amino acid substitutions between alleles (Powis et al. 1993). In humans, four alleles of TAP1 and eight alleles of TAP2 have been identified (Colonna et al. 1992; Powis et al. 1992c, 1993), based on all possible combinations of two dimorphic positions in TAP1 and four in TAP2. Recently, a new TAP allele encoding a large cytoplasmic domain has been found (Cesari et al. 1997). However, thus far no significant or consistent differences between patient and control populations in the frequencies of either TAP1 or TAP2 alleles have been found for a number of MHC-associated diseases, including IDDM (Caillat-Zucman et al. 1993; Lofti et al. 1994; Martinez-Laso et al. 1994; Nakanishi et al. 1994; van Endert et al. 1994b), celiac disease (Djilali-Saiah et al. 1994; Tighe et al. 1994), rheumatoid arthritis (Donn et al. 1994; Marsal et al. 1994; Ploski et al. 1994), multiple sclerosis (Kellar-Wood et al. 1994; Middleton et al. 1994; Spurkland et al. 1994; Moins-Teisserenc et al. 1995), Crohn's disease (Hesresbach et al. 1996), psoriasis (Fakler et al. 1994), and ankylosing spondylitis (Burney et al. 1994). In addition, the extent of linkage disequilibrium observed between TAP and class II alleles in different studies varies and may differ between ethnic groups.

7 Additional Components of the Class I Pathway

A growing body of evidence suggests that there may be more components in the class I pathway than are currently recognized. A mutant human cell line, LCL 721.220, is defective in MHC class I antigen processing and expresses low levels of cell surface MHC class I. Despite the presence of structurally normal class I and TAP proteins in these cells, in contrast to what is seen in wild-type cells, no physical association of these molecules is observed (GRANDEA et al. 1995; SOLHEIM et al 1997). The defect in these cells resides in the gene for tapasin, which apparently mediates TAP-class I interaction (GRANDEA et al. 1995; SADASIVAN et al. 1996; SOLHEIM et al. 1997). Interestingly, the tapasin gene is also MHC linked. However, in mice, it is found approximately 65 kb centromeric of *H-2K* (P. Comber, S. Wenderfer, and J. Monaco, unpublished observation) and hence is not clustered together with the TAP and LMP genes, which are in the class II region, telomeric of *H-2K*.

Another apparent antigen-processing polymorphism affects the repertoire of peptides bound to certain HLA molecules without affecting cell surface MHC class I expression (PAZMANY et al. 1992; ROWLAND-JONES et al. 1993). Although the gene responsible for this effect appears to be MHC linked, no coding changes are found in the MHC class I molecules in these cells and the phenotype does not correlate with any known TAP polymorphisms (PAZMANY et al. 1992; ROWLAND-JONES et al. 1993). This phenomenon may therefore result from polymorphisms in the IFN-γ-inducible proteasome subunits LMP2 or LMP7, tapasin, or other unidentified molecules involved in MHC class I antigen processing. Several uncharacterized proteins that may play a role in the MHC class I antigen-processing pathway have been described in the literature, but their functional role is not known. For example, an iodinated ovalbumin peptide binds a 55-kDa protein and other unidentified high molecular weight components that can be coprecipitated with class I heavy chain from mouse liver microsomes (WANG et al. 1996b). WANG et al. (1996a) visualized 58-kDa and 43-kDa proteins in rat microsomes by cross-linking with iodinated influenza A virus nucleoprotein peptide. In similar experiments, MARUSINA et al. (1997a) identified three intralumenal components that consistently bound radiolabeled, cross-linkable peptide derivatives after they had been transported into microsomes. The first component is gp96, a common intralumenal chaperone previously implicated in MHC class I presentation (SUTO and SRIVASTAVA 1995). The second component comprises a group of proteins, referred to as the p36 complex, which bind only peptides containing an N-linked glycosylation signal sequence; these proteins are therefore likely to be specific for the carbohydrate moiety attached to such peptides and may be responsible for trapping such peptides within the lumen of the ER. The identity of the third component, p60, remains unknown. An additional unknown but potentially interesting protein with a molecular weight of approximately 100 kDa is found on microsomal membranes of certain cell lines defective in MHC class I antigen processing, but not in their wild-type parent lines or in normal cells (MARUSINA et al. 1997a). One major unanswered question in the class I antigen-processing pathway is how peptides get from the site of their generation (the proteasome) to the TAP molecule for transport into the ER, but cytosolic chaperones or the peptide-binding molecules described above may participate in this process. Although immunohistochemistry has localized a subset of proteasome particles to the ER membrane, there is thus far no evidence that any of these can be found physically associated with TAP molecules.

8 Factors Affecting the Generation of Peptides

Only relatively few peptides from the repertoire of those that could potentially be derived from a given protein antigen are capable of binding MHC and being recognized by CTL. We will consider some of the factors that determine which peptides will be presented by MHC class I molecules.

8.1 Peptide Generation

$CD8^+$ T cells are extremely sensitive to the quantity of antigen, and at least one study has claimed that a single peptide-MHC complex on a target cell is sufficient to elicit a CTL response (SYKULEV et al. 1996). Although several peptides from a given protein can bind MHC molecules, some peptides are generated at higher levels or more efficiently than others (SIJTS et al. 1996; ANTÓN et al. 1997; DENG et al. 1997); however, the relative abundance of a given peptide does not always correlate with immunodominance (ANTÓN et al. 1997). Direct quantification has suggested that 35 molecules of the p60 hydrolase of *Listeria* are degraded to yield a single antigenic peptide (VILLANUEVA et al. 1994), although this number can vary by at least an order of magnitude even for different epitopes within the same antigen molecule. Since the first demonstration by TOWNSEND et al. (1988) that enhanced intracellular protein degradation correlates with antigen processing, several groups (GRANT et al. 1995; SIJTS et al. 1996b; ANTÓN et al. 1997; TOBERY and SILICIANO 1997) – with one exception (GOTH et al. 1996) – have confirmed this finding. It is, however, not clear whether MHC class I-binding peptides are derived primarily from intact proteins that are targeted for destruction or are generated from prematurely terminated polypeptides or polypeptides that have misfolded during or after translation (termed defective ribosomal products, DRIP) (YEWDELL et al. 1996). The vast majority of CTL-binding peptides are derived from translation of mRNA encoding these proteins in the proper reading frame (ENGELHARD 1994; RAMMENSEE et al. 1995; RAMMENSEE 1996), although a number of exceptions have been noted (MALARKANNAN et al. 1995a; SHASTRI et al. 1995a; WANG et al. 1996c). Although the role of ubiquitination and degradation of ubiquitinated proteins by 26S proteasomes in MHC class I antigen processing is tenuous, it is clear that 20S proteasomes play a major role (see above discussion). Once generated, peptides have a remarkably short half-life in cells, but the binding to MHC class I molecules prevents their degradation and extends their half-life (MALARKANNAN et al. 1995b).

8.2 Peptide Size

Digestion of ovalbumin-derived peptides 22–44 amino acids long with proteasomes in vitro resulted in the formation of peptides eight to 11 amino acids long (NIEDERMANN et al. 1996). Similarly, the generation of MHC class I-binding peptides eight to 17 amino acids long was dependent on proteasome activity (YANG et al. 1996), suggesting that proteasomal activity is required for the generation of MHC class I-binding peptides. Surprisingly, peptides generated by *Thermoplasma* proteasomes also result in peptides that are seven to ten amino acids in size (WENZEL et al. 1994), suggesting that the immune system has recruited an evolutionarily conserved mechanism to generate peptides and that the peptide-binding specificity of MHC class I molecules was shaped by this preexisting mechanism. Moreover, peptides in the range of eight to 12 amino acids are most efficiently transported by TAP, although longer peptides can also be transported with a lower efficiency (ANDROLEWICZ and CRESSWELL 1994b; HEEMELS and PLOEGH 1994; MOMBURG et al. 1994; SCHUMACHER et al. 1994a; KOOPMAN et al. 1996). Interestingly, the rat TAP2B-type transporter, which is more selective in the carboxyl-terminal amino acid of peptides that it transports, is also more relaxed in terms of peptide length than rat TAP2A (HEEMELS and PLOEGH 1995; KOOPMAN et al. 1996). In some cases, peptides longer than the MHC class I-binding epitope are transported more efficiently than the exact MHC-binding peptide (DICK et al. 1996; OSSENDORP et al. 1996). Several MHC molecules use proline at position 2 of the peptide as an anchor residue, despite the fact that such peptides are poorly transported (NEISIG et al. 1995). Longer peptides, extended at the amino-terminal end, may be used for transporting such proline-containing epitopes, which may undergo subsequent trimming in the ER and binding to MHC class I molecules (ROELSE et al. 1994; SNYDER et al. 1994; ELLIOTT et al. 1995). Peptides that are transported into the ER but do not bind MHC molecules may bind other chaperones (SUTO and SRIVASTAVA 1995) or peptide-binding proteins (MARUSINA et al. 1997). Alternatively, peptides may be recycled back to the cytosol by an ATP-dependent mechanism (ROELSE et al. 1994; SCHUMACHER et al. 1994a) for further processing or degradation.

8.3 Peptide Sequence

Using oligopeptide substrates, a subdominant ovalbumin peptide was preferentially destroyed by proteasomal cleavage, although the dominant ova peptide was generated, suggesting a correlation between proteasome activity and the generation of antigenic epitopes (NIEDERMANN et al. 1995). A single amino acid change in a CTL epitope was shown to enhance destruction of the epitope by proteasomal cleavage (OSSENDORP et al. 1996). Thus there is evidence to suggest that cleavage preferences of proteasomes are important in the generation of CTL epitopes, but the rules governing peptide cleavage by proteasomes are not yet completely understood.

In vitro peptide transport or peptide-binding assays demonstrate differences in species-specific TAP substrate specificity. As discussed above, peptides with hydrophobic residues at the carboxyl terminus are transported more efficiently by mouse

TAP and rat TAP2B, whereas the human and rat TAP2A do not possess a particular preference in this regard (HEEMELS et al. 1993; HEEMELS and PLOEGH 1994; MOMBURG et al. 1994b; SCHUMACHER et al. 1994a, 1994b; NEEFJES et al. 1995). In addition, peptides containing proline at positions 2 and 3 are transported inefficiently (NEISIG et al. 1995). Chimeric TAP molecules from different species demonstrated that transport specificity toward the carboxyl-terminal residue of the peptide transported is dependent on both TAP1 and TAP2, with the latter playing a major role. A peptide-binding motif has been identified for human TAP, and the carboxyl-terminal residue in peptides plays an important role in determining whether peptides are capable of being transported by TAP (VAN ENDERT et al. 1995). After the carboxyl-terminal residue, the next most important position appears to be 3, followed by 1, 2, and 7 (VAN ENDERT et al. 1995). Hydrophobic residues in positions 3 and 5 of the peptide contribute significantly to the binding activity of the human transporter (VAN ENDERT et al. 1995). Small differences in internal amino acid composition and sequence of the peptide may play a role in binding to TAP in some cases (VAN ENDERT et al. 1994a; HEEMELS and PLOEGH 1994; MARUSINA et al. 1997a). For instance, the iodinated cross-linkable peptide SIINYSKL, based on the naturally occurring ova-peptide SIINFEKL, fails to bind to TAP and fails to be transported (MARUSINA et al. 1997a), although iodinated SIINYEKL is transported. Thus single amino acid changes, even at internal positions in peptides, can influence TAP binding and transport. In general, peptide binding to TAP appears to be less stringent than peptide binding to MHC class I, allowing for a wide array of peptides to be transported that are capable of binding various MHC class I molecules.

8.4 Role of Flanking Residues

Earlier studies demonstrated a key role of residues flanking CTL epitopes in antigen processing (DEL VAL et al. 1991; EISENLOHR et al. 1992). In general, T cell responses are affected by varying the carboxyl-terminal flanking residues, but not the amino-terminal residues, of a CTL-specific peptide (SHASTRI et al. 1995b; BERGMANN et al. 1996). In addition, the amino-terminal methionine used to express minigenes is clipped and does not affect binding of these peptides to MHC class I molecules (MALARKANNAN et al. 1995b). Recent evidence suggests that amino-terminal trimming of peptides is independent of proteasomes and may be due to aminopeptidases; however, the cleavage at the carboxyl terminus of peptides is mediated by proteasome (K. Rock, personal communication). Insertion of alanine residues flanking a CTL epitope improved proteasome-mediated generation of an epitope 160-fold, suggesting that flanking residues in the primary structure of a protein may influence proteasome cleavage (EGGERS et al. 1995). This is further supported by the correlation of proteasome activity and the inhibitory or stimulatory effect of flanking residues in the presentation of OVA by MHC K^b molecules (NIEDERMANN et al. 1995). Furthermore, because the efficiency of TAP transport is sensitive to peptide size and sequence, flanking residues may play an important role in the generation of the CTL response both at the level of peptide transport and at the level of peptide generation.

8.5 Subcellular Localization of Protein

Targeting a protein to a subcellular compartment other than the one in which it normally resides often causes rapid degradation and elicits peptides that bind MHC class I molecules (TOWNSEND et al. 1988; TOBERY and SILICIANO 1997). Alternatively, two different epitopes can be generated from a single protein by targeting it to different cellular compartments, suggesting that MHC class I epitopes are affected by the intracellular location of the antigenic protein (YAMAZAKI et al. 1997).

8.6 Peptide Binding Affinity to MHC Class I

Although the affinity of peptide binding to MHC class I molecules is important (JAMESON and BEVAN 1992; LIPFORD et al. 1993; CHEN et al. 1994; SETTE et al. 1994; OLDSTONE et al. 1995), peptides that bind MHC class I molecules with the highest affinity do not necessarily generate the best CTL response (WIPKE et al. 1993; DENG et al. 1997). As noted in the discussion above, it is possible that the peptide is not generated in vivo, may be degraded by proteasomes, or may be transported inefficiently by TAP. In addition, it is also possible that appropriate TCR capable of recognizing this peptide with a given MHC class I molecule are lacking, due to deletion resulting from the necessity for self-tolerance.

9 Viral Modulation of MHC Class I Antigen Processing

Viruses have developed several mechanisms to escape detection from the immune system (SPRIGGS 1996). Because the MHC class I-processing pathway is a major method for the detection and elimination of viruses, many of these have evolved mechanisms to downregulate this pathway and gain a selective advantage for viral propagation. Downregulation of MHC class I often involves a reduction in the transcription of genes encoding MHC class I heavy chains in virally infected or transformed cells (MAUDSLEY and POUND 1991), although other mechanisms of reducing MHC class I surface expression have also been described. The adenovirus 2-encoded protein E3-19K binds and retains MHC class I heavy chains in the ER, thus preventing cell surface expression of MHC class I molecules (BURGERT and KVIST 1985). This mechanism is also operative in mouse cytomegalovirus (CMV)-infected cells, where ER retention of MHC class I molecules has been observed (DEL VAL et al. 1992). On the other hand, human CMV downregulates MHC molecules by rapid destruction of newly synthesized MHC class I molecules (BEERSMA et al. 1993; JONES et al. 1995). The products of human CMV genes US2 and US11, which are responsible for MHC class I downregulation, mediate the rapid translocation of newly synthesized MHC class I heavy chain molecules from the ER to the cytosol, which is followed by their degradation by proteasomes (WIERTZ et al. 1996a,b). These two viral products are not functionally redundant due to differences in their ability to destroy different allelic forms of MHC class I heavy chains (MACHOLD et al. 1997). A third human CMV gene product, US3, inhibits the maturation and egress of MHC class I to the cell surface (AHN et al. 1996b; JONES et al. 1996). CMV also encodes a product, gp34, that complexes with folded MHC class I molecules and expresses them on the cell surface (KLEIJNEN et al. 1997). In addition, human CMV also encodes an MHC-like class I heavy chain that is responsible for protecting human CMV-infected cells from lysis by NK cells (FARRELL et al. 1997; REYBURN et al. 1997), although the exact mechanism by which this occurs is unclear. Therefore, CMV-infected cells are prevented from expressing MHC class I that may elicit a CTL response, but by encoding a MHC-like molecule or products that bind and express cell surface MHC class I molecules, they are also protected from lysis by NK cells.

Cells transformed by adenovirus 12 demonstrate low expression of TAP, LMP2, and LMP7, but not other proteasome subunits, as well as reduced ability to transport newly synthesized MHC class I molecules (ROTEM-YEHUDAR et al. 1996). In addition, adenovirus 12 transformation affects the turnover of β_2-microglobulin as well as its ability to bind MHC class I molecules (MEY-TAL et al. 1997). Interestingly, the hepatitis B viral protein X (HUANG J et al. 1996) and the Tax protein encoded by human T lymphotropic virus (HTLV)-1 (ROUSSET et al. 1996) bind 20S proteasome subunits. These viral proteins may enhance the degradation of some cellular proteins by bridging them to proteasomes (ROUSSET et al. 1996) and/or they may possess unidentified functions. A single amino acid change in a CTL epitope against a virus can influence proteasome cleavage such that the CTL epitope containing the change is destroyed and is consequently not generated and presented to CTL. Viruses may therefore be under selection pressure to alter CTL epitopes to introduce proteasome cleavage sites, thus evading this arm of the immune system (OSSENDORP et al. 1996).

The herpes simplex virus-encoded cytosolic protein ICP47 blocks cell surface MHC class I expression (YORK et al. 1994) by binding and inactivating TAP (FRÜH et al. 1995; HILL et al. 1995). The active site of ICP47 maps to the amino terminus (GALOCHA et al. 1997) and prevents peptide, but not ATP, binding by TAP (AHN et al. 1996a; TOMAZIN et al. 1996), thereby inhibiting the transport of peptides from the cytosol to the ER and, consequently, reducing the assembly and cell surface expression of MHC class I molecules. The affinity of recombinant ICP47 binding to human TAP is in the region of 50 n*M*, which is higher than the binding of the majority of peptides to TAP (AHN et al. 1996a). Interestingly, ICP47 interacts differently with the murine and human TAP peptide-binding site (FRÜH et al. 1995; HILL et al. 1995; AHN et al. 1996a; TOMAZIN et al. 1996). Inhibition of transport of a reporter peptide in insect cell microsomes containing human TAP is achieved at a one to twofold molar excess of ICP47, but requires 60-fold molar excess for 50% inhibition of transport by murine TAP (AHN et al. 1996a). Human CMV also encodes an ER-resident glycoprotein, US6, that inhibits TAP function. US6 inhibits TAP by a mechanism distinct from ICP47 and binds TAP in the lumen of the ER; it reduces peptide transport, but does not prevent peptide binding to TAP (AHN et al. 1997; HENGEL et al. 1997; LEHNER et al. 1997).

Epstein-Barr virus encodes a protein, EBNA-1, which is expressed in high levels in malignant cells; however, EBNA-1-specific CTL are not generated. Recently, it was shown that the glycine-alanine repeats in this protein prevent MHC class I antigen processing and presentation (LEVITSKAYA et al. 1995). Although this motif plays a key role, the mechanism by which it subverts antigen processing is not known. Unlike the reduction in cell surface MHC class I molecules by viruses discussed above, infection by flavivirus results in the upregulation of MHC class I molecules (MÜLLBACHER and LOBIGS 1995); this upregulation is independent of IFN-γ and TAP and is probably due to virally induced immunopathology, although the key mediators have not been identified.

References

Ahn JY, Tanahashi N, Akiyama K, Hisamatsu H, Noda C, Tanaka K, Chung AH, Shimbara N, Willy PJ, Mott JD, Slaughter CA, DeMartino GN (1995) Primary structures of two homologous subunits of PA28, a γ-interferon-inducible protein activator of the 20S proteasome. FEBS Lett 366:37–42

Ahn K, Meyer TH, Uebel S, Sempe P, Djaballah H, Yang Y, Peterson PA, Früh K, Tampe R (1996a) Molecular mechanism and species specificity of TAP inhibition by herpes simplex virus protein ICP47. EMBO J 15:3247–3255

Ahn K, Angulo A, Ghazal P, Peterson PA, Yang Y, Früh K (1996b) Human cytomegalovirus inhibits antigen presentation by a sequential multistep process. Proc Natl Acad Sci USA 93:10990–10995

Ahn K, Gruhler A, Galocha B, Jones TR, Wiertz EJ, Ploegh HL, Peterson PA, Yang Y, Früh K (1997) The ER-luminal domain of the HCMV glycoprotein US6 inhibits peptide translocation by TAP. Immunity 6:613–621

Aki M, Shimbara N, Takashina M, Akiyama K, Kagawa S, Tamura T, Tanahashi N, Yoshimura T, Tanaka K, Ichihara A (1994) Interferon-γ induces different subunit organizations and functional diversity of proteasomes. J Biochem 115:257–269

Akopian TN, Kisselev AF, Goldberg AL (1997) Processive degradation of proteins and other catalytic properties of the proteasome from *Thermoplasma acidophilum*. J Biol Chem 272:1791–1798

Aldrich CJ, DeCloux A, Woods AS, Cotter RJ, Soloski MJ, Forman J (1994a) Identification of a TAP-dependent leader peptide recognized by alloreactive T cells specific for a class Ib antigen. Cell 79:649–658

Aldrich CJ, Ljunggren H-G, van Kaer L, Ashton-Rickardt PG, Tonegawa S, Forman J (1994b) Positive selection of self- and alloreactive $CD8^+$T cells in Tap-1 mutant mice. Proc Natl Acad Sci USA 91:6525–6528

Anderson K, Cresswell P, Gammon M, Hermes J, Williamson A, Zweerink H (1991) Endogenously synthesized peptide with an endoplasmic reticulum signal sequence sensitizes antigen processing mutant cells to class I-restricted cell mediated lysis. J Exp Med 174:489–492

Androlewicz MJ, Anderson KS, Cresswell P (1993) Evidence that transporters associated with antigen processing translocate a major histocompatibility complex class I-binding peptide into the endoplasmic reticulum in an ATP-dependent manner. Proc Natl Acad Sci USA 90:9130–9134

Androlewicz MJ, Ortmann B, van Endert PM, Spies T, Cresswell P (1994a) Characteristics of peptide and major histocompatibility complex class I/β2-microglobulin binding to the transporters associated with antigen processing (TAP1 and TAP2). Proc Natl Acad Sci USA 91:12716–12720

Androlewicz MJ, Cresswell P (1994b) Human transporters associated with antigen processing possess a promiscuous peptide-binding site. Immunity 1:7–14

Antón LC, Yewdell JW, Bennink JR (1997) MHC class I-associated peptides produced from endogenous gene products with vastly different efficiencies. J Immunol 158:2535–2542

Armandola E, Momburg F, Nijenhuis M, Bulbic N, Hämmerling G (1996) A point mutation in the human transporter associated with antigen processing (TAP2) alters the peptide transport specificity. Eur J Immunol 26:1748–1755

Arnold D, Driscoll J, Androlewicz M, Hughes E, Cresswell P, Spies T (1992) Proteasome subunits encoded in the MHC are not generally required for the processing of peptides bound by MHC class I molecules. Nature 360:171–174

Attaya M, Jameson S, Martinez CK, Hermel E, Aldrich C, Forman J, Lindahl KF, Bevan MJ, Monaco JJ (1992) Ham-2 corrects the class I antigen-processing defect in RMA-S cells. Nature 355:647–649

Bacik I, Cox JH, Anderson R, Yewdell JW, Bennink JR (1994) TAP (transporter associated with antigen processing)-independent presentation of endogenously synthesized peptides is enhanced by endoplasmic reticulum insertion sequences located at the amino- but not carboxyl-terminus of the peptide. J Immunol 152:381–387

Beck S, Kelly A, Radley E, Khurshid F, Alderton RP, Trowsdale J (1992) DNA sequence analysis of 66 kb of the human MHC class II region encoding a cluster of genes for antigen processing. J Mol Biol 228:433–441

Beersma MFC, Bijlmakers MJE, Ploegh H (1993) Human cytomegalovirus down-regulate HLA expression by reducing the stability of class I H chains. J Immunol 151:4455–4464

Bennink JR, Yewdell JW, Gerhard W (1982) A viral polymerase involved in recognition of influenza virus-infected cells by a cytotoxic T cell clone. Nature 296:75–76

Bergmann CC, Yao Q, Ho C-K, Buckwold SL (1996) Flanking residues alter antigenicity and immunogenicity of multi-unit CTL epitopes. J Immunol 157:3242–3249

Bevan M (1995) Antigen presentation to cytotoxic T lymphocytes in vivo. J Exp Med 182:639–641

Boehm U, Klamp T, Groot M, Howard JC (1997) Cellular responses to interferon-γ. Annu Rev Immunol 15:749–795

Boes B, Hengel H, Ruppert T, Multhaup G, Koszinowski UH, Kloetzel P-M (1994) IFN-γ stimulation modulates the proteolytic activity and cleavage site preference of 20S mouse proteasomes. J Exp Med 179:901–909

Brannigan JA, Dodson G, Duggleby HJ, Moody PCE, Smith JL, Tomchick DR, Murzin AG (1995) A protein catalytic framework with an N-terminal nucleophile is capable of self-activation. Nature 378:416–419

Brutkiewicz RR, Bennink JR, Yewdell JW, Bendalac A (1995) TAP-independent, β2-microglobulin-dependent surface expression of functional mouse CD1.1. J Exp Med 182:1913–1919

Burgert H-G, Kvist S (1985) An adenovirus type 2 glycoprotein blocks cell surface expression of human histocompatibility class I antigens. Cell 41:987–997

Burney RO, Pile KD, Gibson K, Calin A, Kennedy LG, Sinnott PJ, Powis SH, Wordsworth BP (1994) Analysis of the MHC class II encoded components of the HLA class I antigen processing pathway in ankylosing spondylitis. Ann Rheum Dis 53:58–60

Caillat-Zucman S, Bertin E, Timsit J, Boitard C, Assan R, Bach J-F (1993) Protection from insulin dependent diabetes mellitus is linked to a peptide transporter gene. Eur J Immunol 23:1784–1788

Carreno BM, Solheim J, Harris M, Stroynowski I, Connolly J, Hansen T (1995) Tap associates with a unique class I conformation, whereas calnexin associates with multiple class I forms in mouse and man. J Immunol 155:4726–4733

Cerundolo V, Kelly A, Elliott T, Trowsdale J, Townsend A (1995) Genes encoded in the major histocompatibility complex affecting the generation of peptides for TAP transport. Eur J Immunol 25:554–562

Cerundolo V, Benham A, Braud V, Mukherjee S, Gould K, Macino B, Neefjes J, Townsend A (1997) The proteasome-specific inhibitor lactacystin blocks presentation of cytotoxic T lymphocyte epitopes in human and murine cells. Eur J Immunol 27:336–341

Cesari MM, Dulay SJ, Caillens H, Robert C, Rouch C, Cadet F, Pabion M (1997) A new human transporter associated with antigen processing allele encodes a large C-terminal protein domain. Immunogenetics 45:280–281

Chen P, Hochstrasser M (1996) Autocatalytic subunit processing couples active site formation in the 20S proteasome to completion of assembly. Cell 86:961–972

Chen W, Khilko S, Fecondo J, Margulies DH, McCluskey J (1994) Determinant selection of major histocompatibility complex class I-restricted antigenic peptides is explained by class I-peptide affinity and is strongly influenced by nondominant anchor residues. J Exp Med 180:1471–1483

Colonna M, Bresnahan M, Bahram S, Strominger JL, Spies T (1992) Allelic variants of the human putative peptide transporter involved in antigen processing. Proc Natl Acad Sci USA 89:3932–3936

Coux O, Tanaka K, Goldberg AL (1996) Structure and functions of the 20S and 26S proteasomes. Annu Rev Biochem 65:801–847

Cox JH, Galardy P, Bennink JR, Yewdell JW (1995) Presentation of endogenous and exogenous antigens is not affected by inactivation of E1 ubiquitin-activating enzyme in temperature-sensitive cell lines. J Immunol 154:511–519

Craiu A, Gaczynska M, Akopian T, Gramm CF, Fenteany G, Goldberg AL, Rock KL (1997) Lactacystin and clasto-lactacystin β-lactone modify multiple proteasome β subunits and inhibit intracellular protein degradation and major histocompatibility complex class I antigen presentation. J Biol Chem 272:13437–13445

De la Salle H, Hanau D, Fricker D, Urlacher A, Kelly A, Salamero J, Powis SH, Donato L, Bausinger H, Laforet M, Jeras M, Spehner D, Bieber T, Falkenrodt A, Cazenave J-P, Trowsdale J, Tongio M-M (1994) Homozygous human TAP peptide transporter mutation in HLA class I deficiency. Science 265:237–241

Del Val M, Schlicht H-J, Ruppert T, Reddehase MJ, Koszinowski UH (1991) Efficient processing of an antigenic sequence for presentation by MHC class I molecules depends on its neighboring residues in the protein. Cell 66:1145–1153

Del Val M, Hengel H, Hacker H, Hartlaub U, Ruppert T, Lucin P, Koszinowski UH (1992) Cytomegalovirus prevents antigen presentation by blocking the transport of peptide-loaded major histocompatibility complex class I molecules into the medial-golgi compartment. J Exp Med 176:729–738

Deng GY, Muir A, Maclaren NK, She J-X (1995) Association of *LMP2* and *LMP7* genes within the major histocompatibility complex with insulin-dependent diabetes mellitus: population and family studies. Am J Hum Genet 56:528–534

Deng Y, Yewdell JW, Eisenlohr LC, Bennink JR (1997) MHC affinity, peptide liberation, T cell repertoire, and immunodominance all contribute to the paucity of MHC class-I restricted peptides recognized by antiviral CTL. J Immunol 158:1507–1515

Deverson EV, Gow ER, Coadwell WJ, Monaco JJ, Butcher GW, Howard JC (1990) MHC class II region encoding proteins related to the multidrug resistance family of membrane transporters. Nature 348:738–741

Dick LR, Aldrich C, Jameson SC, Moomaw CR, Pramanik BC, Doyle CK, DeMartino GN, Bevan MJ, Forman JM, Slaughter CA (1994) Proteolytic processing of ovalbumin and β-galactosidase by the proteasome to yield antigenic peptides. J Immunol 152:3884–3894

Dick LR, Cruikshank AA, Destree AT, Grenier L, McCormack TA, Melandri FD, Nunes SL, Palombella VJ, Parent LA, Plamondon L, Stein RL (1997) Mechanistic studies on the inactivation of the proteasome by lactacystin in cultured cells. J Biol Chem 272:182–188

Dick TP, Ruppert T, Groettrup M, Kloetzel PM, Kuehn L, Koszinowski UH, Stevanovic S, Schild H, Rammensee H-G (1996) Coordinated dual cleavages induced by the proteasome regulator PA28 lead to dominant MHC ligands. Cell 86:253–262

Djilali-Saiah I, Caillat-Zucman S, Schmitz J, Chaves-Vieira ML, Bach J-F (1994) Polymorphism of antigen processing (TAP, LMP) and HLA class II genes in celiac disease. Hum Immunol 40:8–16

Donn RP, Davies EJ, Holt PL, Thompson W, Ollier W (1994) Increased frequency of TAP2B in early onset pauciarticular juvenile chronic arthritis. Ann Rheum Dis 53:261–264

Driscoll J, Brown MG, Finley D, Monaco JJ (1993) MHC-linked *LMP* gene products specifically alter peptidase activities of the proteasome. Nature 365:262–264

Dubiel W, Pratt G, Ferrell K, Rechsteiner M (1992) Purification of an 11S regulator of the multicatalytic protease. J Biol Chem 267:22369–22377

Eggers M, Boes-Fabian B, Ruppert T, Kloetzel P-M, Koszinowski UH (1995) The cleavage preference of the proteasome governs the yield of antigenic peptides. J Exp Med 182:1865–1870

Ehring B, Meyer TH, Eckerskorn C, Lottspeich F, Tampe R (1996) Effects of major-histocompatibility-complex-encoded subunits on the peptidase and proteolytic activities of human 20S proteasomes. Eur J Biochem 235:404–415

Eisenlohr LC, Yewdell JW, Bennink JR (1992) Flanking sequences influence the presentation of an endogenously synthesized peptide to cytotoxic T lymphocytes. J Exp Med 175:481–487

Elliott T, Willis A, Ccrundolo V, Townsend A (1995) Processing of major histocompatibility class I restricted antigens in the endoplasmic reticulum. J Exp Med 181:1481–1491

Engelhard VH (1994) Structure of peptides associated with class I and class II MHC molecules. Annu Rev Immunol 12:181–207

Epperson DE, Arnold D, Spies T, Cresswell P, Pober J, Johnson D (1992) Cytokines increase transporter in antigen processing-1 expression more rapidly than HLA class I expression in endothelial cells. J Immunol 149:3297–3301

Fakler J, Schmitt-Egenolf M, Vejbaesya S, Boehncke W-H, Sterry W, Eiermann TH (1994) Analysis of TAP2 and HLA-DP gene polymorphism in psoriasis. Hum Immunol 40:299–302

Falk K, Rotzschke O, Stevanovic S, Jung G, Rammensee H-G (1991) Allele-specific motifs revealed by sequencing of self-peptides eluted from MHC molecules. Nature 351:290–296

Farrell HE, Vally H, Lynch DM, Fleming P, Shellam GR, Scalzo AA, Davis-Poynter NJ (1997) Inhibition of natural killer cells by a cytomegalovirus MHC class I homologue *in vivo*. Nature 386:510–514

Fehling HJ, Swat W, Laplace C, Kühn R, Rajewsky K, Müller U, von Boehmer H (1994) MHC class I expression in mice lacking the proteasome subunit LMP-7. Science 265:1234–1237

Fenteany G, Standaert RF, Lane WS, Choi S, Corey EJ, Schreiber S (1995) Inhibition of proteasome activities and subunit-specific amino terminal threonine modification of lactacystin. Science 268:726–731

Früh K, Yang Y, Arnold D, Chambers J, Wu L, Waters JB, Spies T, Peterson P (1992) Alternative exon usage and processing of the major histocompatibility complex-encoded proteasome subunits. J Biol Chem 267:22131–22140

Früh K, Ahn K, Djaballah H, Sempe P, van Endert PM, Tampe R, Peterson P, Yang Y (1995) A viral inhibitor of peptide transporters for antigen presentation. Nature 375:415–418

Gabathuler R, Reid G, Kolaitis G, Driscoll J, Jefferies WA (1994) Comparison of cell lines deficient in antigen presentation reveals a functional role for TAP-1 alone in antigen processing. J Exp Med 180:1415–1425

Gaczynska M, Rock, KL, Goldberg AL (1993) γ-Interferon and expression of MHC genes regulate peptide hydrolysis by proteasomes. Nature 365:264–267

Gaczynska M, Rock KL, Spies T, Goldberg AL (1994) Peptidase activities of proteasomes are differentially regulated by the major histocompatibility complex-encoded genes for LMP2 and LMP7. Proc Natl Acad Sci USA 91:9213–9217

Galocha B, Hill A, Barnett BC, Dolan A, Raimondi A, Cook RF, Brunner J, McGrach DJ, Ploegh HL (1997) The active site of ICP47, a herpes simplex virus-encoded inhibitor of the major histocompatibility (MHC)-encoded peptide transporter associated with antigen processing (TAP), maps to the NH_2 terminal 35 residues. J Exp Med 185:1565–1572

Gao GF, Tormo J, Gerth UC, Wyer JR, McMichael AJ, Stuart DI, Bell JI, Yvonne Jones E, Jakobsen BK (1997) Crystal structure of the complex between human CD8αa and HLA-A2. Nature 387:630–634

Garboczi DN, Ghosh P, Utz U, Fan QR, Biddison WE, Wiley DC (1996) Structure of the complex between human T-cell receptor, viral peptide and HLA-A2. Nature 384:134–141

Garcia KC, Degano M, Stanfield RL, Brunmark A, Jackson MR, Peterson PA, Teyton L, Wilson IA (1996a) An αβ T cell receptor structure at 2.5 Å and its orientation in the TCR-MHC complex. Science 274:209–219

Garcia KC, Scott CA, Brunark A, Carbone FR, Peterson PA, Wilson IA, Teyton L (1996b) CD8 enhances formation of stable T-cell receptor/MHC class I molecule complexes. Nature 384:577–581

Gaskins HR, Monaco JJ, Leiter EH (1992) Expression of intra-MHC transporter (*Ham*) genes and class I antigens in diabetes-susceptible NOD mice. Science 256:1826–1828

Gileadi U, Higgins CF (1997) Membrane topology of the ATP-binding cassette transporter associated with antigen presentation (Tap1) expressed in Escherichia coli. J Biol Chem 273:11103–11108

Gooding LR, O'Connell KA (1983) Recognition by cytotoxic T lymphocytes of cells expressing fragments of the SV40 tumor antigen. J Immunol 131:2580–2586

Goth S, Nguyen V, Shastri N (1996) Generation of naturally processed peptide/MHC class I complexes is independent of the stability of endogenously synthesized precursors. J Immunol 157:1894–1904

Grandea AG, Androlewicz MJ, Raghbir SA, Geraghty DE, Spies T (1995) Dependence of peptide binding by MHC class I molecules on their interaction with TAP peptide transporters. Science 270:105–108

Grant EP, Michalek MT, Goldberg AL, Rock KL (1995) Rate of antigen degradation by the ubiquitin-proteasome pathway influences MHC class I presentation. J Immunol 155:3750–3758

Gray CW, Slaughter CA, DeMartino GN (1994) PA28 activator protein forms regulatory caps on proteasome stacked rings. J Mol Biol 236:7–15

Groettrup M, Ruppert T, Kuehn L, Seeger M, Standera S, Koszinowski U, Kloetzel P-M (1995) The interferon-γ-inducible 11S regulator (PA28) and the LMP2/LMP7 subunits govern the peptide production by 20S proteasome *in vitro*. J Biol Chem 270:23808–23815

Groettrup M, Kraft R, Kostka S, Standera S, Stohwasser R, Kloetzel P-M (1996a) A third interferon-γ-induced subunit exchange in the 20S proteasome. Eur J Immunol 26:863–869

Groettrup M, Soza A, Eggers M, Kuehn L, Dick TP, Schild H, Rammensee H-G, Koszinowski UH, Kloetzel P-M (1996b) A role of the proteasome regulator PA28α in antigen presentation. Nature 381:166–168

Groll M, Ditze L, Löwe J, Stock D, Bochtler M, Bartunik HD, Huber R (1997) Structure of 20S proteasome from yeast at 2.4Å resolution. Nature 386:463–471

Hammond SA, Bollinger RC, Tobery TW, Siciliano RF (1993) Transporter-dependent processing of HIV-1 envelope protein for recognition by $CD8^+$ T cells. Nature 364:158–161

Hanau D, Fricker D, Bieber T, Esposito-Farese M-E, Bausinger H, Cazenave J-P, Donato L,Tongio M-M, de la Salle H (1994) CD1 expression is not affected by human peptide transporter deficiency. Hum Immunol 41:61–68

Harding C, France J, Song R, Farah JM, Chatterjee S, Iqbal M, Siman R (1995) Novel dipeptide aldehydes are proteasome inhibitors and block the MHC-I antigen processing pathway. J Immunol 155:1767–1775

Heemels M-T, Ploegh H (1994) Substrate specificity of allelic variants of the TAP peptide transporter. Immunity 1:775–784

Heemels M-T, Ploegh H (1995) Generation, translocation, and presentation of MHC class I-restricted peptides. Annu Rev Biochem 64:463–491

Heemels M-T, Schumacher TNM, K Wonigeit, Ploegh HL (1993) Peptide translocation by variants of the transporter associated with antigen processing. Science 262:2059–2061

Henderson RA, Michel H, Sakaguchi K, Shabanowitz J, Appella E, Hunt DF, Engelhard VH (1992) HLA-A2.1-associated peptides from a mutant cell line: a second pathway of antigen presentation. Science 255:1264–1266

Hengel H, Koopman J-O, Flohr T, Muranyi W, Goulmy E, Hämmerling GJ, Koszinowski UH, Momburg F (1997) A viral ER-resident glycoprotein inactivates the MHC-encoded peptide transporter. Immunity 6:623–632

Hesresbach D, Alizadeh M, Bretagne JF, Gautier A, Quillivic F, Lemarchand B, Gosselin M, Genetet B, Semana G (1996) Investigation of the association of major histocompatibility complex genes, including HLA class I, class II and Tap genes, with clinical forms of Crohn's disease. Eur J Immunogen 23:141–151

Hill A, Judovic P, York I, Russ G, Bennink J, Yewdell J, Ploegh H, Johnson D (1995) Herpes simplex virus turns off the TAP to evade host immunity. Nature 375:411–415

Hisamatsu H, Shimbara N, Saito Y, Kristensen P, Hendil KB, Fujiwara T, Takahashi N, Tamura T, Ichihara A, Tanaka K (1996) Newly pair of proteasomal subunits regulated reciprocally by interferon γ. J Exp Med 183:1807–1816

Holcombe HR, Castano AR, Cheroutre H, Teitell M, Maher JK, Peterson PA, Kronenberg M (1995) Nonclassical behavior of the thymus leukemia antigen: peptide transporter-independent expression of a nonclassical class I molecule. J Exp Med 181:1433–1443

Huang J, Kwong J, Sun EC-Y, Liang TJ (1996) Proteasome complex as a potential cellular target of hepatitis B virus X protein. J Virol 70:5582–5591

Hughes EA, Ortmann B, Surman M, Cresswell P (1996) The protease inhibitor, N-acetyl-L-leucyl-L-leucyl-L-norleucinal, decreases the pool of major histocompatibility complex class I-binding peptides and inhibits peptide trimming in the endoplasmic reticulum. J Exp Med 183:1569–1578

Hughes EA, Hammond C, Cresswell P (1997) Misfolded major histocompatibility complex class I heavy chains are translocated into the cytoplasm and degraded by the proteasome. Proc Natl Acad Sci USA 94:1896–1901

Jameson SC, Bevan MJ (1992) Dissection of major histocompatibility complex (MHC) and T cell receptor contact residues in a Kb-restricted ovalbumin peptide and an assessment of the predictive power of MHC-binding motifs. Eur J Immunol 22:2663–2667

Jiang H, Monaco JJ (1997) Sequence and expression of mouse proteasome activator PA28 and the related autoantigen, Ki. Immunogenetics 46:93–98

Joly E, Deverson EV, Coadwell WJ, Günther E, Howard JC, Butcher GW (1994) The distribution of Tap2 alleles among laboratory rat RT1 haplotypes. Immunogenetics 40:45–53

Jondal M, Schirmbeck R, Reimann J (1996) MHC class I-restricted CTL responses to exogenous antigens. Immunity 5:295–302

Jones TR, Hanson LK, Sun L, Slater JS, Stenberrg RM, Campbell AE (1995) Multiple independent loci within the human cytomegalovirus unique short region down-regulate expression of major histocompatibility complex class I heavy chains. J Virol 69:4830–4841

Jones TR, Wiertz EJ, Sun L, Fish EN, Nelson JA, Ploegh H (1996) Human cytomegalovirus US3 impairs transport and maturation of major histocompatibility complex class I heavy chains. Proc Natl Acad Sci USA 93:11327–11333

Joyce S, Kuzushima K, Kepecs G, Angeletti RH, Nathenson SG (1994) Characterization of an incompletely assembled major histocompatibility class I molecule (H-2K^b) associated with unusually long peptides: implications for antigen processing and presentation. Proc Natl Acad Sci USA 91:4145–4149

Kandil E, Namikawa C, Nonaka M, Greenberg AS, Flajnik MF, Ishibashi T, Kasahara M (1996) Isolation of low molecular mass polypeptide complementary DNA clones from primitive vertebrates:implications for the origin of MHC class I restricted antigen presentation. J Immunol 156:4245–4253

Kellar-Wood HF, Powis SH, Gray J, Compson DAS (1994) MHC-encoded TAP1 and TAP2 dimorphisms in multiple sclerosis. Tissue Antigens 43:129–132

Kelly A, Powis S, Kerr L-A, Mockridge I, Elliot T, Bastin J, Uchanska-Ziegler B, Ziegler A, Trowsdale J, Townsend A (1992) Assembly and function of the two ABC transporter proteins encoded in the human major histocompatibility complex. Nature 355:641–644

Khanna R, Burrows SR, Argaet V, Moss D (1994) Endoplasmic reticulum signal sequence facilitated transport of peptide epitopes, restores immunodeficiency of an antigen processing defective tumor cell line. Int Immunol 6:639–645

Kim T-G, Lee Y-H, Choi H-B, Han H (1996) Two newly discovered alleles of major histocompatibility complex-encoded LMP7 in Korean populations. Immunogenetics 46:61–64

Kishi F, Suminami Y, Monaco JJ (1993) Genomic organization of the mouse *Lmp-2* gene and characteristic structure of its promoter. Gene 133:243–248

Kleijmeer MJ, Kelly A, Geuze HJ, Slot JW, Townsend A, Trowsdale J (1992) Location of MHC- encoded transporters in the cytoplasmic reticulum and cis-Golgi. Nature 357:342–344

Kleijnen MF, Huppa JB, Lucin P, Mukherjee S, Farrell H, Capmbell AE, Koszinowski, Hill AB, Ploegh HL (1997) A mouse cytomegalovirus glycoprotein, gp34, forms a complex with folded class I MHC molecules in the ER which is not retained but is transported to the cell surface. EMBO J 16:685–694

Koopman JO, Post M, Neefjes JJ, Hämmerling GJ, Momburg F (1996) Translocation of long peptides by transporters associated with antigen processing (TAP). Eur J Immunol 26:1720–1728

Kopp F, Dahlmann B, Hendil KB (1993) Evidence indicating that the human proteasome is a complex dimer. J Mol Biol 229:14–19

Kopp F, Kristensen P, Hendil KB, Johnsen A, Sobek A, Dahlmann B (1995) The human proteasome subunit HsN3 is located in the inner rings of the complex dimer. J Mol Biol 248:264–272

Kopp F, Hendil KB, Dahlmann B, Kristensen P, Sobek A, Uerkvitz W (1997) Subunit arrangement in the human 20S proteasome. Proc Natl Acad Sci USA 94:2939–2944

Kuckelkorn U, Frentzel S, Kraft R, Kostka S, Groettrup M, Kloetzel P-M (1995) Incorporation of major histocompatibility complex-encoded subunits LMP-2 and LMP-7 changes the quality of the 20S proteasome polypeptide processing products independent of interferon-γ. Eur J Immunol 25:2605–2611

Lanier LL (1997) Natural killer cells: from no receptors to too many. Immunity 6:371–378

Larsen F, Solheim J, Kristensen T, Kolsto A-B, Prydz H (1993) A tight cluster of five unrelated human genes on chromosome 16q22.1. Hum Mol Genet 2:1589–1595

Lee N, Malacko AR, Ishitani A, Chen M, Bajorath J, Marquardt H, Geraghty D (1995) The membrane-bound and soluble forms of HLA-G bind identical sets of endogenous peptides but differ with respect to TAP association. Immunity 3:591–600

Lee SP, Thomas WA, Blake NW, Rickinson AB (1996) Transporter (TAP)-independent processing of a multiple membrane-spanning protein, the Epstein-Barr virus latent membrane protein 2. Eur J Immunol 26:1875–1885

Lehner PJ, Kartunen JT, Wilkinson GWG, Russell P (1997) The human cytomegalovirus US6 glycoprotein inhibits transporter associated with antigen processing-dependent peptide translocation. Proc Natl Acad Sci USA 94:6904–6909

Levitskaya J, Coram M, Levitsky V, Imreh S, Steigerwald-Mullen PM, Klein G, Kurilla MG, Masucci MG (1995) Inhibition of antigen processing by the internal repeat region of the Epstein-Barr nuclear antigen-1. Nature 375:685–688.

Lewis JW, Neisig A, Neefjes J, Elliot T (1996) Point mutations in the α2 domain of HLA-A2.1 define a functionally relevant interaction with TAP. Curr Biol 6:873–883

Lipford GB, Hoffman M, Wagner H, Heeg K (1993) Primary *in vivo* responses to ovalbumin: probing the predictive value of the K^b binding motif. J Immunol 150:1212–1222

Ljunggren H-G, Stam NJ, Öhlén C, Neefjes JJ, Höglund P, Heemels M-T, Bastin J, Schumacher TNM, Townsend A, Kärre K, Ploegh HL (1990) Empty MHC class I molecules come out in the cold. Nature 346:476–480

Ljunggren H-G, van Kaer L, Ploegh HL, Tonegawa S (1994) Altered natural killer cell repertoire in Tap-1 mutant mice. Proc Natl Acad Sci USA 91:6520–6524

Lobigs M, Rothenfluh H, Blanden RV, Mullbacher A (1995) Polymorhic peptide transporters in MHC class I monomorphic Syrian hamster. Immunogenetics 42:398–407

Lofti M, Sastry A, Ye M, Van der Meulen J, Dosch HM, Singal DP (1994) HLA-DQ and TAP2 genes in patients with insulin-dependent diabetes mellitus. Immunol Lett 41:201–204

Löwe J, Stock D, Jap B, Zwickl P, Baumeister W, Huber R (1995) Crystal structure of the 20S proteasome from the archaeon *T. acidophilus* at 3.4 Å resolution. Science 268:533–539

Ma C-P, Slaughter CA, DeMartino GN (1992) Identification, purification, and characterization of a protein activator (PA28) of the 20S proteasome (macropain). J Biol Chem 267:10515–10523

Machold RP, Wiertz EJ, Jones TR, Ploegh HL (1997) The HCMV gene products US11 and US2 differ in their ability to attack allelic forms of murine histocompatibility complex (MHC) class I heavy chains. J Exp Med 185:363–366

Maksymowych WP, Russell AS (1995) Polymorphism in the LMP2 gene influences the relative risk for acute anterior uveitis in unselected patients with ankylosing spondylitis. Clin Invest Med 18:42–46

Maksymowych WP, Wessler A, Schmitt-Engenolf M, Suarez-Almazor M, Ritzel G, Von Borstel RC, Pazderka F, Russell AS (1994) Polymorphism in an HLA linked proteasome gene influences phenotypic expression of disease in HLA-B27 positive individuals. J Rheumatol 21:665–669

Maksymowych WP, Tao S, Luong M, Suarez-Almazor M, Nelson R, Pazderka F, Russell AS (1995) Polymorphism in the LMP2 and LMP7 genes and adult rheumatoid arthritis: no relationship with disease susceptibility or outcome. Tissue Antigens 46:136–139

Malarkanan S, Afrarian M, Shastri N (1995a) A rare cryptic translation product is presented by K^b MHC class I molecule to alloreactive T-cells. J Exp Med 182:1739–1750

Malarkanan S, Goth S, Buchholz DR, Shastri N (1995b) The role of MHC class I molecules in the generation of endogenous peptide/MHC complexes. J Immunol 154:585–598

Marsal S, Hall MA, Panayi GS, Lanchbury JS (1994) Association of TAP2 polymorphism with rheumatoid arthritis is secondary to allelic association with HLA-DRB1. Arthritis Rheum 37:504–513

Martinez-Laso J, Martin-Villa JM, Alvarez M, Martinez-Quiles N, Lledo G, Arnaiz-Villena A (1994) Susceptibility to insulin-dependent diabetes mellitus and short cytoplasmic ATP-binding domain TAP2 01 alleles. Tissue Antigens 44:184–188

Marusina K, Reid G, Gabathuler R, Jefferies W, Monaco JJ (1997a) Novel peptide-binding proteins and peptide transport in normal and TAP-deficient microsomes. Biochemistry 36:856–863

Marusina K, Iyer M, Monaco JJ (1997b) Allelic variation in the mouse *Tap-1* and *Tap-2* transporter genes. J Immunol 158:5251–5256

Maudsley DJ, Pound JD (1991) Modulation of MHC antigen expression by viruses and oncogenes. Immunol Today 12:429–431

Meyer TH, van Endert PM, Uebel S, Ehring B, Tampé (1994) Functional expression and purification of the ABC transporter complex associated with antigen processing (TAP) in insect cells. FEBS Lett 351:443–447

Mey-Tal SV, Schechter C, Ehrlich R (1997) Synthesis and turnover of β2-microglobulin in Ad12-transformed cells defective in assembly and transport of class I major histocompatibility complex molecules. J Biol Chem 272:353–361

Michalek MT, Grant EP, Gramm C, Goldberg AL, Rock KL (1993) A role for the ubiquitin-dependent proteolytic pathway in MHC class I-restricted antigen presentation. Nature 363:552–554

Middleton D, Megaw G, Cullen C, Hawkins S, Darke C, Savage D (1994) TAP1 and TAP2 polymorphism in multiple sclerosis patients. Hum Immunol 40:131–134

Min W, Pober JS, Johnson DR (1996) Kinetically coordinated induction of TAP1 and HLA class I by IFN-γ. J Immunol 156:3174–3183

Moins-Teisserenc H, Semana G, Alizadeh M, Loiseau P, Bobrynina V, Deschamps I, Edan G, Birebent B, Genetet B, Sabouraud O, Charron D (1995) TAP2 gene polymorphism contributes to genetic susceptibility to multiple sclerosis. Hum Immunol 42:195–202

Momburg F, Roelse J, Howard JC, Butcher GW, Hämmerling G, Neefjes JJ (1994a) Selectivity of MHC-encoded peptide transporters from human, mouse and rat. Nature 367:648–651

Momburg F, Roelse J, Hämmerling GJ, Neefjes (1994b) Peptide size selection by the major histocompatibility complex-encoded peptide transporter. J Exp Med 179:1613–1623

Momburg F, Armandola EA, Post M, Hämmerling GJ (1996) Residues in TAP2 controlling substrate specificity. J Immunol 156:1756–1763

Monaco JJ (1992) A molecular model of MHC class I-restricted antigen processing. Immunol Today 13:173–79

Monaco JJ, McDevitt HO (1986) The LMP antigens: a stable MHC-controlled multisubunit protein complex. Hum Immunol 15:416–426

Monaco JJ, Nandi D (1995) The genetics of proteasomes and antigen processing. Annu Rev Genet 29:729–754

Monaco JJ, Cho S, Attaya M (1990) Transport protein genes in the murine MHC: possible implications for antigen processing. Science 250:1723–1726

Moore MW, Carbone FR, Bevan MJ (1988) Introduction of soluble protein into the class I pathway of antigen processing and presentation. Cell 54:777–785

Morrison LA, Lukacher AE, Braciale VL, Fan DP, Braciale TJ (1986) Differences in antigen presentation to MHC class I- and class II-restricted influenza virus-specific cytolytic T lymphocyte clones. J Exp Med 163:903–921

Müllbacher A, Lobigs M (1995) Up-regulation of MHC class I by flavivirus-induced peptide translocation into the endoplasmic reticulum. Immunity 3:207–214

Muller K, Ebensperger C, Tampe R (1994) Nucleotide binding to the hydrophilic C terminal domain of the transporter associated with antigen processing (TAP). J Biol Chem 19:14032–14037

Murakami Y, Matsufuji S, Kameji T, Hayashi S, Igarashi K, Tamura T, Tanaka K, Ichihara A (1992) Ornithine decarboxylase is degraded by 26S proteasomes without ubiquitination. Nature 360:597–599

Nakanishi K, Kobayashi T, Murase T, Kosaka K (1994) Lack of association of the transporter associated with antigen processing with Japanese insulin-dependent diabetes mellitus. Metabolism 43:1013–1017

Nandi D, Jiang H, Monaco JJ (1996a) Identification of MECL-1 (LMP-10) as the third IFNγ-inducible proteasome subunit. J Immunol 156:2361–2364

Nandi D, Iyer MN, Monaco JJ (1996b) Molecular and serological analysis of polymorphisms in the murine major histocompatibility-encoded proteasome subunits, LMP-2 and LMP-7. Exp Clin Immunogenet 13:20–29

Nandi D, Woodward E, Ginsburg D, Monaco JJ (1997) Intermediates in the formation of mouse 20S proteasomes: implications for the assembly of precursor β subunits. EMBO J 16:5363–5375

Neefjes JJ, Momburg F, Hämmerling GJ (1993) Selective and ATP-dependent translocation of peptides by the MHC-encoded transporter. Science 261:769–771

Neefjes JJ, Gottfried E, Roelse J, Gromme M, Obst R, Hämmerling GJ, Momburg F (1995) Analysis of the fine specificity of rat, mouse and human TAP peptide transporters. Eur J Immunol 25:1133–1136

Neisig A, Roelse J, Sijts AJ, Ossendorp F, Feltkamp MCW, Kast WM, Melief CJM, Neefjes JJ (1995) Major differences in transporter associated with antigen presentation (TAP)-dependent translocation of MHC class I presentable peptides and the effect of flanking sequences. J Immunol 154:1273–1279

Neisig A, Wubbolts R, Zang X, Melief C, Neefjes J (1996) Allele-specific differences in the interaction of MHC class I molecules with transporters associated with antigen processing. J Immunol 156:3196–3206

Niedermann G, Butz S, Ihlenfeldt G, Grimm R, Lucchiari M, Hoschutzky, Jung G, Maier B, Eichmann K (1995) Contribution of proteasome-mediated proteolysis to the hierarchy of epitopes presented by major histocompatibility complex class I molecules. Immunity 2:289–299

Niedermann G, King G, Butz S, Birsner U, Grimm R, Shabanowitz J, Hunt DF, Eichmann K (1996) The proteolytic fragments generated by vertebrate proteasomes: structural relationships to major histocompatibility complex class I binding peptides. Proc Natl Acad Sci USA 93:8572–8577

Nijenhuis M, Hämmerling GJ (1996) Multiple regions of the transporter associated with antigen processing (TAP) contribute to its peptide binding site. J Immunol 157:5467–5477

Nijenhuis M, Schmitt S, Armandola E, Obst R, Brunner J, Hämmerling G (1996) Identification of a contact region of the transporter associated with antigen processing. J Immunol 156:2186–2195

Oldstone MBA, Lewicki H, Borst P, Hudrisier D, Gairin J (1995) Discriminated selection among viral peptides with the appropriate anchor residues: implications for the size of the cytotoxic T-lymphocyte repertoire and control of viral infection. J Virol 69:7423–7429

Orlowski M, Cardozo C, Michaud C (1993) Evidence for the presence of five distinct proteolytic components in the pituitary multicatalytic proteinase complex. Properties of two components cleaving bonds on the carboxyl side of branched chain and small neutral amino acids. Biochemistry 32:1563–1572

Ortiz-Navarette V, Hämmerling GJ (1991) Surface appearance and instability of empty H-2 class I molecules under physiological conditions. Proc Natl Acad Sci USA 88:3594–3597

Ortmann B, Androlewicz MJ, Cresswell P (1994) MHC class I/β2-microglobulin complexes associate with TAP transporters before peptide binding. Nature 368:864–867

Ossendorp F, Eggers M, Neisig A, Ruppert T, Groettrup M, Sijts A, Mengedé E, Kloetzel P-M, Neefjes J, Koszinowski U, Melief C (1996) A single residue exchange within a viral CTL epitope alters proteasome-mediated degradation resulting in lack of antigen presentation. Immunity 5:115–124

Parham P, Ohta T (1996) Population biology of antigen presentation by MHC class I molecules. Science 272:67–74

Pazmany L, Rowland-Jones S, Heut S, Hill A, Sutton J, Murray R, Brooks J, McMichael A (1992) Genetic modulation of antigen presentation by HLA-B27 molecules. J Exp Med 175:361–369

Peace-Brewer A, Tussey L, Matsui M, Li G, Quinn DG, Frelinger JA (1996) A point mutation in HLA-A*201 results in failure to bind the TAP complex and to present virus derived peptides to CTL. Immunity 4:505–514

Pearce RB, Trigler L, Svaas EK, Peterson CM (1993) Polymorphism in the mouse *Tap-1* gene: association with abnormal $CD8^+$ T cell development in the nonobese nondiabetic mouse. J Immunol 151:5338–5347

Peters J-M, Cejka Z, Robin Harris J, Kleinschmidt JA, Baumeister W (1993) Structural features of the 26S proteasome complex. J Mol Biol 234:932–937

Ploski R, Undlien DE, Vinjie O, Forre O, Thorsby E, Ronningen K (1994) Polymorphism of human major histocompatibility complex-encoded transporter associated with antigen processing (TAP) genes and susceptibility to Juvenile rheumatoid arthritis. Hum Immunol 39:54–60

Powis SJ, Townsend ARM, Deverson EV, Bastin J, Butcher GW, Howard JC (1991a) Restoration of antigen presentation to the mutant cell line RMA-S by an MHC-linked transporter. Nature 354:528–531

Powis SJ, Howard JC, Butcher GW (1991b) The major histocompatibility class II-linked *cim* locus controls the kinetics of intracellular transport of a classical class I molecule. J Exp Med 173:913–921

Powis SJ, Deverson EV, Coadwell WJ, Cruela A, Huskisson NS (1992a) Effect of polymorphism of an MHC linked transporter on the peptides assembled in a class I molecule. Nature 357:211–215

Powis S, Mockridge I, Kelly A, Kerr L-A, Glynne R, Gilead U, Beck S, Trowsdale J (1992b) Polymorphism in a second ABC transporter gene located within the class II region of the human major histocompatibility complex. Proc Natl Acad Sci USA 89:1463–1467

Powis S, Tonks S, Mockridge I, Kelly A, Bodmer J, Trowsdale J (1993) Alleles and haplotypes of the MHC-encoded ABC transporters TAP1 and TAP2. Immunogenetics 37:373–380

Powis SJ, Young L, Joly E, Barker P, Richardson L, Brandt RP, Melief C, Howard J, Butcher G (1996) The rat cim effect: TAP allele-dependent changes in a class I MHC anchor motif and evidence against C-terminal trimming of peptides in the ER. Immunity 4:159–165

Pryhuber KG, Murray KJ, Donnelly P, Passo MH, Maksymowych WP, Glass DN, Giannini EH, Colbert RA (1996) Polymorphism in the LMP2 gene influences disease susceptibility and severity in HLA-B27 associated juvenile rheumatoid arthritis. J Rheumatol 23:747–752

Pühler G, Pitzer F, Zwickl P, Baumeister W (1994) Proteasomes: multisubunit proteinases common to *Thermoplasma* and eucaryotes. Syst Appl Microbiol 16:734–741

Rammensee H-G (1996) Antigen presentation – recent developments. Int Arch Allergy Immunol 110:299–307

Rammensee H-G, Friede T, Stevanovic S, (1995) MHC ligands and peptide motifs. First listing. Immunogenetics 41:178–228

Realini C, Dubiel W, Pratt G, Ferrell K, Rechsteiner M (1994) Molecular cloning and expression of a γ-interferon-inducible activator of the multicatalytic protease. J Biol Chem 269:20727–20732.

Reyburn HT, Mandelboim O, Valés-Gómez M, Davis DM, Pazmany L, Strominger J (1997) The class I MHC homologue of human cytomegalovirus inhibits attack by natural killer cells. Nature 386:514–517

Rock KL (1996) A new foreign policy: MHC class I molecules monitor the outside world. Immunol Today 17:131–137

Rock KL, Gramm C, Benacerraf B (1991) Low temperature and peptide favor the formation of class I heterodimers on RMA-S cells at the cell surface. Proc Natl Acad Sci USA 88:4200–4204

Rock KL, Gramm C, Rothstein L, Clark K, Stein R, Dick L, Hwang D, Goldberg AL (1994) Inhibitors of the proteasome block the degradation of most cell proteins and the generation of peptides presented on MHC class I molecules. Cell 78:761–771

Roelse J, Gromme M, Momburg F, Hämmerling G, Neefjes J (1994) Trimming of TAP-translocated peptides in the endoplasmic reticulum and in the cytosol during recycling. J Exp Med 180:1591–1597

Rotem-Yehudar R, Groettrup M, Soza A, Kloetzel PM, Ehrlich R (1996) LMP-associated proteolytic activities and TAP-dependent peptide transport for class I MHC molecules are suppressed in cell lines transformed by highly oncogenic adenovirus 12. J Exp Med 183:499–514

Rousset R, Desbois C, Bantignies F, Jalinot P (1996) Effects on NFκB1/p105 processing of the interaction between the HTLV-1 transactivator Tax and the proteasome. Nature 381:328–331

Rowland-Jones SL, Powis SH, Sutton J, Mockridge I, Gotch FM, Murray N, Hill AB, Rosenberg WM, Trowsdale J, McMichael AJ (1993) An antigen processing polymorphism revealed by HLA-B8-restricted cytotoxic T lymphocytes which does not correlate with TAP gene polymorphism. Eur J Immunol 23:1999–2004

Sadasivan B, Lehner PJ, Ortmann B, Spies T, Cresswell P (1996) Roles for calreticulin and a novel glycoprotein, tapasin, in the interaction of MHC class I molecules with TAP. Immunity 5:103–114

Sandberg JK, Chambers BJ, Van Kaer L, Karre K, Ljunggren H-G (1996) TAP1-deficient mice select a $CD8^+$ T cell repertoire that displays both diversity and peptide specificity. Eur J Immunol 26:288–293

Schmidtke G, Kraft R, Kostka S, Henklein P, Frömmel C, Löwe J, Huber R, Kloetzel P-M, Schmidt M (1996) Analysis of mammalian 20S proteasome biogenesis: the maturation of β subunits is an ordered two-step mechanism involving autocatalysis. EMBO J 15:6887–6898

Schumacher TNM, Heemels M-T, Neefjes JJ, Kast WM, Melief CJM, Ploegh H (1990) Direct binding of peptides to empty MHC class I molecules on intact cells and in vitro. Cell 62:563–567

Schumacher TNM, Kantesaria DV, Heemels M-T, Ashton-Rickardt PG, Shepherd JC, Früh K, Yang Y, Peterson P, Tonegawa S, Ploegh H (1994a) Peptide length and specificity of the mouse TAP1/TAP2 translocator. J Exp Med 179:533–540

Schumacher TNM, Kantesaria D, Serreze DV, Roopenian DC, Ploegh H (1994b) Transporters from $H\text{-}2^b$, $H\text{-}2^d$, $H\text{-}2^s$, $H\text{-}2^k$, and $H\text{-}2^{g7}$ (NOD/Lt) haplotype translocate similar sets of peptides. Proc Natl Acad Sci USA 91:13004–13008

Seemüller E, Lupas A, Stock D, Löwe J, Huber R, Baumeister W (1995) Proteasome from *Thermoplasma acidophilum*: a threonine protease. Science 268:579–582

Seemüller E, Lupas A, Baumeister W (1996) Autocatalytic processing of the 20S proteasome. Nature 382:468–470

Sette A, Vitiello A, Reherman B, Fowler P, Nayersina R, Kast WA, Melief CJM, Oseroff C, Yuan L, Ruppert J, Sidney J, del Guercio M-F, Southwood S, Kubo RT, Chesnut RW, Gret HM, Chisari FV (1994) The relationship between class I binding affinity and immunogenicity of potential cytotoxic T cell epitopes. J Immunol 153:5586–5592

Shastri N, Nguyen V, Gonzalez F (1995a) Major histocompatibility class I molecules can present cryptic translation products to T-cells. J Biol Chem 270:1088–1091

Shastri N, Serwold T, Gonzalez F (1995b) Presentation of endogenous peptide/MHC class I complexes is profoundly influenced by specific C-terminal flanking residues. J Immunol 155:4339–4346

Shepherd JC, Schumacher TNM, Ashton-Rickardt PG, Imaeda S, Ploegh HL, Janeway Jr CA, Tonegawa S (1993) TAP1-dependent peptide translocation in vitro is ATP dependent and peptide selective. Cell 74:577–584

Shimizu Y, DeMars R (1989) Production of human cells expressing individual transferred HLA-A, -B, -C genes using a HLA-A, -B, -C null human cell line. J Immunol 142:3320–3328

Sijts AJ, Neisig A, Neefjes J, Pamer EG (1996a) Two Listeria monocytogenes CTL epitopes are processed from the same antigen with different efficiencies. J Immunol 156:685–692

Sijts AJ, Villanueva MS, Pamer EG (1996b) CTL epitope generation is tightly linked to cellular proteolysis of a *Listeria monocytogenous* antigen. J Immunol 156:1497–1503

Siliciano RF, Soloski MJ (1995) MHC class I-restricted processing of transmembrane proteins. J Immunol 155:2–5

Sinha AA, Lopez MT, McDevitt HO (1990) Autoimmune diseases: the failure of self tolerance. Science 248:1380–1388

Snyder HL, Yewdell JW, Bennink JR (1994) Trimming of antigenic peptides in an early secretory compartment. J Exp Med 180:2389–2394

Solheim JC, Harris, MR, Kindle CS, Hansen TH (1997) Prominence of β2-microglobulin, class I heavy chain conformation, and Tapasin in the interactions of class I heavy chain with calreticulin and the transporter associated with antigen processing. J Immunol 158:2236–2241

Spies T, DeMars R (1991) Restored expression of major histocompatibility class I molecules by gene transfer of a putative peptide transporter. Nature 351:323–324

Spies T, Bresnahan M, Bahram S, Arnold D, Blanck G, Mellins E, Pious D, DeMars R (1990) A gene in the human major histocompatibility complex class II region controlling the class I antigen processing pathway. Nature 348:744–747

Spies T, Cerundolo V, Colonna M, Cresswell P, Townsend A, DeMars R (1992) Presentation of viral antigen by MHC class I molecules is dependent on putative peptide transporter heterodimer. Nature 355:644–646

Spriggs M (1996) One step ahead of the game: viral immunomodulatory molecules. Annu Rev Immunol 14:101–130

Spurkland A, Knutsen I, Undlien DE, Vartdal F (1994) No association of multiple sclerosis to alleles at the TAP2 locus. Hum Immunol 39:299–301

Stein RL, Melandri F, Dick L (1996) Kinetic characterization of the chymotryptic activity of the 20S proteasome. Biochemistry 35:3899–3908

Stohwasser R, Kuckelkorn U, Kraft R, Kostka S, Kloetzel P-M (1996) 20S proteasome from LMP7 knock out mice reveals altered proteolytic activities and cleavage site preferences. FEBS Lett 383:109–113

Stohwasser R, Standera S, Peters I, Kloetzel P-M, Groettrup M (1997) Molecular cloning of the mouse proteasome subunits MC14 and MECL-1: reciprocally regulated tissue expression of interferon-γ-modulated proteasome subunits. Eur J Immunol 27:1182–1187

Suh W-K, Cohen-Doyle MF, Früh K, Wang K, Peterson PA, Williams D (1994) Interaction of MHC class I molecules with the transporter associated with antigen processing. Science 264:1322–1325

Suh W-K, Mitchell EK, Yang Y, Peterson PA, Waneck GL, Williams DB (1996) MHC class I molecules form ternary complexes with calnexin and TAP and undergo peptide-regulated interaction with TAP via their extracellular domains. J Exp Med 184:337–348

Suto R, Srivastava PK (1995) A mechanism for the specific immunogenicity of heat-shock protein chaperoned peptides. Science 269:1585–1588

Sykulev Y, Joo M, Vturina I, Tsomides TT, Eisen H (1996) Evidence that a single peptide-MHC complex on a target cell can elicit a cytolytic T cell response. Immunity 4:565–571

Tabaczewski P, Stroynowski I (1994) Expression of secreted and glycosylphosphatidylinositol-bound Qa-2 molecules is dependent on functional TAP-2 peptide transporter. J Immunol 152:5268–5274

Tamura T, Nagy I, Lupas A, Lottspeich F, Cejka Z, Schoofs G, Tanaka K, De Mot R, Baumeister W (1995) The first characterization of a eubacterial proteasome: the 20S complex of *Rhodococcus*. Curr Biol 5:766–774

Tanaka K, Tsurumi C (1997) The 26S proteasome:subunits and functions. Mol Biol Reports 24:3–11

Tighe MR, Hall MA, Cardi E, Ashkenazi A, Lanchbury JS, Ciclitira PJ (1994) Associations between alleles of the major histocompatibility complex-encoded ABC transporter gene TAP2, HLA class II alleles, and celiac disease susceptibility. Hum Immunol 39:9–16

Tobery TW, Siliciano RF (1997) Targeting of HIV-1 antigens for rapid intracellular degradation enhances cytotoxic T lymphocyte (CTL) recognition and the induction of *de novo* CTL responses *in vivo* after immunization. J Exp Med 185:909–920

Tomazin R, Hill AB, Jugovic P, York I, van Endert P, Ploegh HL, Andrews DW, Johnson DC (1996) Stable binding of the herpes simplex virus ICP47 protein to the peptide binding site of TAP. EMBO J 15:3256–3266

Townsend A, Bastin J, Gould K, Brownlee G, Andrew M, Coupar B, Boyle D, Chan S, Smith G (1988) Defective presentation to class I-restricted cytotoxic T lymphocytes in vaccinia-infected cells is overcome by enhanced degradation of antigen. J Exp Med 168:1211–1224

Townsend ARM, Bastin J, Gould K, Brownlee GG (1986) Cytotoxic T lymphocytes recognize influenza haemagglutinin that lacks a signal sequence. Nature 324:575–577

Trowsdale J (1996) Molecular genetics of HLA class I and class II regions. In: Browning MJ, McMichael AJ (eds) HLA and MHC:genes, molecules and function. Bios, Oxford

Trowsdale J, Hanson I, Mockridge I, Beck S, Townsend A, Kelly A (1990) Sequences encoded in the class II region of the MHC related to the 'ABC' superfamily of transporters. Nature 348:741–744

Uebel S, Meyer TH, Kraas W, Kienle S, Jung G, Wiesmüller, Tampe R (1995) Requirements for peptide binding to the human transporter associated with antigen processing revealed by peptide scans and complex peptide libraries. J Biol Chem 270:18512–18516

Urban RG, Chicz RM, Lane WS, Strominger JL, Rehm A, Kenter MJH, Uytdehaag FGCM, Ploegh H, Uchanska-Ziegler B, Zieglar A (1994) A subset of HLA-B27 molecules contains peptides much longer than nonamers. Proc Natl Acad Sci USA 91:1534–1538

Ustrell V, Pratt G, Rechsteiner M (1995a) Effects of interferon and major histocompatibility complex-encoded subunits on peptidase activities of human multicatalytic proteases. Proc Natl Acad Sci USA 92:584–588

Ustrell V, Realini C, Pratt G, Rechsteiner M (1995b) Human lymphoblast and erythrocyte multicatalytic proteases: differential peptidase activities and responses to the 11S regulator. FEBS Lett 376:155–158

van Endert PM, Tampe R, Meyer TH, Tisch R, Bach J-F, McDevitt HO (1994a) A sequential model for peptide binding and transport by the transporters associated with antigen processing. Immunity 1:491–500

van Endert PM, Liblau RS, Patel SD, Fugger L, Lopez T, Pociot F, Nerup J, McDevitt HO (1994b) Major histocompatibility complex-encoded antigen processing gene polymorphism in IDDM. Diabetes 43:110–117

van Endert PM, Riganelli D, Greco G, Fleischhauer K, Sidney J, Sette A, Bach J-F (1995) The peptide binding motif for the human transporter associated with antigen processing. J Exp Med 182:1883–1895

Van Kaer L, Ashton-Rickardt PG, Ploegh HL, Tonegawa S (1992) TAP1 mutant mice are deficient in antigen presentation, surface class I molecules, and $CD4^{-}8^{+}$ T cells. Cell 71:1205–1214

Van Kaer L, Ashton-Rickardt PG, Eichelberger M, Gaczynska M, Nagashima N, Rock KL, Goldberg AL, Doherty PC, Tonegawa S (1994) Altered peptidase and viral-specific T cell response in *Lmp2* mutant mice. Immunity 1:533–541

Van Santen HM, Woolsey A, Ashton-Rickardt PG, van Kaer L, Baas EJ, Berns A, Tonegawa S, Ploegh H (1995) Increase in positive selection of $CD8^{+}$ T cells in TAP1 mutant mice by human β2-microglobulin transgene. J Exp Med 181:787–792

Varshavsky A (1996) The N-end rule: functions, mysteries, uses. Proc Natl Acad Sci USA 93:12142–12149

Villanueva MA, Fischer P, Feen K, Pamer EG (1994) Efficiency of MHC class I antigen processing, a quantitative analysis. Immunity 1:479–489

Wang K, Früh K, Peterson PA, Yang Y (1994) Nucleotide binding of the C-terminal domains of the major histocompatibility complex-encoded transporter expressed in *Drosophila melanogaster* cells. FEBS Lett 350:337–341

Wang P, Gyllner G, Kvist S (1996a) Selection and binding of peptides to human transporters associated with antigen processing and rat cim-a and -b. J Immunol 157:213–220

Wang P, Raynoschek C, Svensson K, Ljunggren H-G (1996b) Binding of H-2Kb-specific peptides to TAP and major histocompatibility complex class I in microsomes from wild type, TAP1, and β2-microglobulin mutant mice. J Biol Chem 271:24830–24835

Wang R-F, Parkhurst MR, Kawakami Y, Robbins PF, Rosenberg SA (1996c) Utilization of an alternative open reading frame of a normal gene in generating a novel human cancer antigen. J Exp Med 183:1131–1140

Wei ML, Cresswell P (1992) HLA-A2 molecules in an antigen-processing mutant cell contain signal sequence-derived peptides. Nature 356:443–446

Wenzel T, Eckerskorn C, Lottspeich F, Baumeister W (1994) Existence of a molecular ruler in proteasomes suggested by analysis of degradation products. FEBS Lett 349:205–209

White LC, Wright KL, Felix NJ, Ruffner H, Reis LFL, Pine R, Ting JP-Y (1996) Regulation of LMP2 and TAP1 genes by IRF-1 explains the paucity of CD8+ T cells in IRF-1-/- mice. Immunity 5:365–376

Wiertz EJ, Jones TR, Sun L, Bogyo M, Geuze H, Ploegh HL (1996a) The human cytomegalovirus US11 gene product dislocates MHC class I heavy chains from the endoplasmic reticulum to the cytosol. Cell 84:769–779

Wiertz EJ, Tortorella D, Bogyo M, Yu J, Mothes W, Jones TR, Rapoport TA, Ploegh HL (1996b) Sec61-mediated transfer of a membrane protein from the endoplasmic reticulum to the proteasome for destruction. Nature 384:432–438

Wipke BT, Jameson SC, Bevan MJ, Pamer EG (1993) Variable binding affinities of listeriolysin 0 peptides for the $H\text{-}2K^{d}$ class I molecule. Eur J Immunol 23:2005–2010

Wright KL, White LC, Kelly A, Beck S, Trowsdale J, Ting JP (1995) Coordinate regulation of the human TAP1 and LMP2 genes from a shared bidirectional promotor. J Exp Med 181:1459–1471

Yamazaki H, Tanaka M, Nagoya M, Fujimaki H, Sato K, Yago T, Nagata T, Minami M (1997) Epitope selection in major histocompatibility complex class I-mediated pathway is affected by the intracellular localization of an antigen. Eur J Immunol 27:347–353

Yang B, Braciale TJ (1995) Characteristics of ATP-dependent peptide transport in isolated microsomes. J Immunol 155:3889–3896

Yang B, Hahn YS, Hahn CS, Braciale TJ (1996) The requirements for proteasome activity in class I major histocompatibility complex antigen presentation is dictated by the length of preprocessed antigen. J Exp Med 183:1545–1552

Yewdell JW, Bennink JR, Hosaka Y (1988) Cells process exogenous proteins for recognition by cytotoxic T lymphocytes. Science 239:637–640

Yewdell JW, Antón LC, Bennink JR (1996) Defective ribosomal products (Drips), a major sources of antigenic peptides for MHC class I molecules. J Immunol 157:1823–1826

York IA, Rock KL (1996) Antigen processing and presentation by the class I major histocompatibility complex. Annu Rev Immunol 14:369–396

York IA, Roop C, Andrews DW, Riddell SR, Graham FL, Johnson DC (1994) A cytosolic herpes simplex virus protein inhibits antigen presentation to $CD8^+$ T lymphocytes. Cell 77:525–535

Yoshimura T, Kameyama K, Takagi T, Ikai A, Tokunaga F, Koide T, Tanahashi N, Tamura T, Cejka Z, Baumeister W, Tanaka K, Ichihara A (1993) Molecular characterization of the 26S proteasome complex from rat liver. J Struct Biol 111:200–211

Zanelli E, Zhou P, Cao H, Smart MK, David CS (1995) Genetic polymorphism of the mouse major histocompatibility complex-associated proteasome subunit *Lmp7*. Immunogenetics 41:251–254

Zhou N, Glass R, Momburg F, Hämmerling GJ, Jondal M, Ljunggren H-G (1993a) TAP2-defective RMA-S cells present Sendai virus antigen to cytotoxic T lymphocytes. Eur J Immunol 23:1796–1801

Zhou P, Cao H, Smart M, Davis C (1993b) Molecular basis of genetic polymorphism in major histocompatibility complex-linked proteasome gene (*Lmp-2*). Proc Natl Acad Sci USA 90:2681–2684

Zijlstra M, Bix M, Simister NE, Loring JM, Raulet DH, Jaenisch R (1990) β2-Microglobulin deficient mice lack $CD4^-8^+$ cytolytic T cells. Nature 344:742–746

Zwickl P, Kleinz J, Baumeister W (1994) Critical elements in proteasome assembly. Nature Struct Biol 1:765–770

Molecular Requirements for Assembly and Intracellular Transport of Class I Major Histocompatibility Complex Molecules

ANNE M. FOURIE and YOUNG YANG

1 Introduction

The major function of the immune system is to detect and eliminate invading pathogens from the cellular host (PAUL 1993). The presentation of peptides derived from intracellular antigens by class I major histocompatibility complex (MHC) molecules plays a central role in the cellular immune response, since specialized immune $CD8^+$ T lymphocytes recognize non-self peptides bound to class I molecules on the surface of infected cells (YANG et al 1996). Thus class I molecules are peptide-binding proteins that provide an extracellular representation of intracellular

The R. W. Johnson Pharmaceutical Research Institute, 3535 General Atomics Ct., Suite 100, San Diego, CA 92121, USA antigen content for inspection by the $CD8^+$ T lymphocytes through interactions of their T cell receptors (TCR) with the class I MHC-peptide complex.

Class I MHC antigen presentation is the first line of defense against viral infections and malignant transformations. Since the assembly of the trimeric class I complex of a transmembrane heavy chain, a soluble light-chain β_2-microglobulin, and a peptide of eight to ten amino acids takes place in the endoplasmic reticulum (ER), distinct biochemistry and cell biology strategies for antigenic peptide acquisition by class I molecules have evolved. The peptides are generated in the cytosol by the proteasome, a multisubunit and multicatalytic protease complex (COUX et al 1996), from intracellular antigens, either self or non-self (e.g., viral antigens), which are ubiquitinated by the ubiquitination system (HOCHSTRASSER 1996a,b). The peptides of eight to ten amino acids are then translocated across the ER membrane by the transporter associated with antigen processing (TAP) (VAN ENDERT 1996), where they bind class I molecules, stabilizing the class I heterodimers. Only after the assembly of the trimolecular complex can these complexes be transported to the cell surface, via the exocytic pathway, for inspection by $CD8^+$ cytotoxic T lymphocytes. This chapter describes our current molecular understanding of the mechanisms of antigen processing and presentation by class I molecules, with particular emphasis on the requirements for the assembly and intracellular transport of class I molecules, and the nature of the peptides they present.

2 Molecular Structure of Class I MHC Molecules

The MHC has been extensively characterized in mouse and humans and has been mapped to chromosomes 6 and 17, respectively (PAUL 1993). Three major gene families, designated as class I, class II, and class III, have been identified in the MHC region. The human class I gene family consists of a minimum of 36 open reading frames (CAMPBELL et al. 1993; KRISHNAN et al. 1995), including the well-characterized HLA-A, -B, and -C heavy chains, which are currently known to have over 250 alleles on the basis of nucleotide sequence (PARHAM et al. 1996). The class I heavy chains with a molecular mass of approximately 45 kDa are expressed as transmembrane glycoproteins on the surface of almost all nucleated cells in noncovalent association with β_2-microglobulin, a non-MHC-encoded light chain with a molecular mass of approximately 12 kDa (YANG et al. 1996). The completion of class I MHC assembly and its transport to the cell surface requires a third component, the antigenic peptide of approximately eight to ten amino acids (BJORKMAN et al. 1987).

The molecular organization of the genes for class I heavy chains reflects their domain structure, with a separate exon encoding each domain (PAUL 1993). A transmembrane domain of approximately 25 amino acids connects three N-terminal extracellular domains of approximately 90 amino acids each, designated α_1, α_2, and α_3, to a cytoplasmic tail of approximately 30 amino acids at the carboxyl end. The α_3-domain of the class I heavy chain has an immunoglobulin-like structure and displays a high degree of amino acid sequence homology to the constant region of

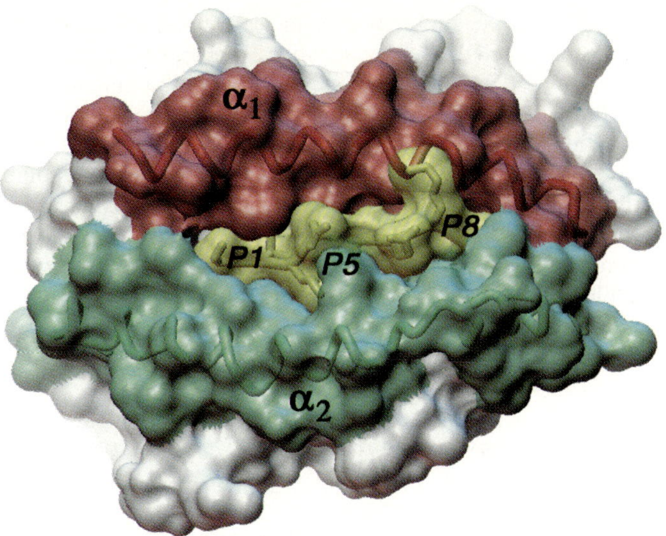

Fig. 1. Molecular model of a class I–peptide complex presenting an octamer peptide. The OVA-8 peptide (*P1–P8*) complexed with the class I molecule H-2Kb is shown in *yellow*, while the α_1 and α_2 domains of the heavy chain are shown in *red* and *green*, respectively

immunoglobulin molecules (PAUL 1993). As revealed from the three-dimensional crystal structures of class I molecules (YOUNG et al. 1995), the α_1- and α_2-domains form a β-sheet platform spanned by two α-helices that together form a peptide-binding pocket (Fig. 1) and also make contacts with both the α_3-domain and β_2-microglobulin.

The peptide-binding groove of the class I molecule tethers the amino- and carboxyl-termini of peptides through a network of hydrogen bonds between several conserved residues at each end of the groove and the peptide (YOUNG et al. 1995). The binding of peptides to class I molecules is mainly dependent on the backbone of the peptide, with the exception of one or two anchor residues that fit into structurally complementary pockets of the groove. Since almost all the polymorphic amino acid residues within the family of class I molecules are clustered around the peptide-binding groove, the shape of the groove and the spectrum of peptides bound are accordingly altered for each particular class I allele.

3 Characteristics of the Antigenic Peptides Bound by Class I MHC Molecules

As described above, the bound peptide forms an integral part of the final class I quarternary structure. Furthermore, the peptide represents the element that, in the context of a particular class I molecule, elicits an effector response upon recognition by the antigen-specific receptors on $CD8^+$ T lymphocytes. To elucidate this process, detailed sequence analysis of class I MHC-binding peptides within protein antigens is a prerequisite for predicting T cell epitopes in uncharacterized antigens. Extensive progress towards achieving this goal has been made in recent years. A list of all known MHC ligands (RAMMENSEE 1996; RAMMENSEE et al. 1993, 1995), as well as other information on peptide binding specificity to class I molecules, has been compiled, and a number of accessible databases and internet web sites have been established (e.g., MHCPEP: http://wehih.wehi.edu.au/mhcpep/; Parker/BIMAS: http://www_bimas.dcrt.hih.gov/molbio/hla_bind/hla_motif_search_info.html; Epimatrix: http: //www.brown.edu/Research/TB-HIV_Lab/Epimatrix/epimatrix.html; Immunogenetics HLA Database at EBI: http://www.ebi.ac.uk/imgt/docs/HLA-DB.html).

3.1 Motifs of Endogenous Peptides Bound by Class I MHC Molecules

The earliest approach to the identification of class I MHC-associated peptides was acid extraction of peptides from whole cells, followed by fractionation using gel filtration and high-performance liquid chromatography (HPLC), and testing of individual fractions for recognition by class I MHC-restricted T cells (ROETZSCHKE et al. 1990). This procedure was limited to the identification of peptides for which specific $CD8^+$ cytotoxic T cells were available. A significant breakthrough in characterization of class I MHC-binding peptide motifs was the approach of pooled sequencing of peptides eluted from purified class I molecules by Edman degradation (FALK et al. 1991) or separation of individual eluted peptides by HPLC (VAN BLEEK et al. 1990) and their identification by mass spectrometry (HUNT et al. 1992).

The results of these analytical techniques indicated that the peptides presented by class I molecules follow stringent rules which are different for each individual class I allele (see example of murine class I Kb molecule in Table 1). The peptide length was usually restricted to eight to ten amino acids (FALK et al. 1991), because the peptide-binding pocket of the class I molecule is closed at both ends, and the N- and C-termini of the peptides contribute significantly to the binding affinity (MATSUMURA et al. 1992a). The allele-specific interactions are characterized by so-called anchor positions, which result from anchoring peptide side chains in complementary pockets of the class I peptide-binding groove. The type of anchor residue or residues required for binding is determined by the nature of the side chains in the binding pocket of a particular class I allele-binding groove (MADDEN et al. 1992). The sum of allele-specific peptide-class I MHC interaction requirements defines the binding motif for a given class I molecule, characterized by the number, spacing, and specificity of anchors, and also more degenerate preferences at non-anchor positions within the peptide. Most of the peptide motifs determined by pooled sequencing of peptides naturally presented by class I molecules have anchor residues in position P2, P3, or P5/P7 (MADDEN et al. 1992; RAMMENSEE et al. 1995; RAMMENSEE 1996) in addition to a C-terminal anchor which can be hydrophobic or charged depending on the particular class I allele.

Table 1. Predictions for peptide binding motif of murine class I Kb molecule

Type of motif	Analysis method	Motif information	Position P1	P2	P3	P4	P5	P6	P7	P8
Simple (FALK et al. 1991)	Sequencing of endogenous peptides eluted from class I	Dominant anchors	–	–	–	–	F, Y	–	–	L
		Strong anchors	–	–	Y	–	–	–	–	M
Expanded (UDUKA 1995)	In vitro peptide-binding studies using octapeptide libraries	Favored residues	I	V, S, T, N	Y, F	–	Y	–	P, S, N, Q, H, R	L, V, M, I
		Disfavored residues	P, D	D, E, K	G, A, D, E, K	–	D, E, K, G, I, M, N, Q, R	W, D, E	W, D	N, G, P, Y, W, D, K, R, E, S
Complex (BRUSIC et al. 1994)	Artificial neural network prediction	Sequence prediction not based on motif	X	X	X	X	X	X	X	X

3.2 In Vitro Synthetic Peptide-Binding Studies

To gain insight into the complexity of class I MHC-peptide binding specificity, synthetic peptide libraries and combinatorial peptide libraries were used in class I peptide-binding assays. The peptide libraries utilized typically contain peptides of constant length but heterogeneous sequences, but can in some cases contain distinct sequences or a fixed amino acid at a certain position with a mixture of amino acids at other positions. The influence of amino acids other than the anchor residues has been studied using libraries of synthetic peptides in quantitative assays of binding to the human class I HLA-A2.1 molecule (RUPPERT et al. 1993). In addition to confirming the importance of major anchor residues at position 2 (with preference for L or M) and position 9 or 10 (with preferences for V, I, or L), this study also showed that amino acids in other positions had a significant impact on binding affinity for HLA-A2.1. When nonamer and decamer peptide libraries (based on viral and tumor cytotoxic T lymphocyte epitopes) were used, containing L or M in position 2 and V, I ,or L at the C terminus (position 9 or 10), it was found that only 34% of the nonamer peptides (K_d, t n*M*) and only 23% of the decamer peptides bound with high affinity. The high percentage of weak and non-binding peptides in the libraries indicated that factors other than size and anchor residues are critical in determining the affinity of a peptide for class I MHC binding. Prominent roles for the amino acids in positions 1, 3, and 7, corresponding to "secondary" pockets in HLA-A2.1 predicted from X-ray crystallography (SAPER et al. 1991), were also demonstrated. The resulting "expanded" motif increased predictability of A2.1-binding epitopes to 70% as compared to 20%–30% based on a simple peptide motif defined only by length and anchor residues.

Based on the assumption that the binding of a peptide could be interpreted as the sum of the independent contributions of each residue position and the total score would be proportional to the binding energy of the peptide, a panel of nonamer peptides of unique sequence was used in which each peptide differed by only one amino acid from at least one other peptide in order to determine the contributions of each position in a given peptide to binding to HLA-A2.1 (PARKER et al. 1994). It was found that there was sequence-dependent variability of the peptide dissociation rate that could not be predicted from the individual scores, indicating interdependent interaction of amino acids within the peptide. However, the known A2.1-binding peptides were in the top 2% of predicted binders in each case, suggesting that the individual side chains bind somewhat independently of one another.

Similarly, the binding requirements for the murine class I Kb molecule have been elucidated using peptide libraries (Table 1; UDAKA et al. 1995a). This study showed that all amino acid positions in Kb-restricted octapeptides exhibited positive or negative effects on Kb-peptide binding as well as recognition of the resulting complexes by cytotoxic T cells. In addition, a strong interdependence was observed for the effects of individual residues in the epitope-derived peptides. To assess the contribution of each amino acid relative to a mixture of amino acids at each position and to eliminate interdependent interactions, octamer peptide libraries were used with one amino acid fixed at a particular position, while the other positions had all amino acids except cysteine equally represented (UDAKA et al. 1995b). It was concluded that, although major anchor residues have a primary influence on peptide binding to class I molecules, all amino acids in the peptide contribute to the overall binding affinity.

3.3 Screening of Bacteriophage Peptide Display Libraries

The peptide motifs of class II MHC molecules have been determined using peptide libraries displayed on bacteriophage (HAMMER et al. 1992). However, free amino and carboxyl termini provide substantial binding energy for peptide-class I MHC interactions and are normally required for stable binding, making phage peptide display less suitable for analysis of class I MHC-binding peptides. Nevertheless, the N-terminal residue appears more critical for binding to class I molecules than the C terminus, and additional residues at the C terminus are less disruptive for binding (MATSUMURA et al. 1992b). Therefore, a bacteriophage peptide display library of random 15-amino acid peptides expressed at the N terminus of gene VIII (SMITH et al. 1995) of M13 phage has been used to identify sequences of class I MHC-binding peptides (M. Milik et al., unpublished data). As expected, most of the "strong" binders contained the anchor residues reported by previous studies. In addition, the information obtained from binding of random peptides displayed on phages to class I molecules is being used as a database for "training" a neural network (see Sect. 3.4) to predict class I MHC-binding peptides within uncharacterized antigens.

3.4 Algorithms and Artificial Neural Networks for Prediction of Class I MHC-Binding Peptides

Prediction of class I MHC-binding peptides based only on a simple motif defined by length and primary anchor residues is only effective in approximately 30% of cases (RUPPERT et al. 1993). Expanding the prediction to include contributions by secondary anchors can only potentially increase the predictability up to 70% (RUPPERT et al. 1993). It became clear that utilization of all the available information generated from detailed analysis of peptide binding to class I molecules is essential to make correct predictions of class I MHC-binding peptides, because all positions in a peptide contribute either positively or negatively to the overall binding affinity. Thus, given a database of peptides that bind with known affinity to a class I molecule, the relative frequency of each amino acid at each position can be deduced and programmed in a matrix. Such a matrix (named "Epimatrix") has been developed, based on a database of peptides derived from sequencing of peptides eluted from class I molecules (JESDALE et al. 1996). Comparing the predictions for class I MHC-binding peptides within the sequences of 67 proteins with those for the known class I ligands (RAMMENSEE 1995, 1996; RAMMENSEE et al. 1993, 1995), Epimatrix correctly identified the published MHC ligand within the top two scoring peptides in 11 of 13 cases or in 55 of 67 cases. Thus Epimatrix permitted the selection of 67 class I MHC-binding peptides from over 33 000 possible decapeptides.

Mathematical matrix-based algorithms such as Epimatrix are based solely on independent side chain values, whereas artificial neural networks can capture subtle relationships within binding data and generalize them in order to classify nonlinearly separable data. Thus artificial neural networks are able to predict sequences which have a certain quality, such as high-affinity binding to class I molecules. Such a neural network, trained on phage display results, has been employed to predict the class I MHC-binding peptides with very good correlations between the immunodominant epitopes and the high-affinity peptides (M. Milik et al. unpublished data). In addition, prediction by the neural network was accurate in approximately 70% of cases. Similarly, predictions of peptide binding for both HLA-A2 and H2-Kb have been reported using another artificial neural network "trained" with published data drawn from entries in the MHCPEP database (Brusic et al. 1996). Overall, the predictive ability using the artificial neural network was 78% for HLA-A2.1 and 88% for H-2Kb. Artificial neural network methodology has many advantages over other approaches for predicting class I MHC-binding peptides, particularly as new data can easily be incorporated. In addition, the flexibility of artificial neural networks allows different applications of the same data, e.g., for prediction of all possible binders for vaccine design and prediction of high-affinity peptides for drug design (Brusic et al. 1994). Thus, at present, artificial neural networks appear to be the method of choice for predicting class I MHC-binding peptides.

4 Generation and Translocation of Antigenic Peptides

4.1 Peptide Generator: The Proteasome

In the ER of the antigen-presenting cells, class I molecules acquire peptides that are derived from intracellular proteins (Yang et al. 1996). These peptides are presumably generated by cytosolic proteases. The first indication that the proteasome, the major nonlysosomal protease in cells (Orlowski 1993), was responsible for class I MHC antigen degradation came from studies on ubiquitin-mediated protein degradation. It was found that ubiquitination increased the efficiency of antigenic peptide presentation (Townsend et al. 1988) and that cells defective in an ubiquitin-activating enzyme E1 were incapable of presenting antigens by class I molecules (Michaelek et al. 1993). Since ubiquitinated protein substrates are degraded by proteasomes (Coux et al. 1996), it was concluded that ubiquitin- and proteasome-mediated proteolysis plays a crucial role in class I MHC antigen processing. The discovery that two interferon (IFN)-inducible and MHC-encoded gene products, LMP2 and LMP7 (Fruh et al. 1992), were indeed integral subunits of the proteasome (Fruh et al. 1994), in conjunction with the observation that inhibitors of the proteasome blocked class I MHC antigen presentation (Vinitsky et al. 1994), placed the proteasome at the center of the mechanism for processing antigen for presentation by class I molecules.

The proteasome, a multisubunit and multicatalytic proteinase complex found in eukaryotes, prokaryotes, and archaeobacteria, is an essential part of the proteolytic machinery responsible for ubiquitin-mediated protein degradation (RECHSTEINER et al. 1993). The proteasome has a molecular mass of approximately 700 kDa and consists of a highly conserved cylindrical structure with sevenfold symmetry. X-ray crystallographic analysis of the proteasome from the archaeobacterium *Thermoplasma acidophilum* (LÖWE et al. 1995) revealed that the particle consists of two outer rings of α-subunits and two inner rings of β-subunits with a stoichiometry of 7α7β7β7α. By contrast, the proteasome of eukaryotes is composed of a family of 14 different, but homologous subunits of molecular masses between 20 and 35 kDa (FRUH et al. 1992, 1994) which can be classified into α-type and β-type subunits based on their homology with the α- and β-subunits of the *T. acidophilum* proteasome (PETERS et al. 1993). The highly conserved α-type subunits are thought to be involved in proteasome structure and are proteolytically inactive, whereas β-type subunits are more divergent in their primary structure and appear to be directly involved in enzymatic activity (SEEMÜLLER et al. 1995). A threonine residue crucial for the structure and activity of the proteasome catalytic sites (SEEMÜLLER et al. 1996) is only exposed after cleavage of the amino-terminal portion of six β-type subunits (SEEMÜLLER et al. 1996), including IFN-inducible LMP (SEEMÜLLER et al. 1995) and MECL-1 (GROETTRUP et al. 1996a). Since all the catalytic threonine-containing subunits are affected by IFN treatment, it seems that the whole catalytic spectrum of the proteasome is modified by this process.

Upon IFN upregulation, such IFN-inducible β-type subunits become integral subunits of newly assembled proteasomes (FRUH et al. 1997). For example, during this process of assembly, both LMP2 and LMP7 are autocatalytically processed and displace the housekeeping β-type subunits 2 and 10 (YANG et al. 1992b)(or delta and MB-1 (BELICH et al. 1994), respectively. Thus IFN induction permanently changes the proteasomal subunit organization (YANG et al. 1996). Biochemical studies with proteasome inhibitors (ROCK et al. 1994; VINITSKY et al. 1994) have provided evidence that the proteasome is responsible for the generation of class I MHC-binding peptides, since the inhibitors prevent antigen presentation by class I molecules. Indeed, in vitro enzymatic studies of isolated proteasomes have demonstrated that the altered molecular organization of the proteasome induced by IFN is responsible for its functional changes in the catalytic activity and ultimately in antigen processing (DRISCOLL et al. 1993; AKI et al. 1994; GACZYNSKA et al. 1993, 1994, 1996). Moreover, in vivo evidence obtained from analysis of LMP2 or LMP7 knockout mice (FEHLING et al. 1994; VAN KAER et al. 1994) indicates that IFN-induced proteasome subunits play a major role in proteasome-mediated antigen processing, since both LMP2 and LMP7 knockout mice are deficient in the generation of a subset of antigenic peptides. Thus proteasome subunit exchange could be a fundamental mechanism for modulating proteasome activities by cytokines during immune responses (FRUH et al. 1997).

In eukaryotes, proteasome activities are modulated by specific regulatory proteins that form complexes with proteasomes (YANG et al. 1996). Two regulatory complexes, the ATPase complex and PA28, have been studied to some extent. The ATPase complex associates with the 20S proteasome in an ATP-dependent manner, resulting in the 26S proteasome (RECHSTEINER et al. 1993). This 26S proteasome is involved in the degradation of protein substrates in a ubiquitin-dependent manner

(RECHSTEINER et al. 1993). The proteasome regulator PA28 has been shown to associate with the 20S proteasome in vitro (CHU-PING et al. 1992, 1993) and in vivo (YANG et al. 1995) in an ATP-independent manner. Association of these regulatory complexes appears to be reversible and regulated by phosphorylation (YANG et al. 1995). It is conceivable that evolutionary divergence of these regulatory complexes is coupled with their functional specializations and that regulatory mechanisms exist that render antigenic peptides more likely to become available to class I molecules during immune responses. Antigen degradation could occur in two steps, namely, initial degradation of whole antigen into intermediate-sized fragments by 26S proteasomal complexes followed by degradation of these fragments by the PA28–20S proteasomal complexes to produce peptides of eight to ten residues in length (YANG et al. 1996). Indeed, in vitro kinetic studies on the influence of PA28 on peptide cleavage and specificity of the proteasome (CHU-PING et al. 1992) indicate that PA28 changes the cleavage behavior of the proteasome in a characteristic, qualitative and quantitative manner. In the absence of PA28, the proteasome digests substrates by consecutive and independent single cleavages. Upon association with PA28, products generated by two flanking cleavages appear immediately as main products, while the generation of single-cleavage products is strongly reduced. Since this PA28-induced, coordinated double-cleavage mechanism appears to optimize the generation of dominant T cell epitopes (GROETTRUP et al. 1996b), the regulation of PA28 expression by IFN plays an essential role in proteasome-mediated antigen processing (YANG et al. 1996).

4.2 Transporter Associated with Antigen Processing

Within the ER, nascent class I molecules acquire peptides derived from cytosolic proteins that are degraded by the proteasome (YANG et al. 1996). Thus it was obvious that a translocation machinery is required to transfer peptides from the cytoplasm into the lumen of the ER (TOWNSEND et al. 1989). The first evidence for one or more transporters involved in the class I MHC-binding peptide translocation came from the observation that the two MHC-encoded genes, termed transporters associated with antigen processing (TAP), displayed a high homology with a family of genes encoding proteins known as traffic ATPases or ATP-binding cassette transporters (SPIES et al. 1990), suggesting that TAP function as ATP-dependent transporters (YANG et al. 1996). Furthermore, several lines of evidence suggested that TAP is the transporter that supplies peptides to class I molecules: (a) the expression of TAP1 and TAP2 is strongly enhanced by IFN (YANG et al. 1992a); (b) electron microscope data show that the heterodimer of TAP1 and TAP2 has a cellular localization restricted to the membranes of ER and the pre-Golgi compartments (KLEIJMER et al. 1992); (c) it has been observed that the cell surface expression of class I molecules is abolished in TAP-deficient cells (LJUNGGREN et al. 1990).

Indeed, it has been demonstrated that both the full-length TAP proteins and the soluble carboxyl-terminal domains of TAP1 or TAP2 bind ATP (WANG et al. 1994). The relative potencies of the nucleotides in preventing azido-ATP binding to TAP are in the order of ATP>GTP>CTP>ITP>UTP for both TAP1 and TAP2 (WANG et al. 1994), suggesting that ATP is indeed the natural substrate for the TAP. Direct evidence that TAP translocate peptides in an ATP-dependent manner was provided by analysis of peptide import into microsomes or into the ER of streptolysin O-permeabilized cells (ANDROLEWICZ et al. 1993). In this TAP-mediated peptide transport assay, it was found that, in the presence of a nonhydrolyzable ATP analogue, no TAP activity could be detected and that TAP displayed a significant preference for peptides of eight to 13 amino acids in the presence of ATP (MOMBURG et al. 1994a; SCHUMACHER et al. 1994; HEEMELS et al. 1995). The upper end of this size limit seems to be less stringent than the lower limit, but generally transport rates gradually decrease with peptide length over 13 amino acid residues (ANDROLEWICZ et al. 1994). This length specificity correlates well with the known length requirements for peptide binding to class I molecules (HEEMELS et al. 1995). In addition to the length restriction, TAP appears to display some sequence specificity in the peptides transported. Human and rat TAP2a prefer both hydrophobic and basic amino acid residues at the carboxyl-terminal position, whereas mouse and rat TAP2u are more selective for peptides with hydrophobic carboxyl termini (MOMBURG et al. 1994a,c; SCHUMACHER et al. 1994; HEEMELS et al. 1995). These specializations match the different peptide-binding specificities of the class I molecules of human, mouse, and the two rat haplotypes. The contribution of other positions to the quality of peptide binding and/or transport has been discussed in a number of reports (DAVENPORT et al. 1996), but its biological significance remains to be addressed. Interestingly, artificial neural networks, similar to those used for class I MHC-binding peptides (see Sect. 3.4), have recently been used to predict peptide-binding affinity for TAP. Very good correlation between predicted and experimentally determined binding affinities was obtained (V. Brusic and P. Van Endert, unpublished data)

Very little is known about the mechanism by which peptides are translocated by TAP except for the peptide-binding and ATP-hydrolysis requirements. While ATP binding, but not its hydrolysis, has been demonstrated for TAP1 or TAP2 separately (WANG et al. 1994), peptide binding requires the presence of a TAP1/2 heterodimer (VAN ENDERT 1996). In addition, it has been shown that ATP hydrolysis is not required for peptide binding (MÜLLER et al. 1994; RUSS et al. 1995). The observation that the peptide binding step is an ATP-independent process led to a model in which TAP initially exists as an "inactive" complex with high substrate affinities, but low affinity for ATP (VAN ENDERT et al. 1994). It is likely that peptide binding induces a conformational change in TAP which increases the affinity for ATP. ATP binding and hydrolysis may then result in a conformational change in TAP that facilitates transfer of the bound peptide from the binding site to the translocation site or sites. The peptide could then be released and translocated across the ER membrane by comsuming the energy provided by ATP hydrolysis. This model, however, is debatable, since others report that ATP binding is not correlated with any changes in peptide affinity (UEBEL et al. 1995).

TAP seems to have coevolved with the class I molecules to provide the latter with the best-fitting peptides, since TAP and class I molecules physically interact during the peptide-loading process (see Sect. 4.2) and since TAP transports peptides of essentially eight to 13 amino acids (MOMBURG et al. 1996; VAN ENDERT 1996), whereas the class I peptide groove principally accommodates peptides of eight to ten amino acids. It is estimated that TAP-dependent peptide transport is very efficient, as approximately 20 000 peptides are translocated every minute per cell, far more than are needed to supply class I molecules, which have a much lower rate of synthesis (about 100 molecules per minute) (MOMBURG et al. 1994a). In addition, the K_m of peptide transport for TAP has been estimated to be as high as 661 n*M*, with a maximum velocity of 2.9 fmol/min per μg microsomal protein (YANG and BRACIALE 1995). Thus it is reasonable to assume that overexpression of a non-self protein, e.g., a viral antigen expressed during viral infection, would lead to rapid and preferential display of the resulting antigenic peptides on most class I molecules.

4.3 Rules for Peptides Presentable In Vivo by Class I Molecules and Biological Relevance

An efficient immune response requires that a given class I molecule be capable of binding a very large number and variety of peptides for presentation to $CD8^+$ T lymphocytes, since many infectious agents evolve rapidly (PAUL 1993). As a result, a great number of different peptides may bind to a particular class I molecule with different affinities, consistent with the estimation that a given class I molecule is able to bind over 10 000 different peptides (ENGELHARD et al. 1993). While a given class I molecule is capable of binding equally well to many peptides of distinct primary sequence, a peptide shown to bind to a particular class I molecule can often associate with many other, if not all, class I molecules in a manner which is quantitatively different and generally isotype specific. Moreover, a family of no more than six different class I molecules is expressed in the antigen-presenting cell of a given individual (PAUL 1993), each member having a different peptide-binding specificity.

Elution and sequencing of naturally processed peptides presented in vivo by class I molecules have revealed that the peptide length is typically eight to ten amino acids, which is in good agreement with the length requirements for peptide transport by TAP and the sizes of proteolytic products generated by the proteasome. The requirement for the C-terminal anchor to be either hydrophobic or charged may reflect the specificity of antigen processing and peptide transport rather than class I MHC-binding specificity. As discussed in Sect. 4.1, proteasomes cleave preferentially after hydrophobic or charged residues (DRISCOLL et al. 1993; GACZYNSKA et al. 1993;BOES etal. 1994; FEHLING et al. 1994) and would thus generate peptides with C termini corresponding to the requirements for class I MHC-binding. The C-terminal amino acid also influences the efficiency of peptide transport by TAP (MOMBURG et al. 1994a–c; SCHUMACHER et al. 1994). It seems that the selection of peptides presented by class I molecules is under strict stepwise restrictions from proteasomes, TAP, and class I molecules, with each step further selecting a limited subset of peptides for class I molecules. Since the degree of polymorphism in the LMP and TAP genes is far more limited than the variability in different class I MHC peptide-binding grooves, their specificity for peptide sequences should accordingly be less stringent than the related class I MHC peptide anchor residue requirements. Thus the major role in shaping the immune response appears to be contributed by class I MHC-binding specificity with some influence of the selectivity of the LMP proteasome or TAP superimposed.

Apart from the known major and secondary anchor preferences, certain amino acid residues appear to be favored or disfavored at certain positions in vivo. Some closely related class I molecules exhibit fine specificity differences, e.g., HLA-A31- and HLA-A33-bound peptides have similar anchors but different preferences at P1 position (FALK et al. 1994). These fine differences may be related to the linkage of disease susceptibility to different alleles (FALK et al. 1995a–c). In addition, the preferences for non-anchor residues in natural ligands may also be related to specificity of antigen processing in vivo, such as the proteolytic specificity of the proteasome, TAP specificity, the peptide shuttling specificity of chaperones (FALK et al. 1995a–c; SRIVASTAVA 1993, 1996; SRIVASTAVA et al. 1994), or the specificity of the interaction with the TCR (DAVIS et al. 1993, 1996).

The peptide structure in the context of a particular class I molecule seems to be the primary focus of T cell recognition, and TCR dedicated to this task have evolved in higher eukaryotes. Through the interaction between the TCR and the class I MHC-peptide complex, T cells are able to sense fine differences in ligands (UDAKA 1996), although most of the peptide (66%) is buried in the binding groove and hence contributes little to the surface exposed for recognition by the TCR (FREMONT et al. 1992). Single substitutions in exposed peptide residues can be sufficient to abolish T cell recognition, even though the alteration may not significantly influence class I MHC-peptide binding (MARYANSKI et al. 1990). Studies using a complete replacement set of peptide analogues for T cell antigens have identified peptide residues that interact with the class I molecule and/or the TCR and have defined the biochemical character of class I and TCR contact residues (reviewed in JORGENSEN et al. 1992; DAVIS et al. 1996). Residues 1–3, 5, and 8 should only indirectly affect interaction with the T cell, while amino acids in position 7, 4, 6, and possibly 2 should contribute most to interaction between the class I MHC-peptide complex and the TCR. Both main-chain and side-chain variations in conformation of the class I molecule take place upon peptide binding (FREMONT et al. 1992), with the most profound changes being in the region of interaction with the TCR. Thus even minor alterations in class I conformation may be sufficient to prevent association with the TCR. The interaction of the TCR with the class I MHC-peptide complex is therefore exquisitely sensitive to peptide sequence differences, either directly or indirectly through conformational changes in the class I molecule to which it is bound. Thus the selectivity of the $CD8^+$ T lymphocytes appears to result from the fine specificity with which TCR are engaged to recognize peptides presented in the context of class I molecules. The crystallographic structure of the TCR-class I MHC complex (Fig. 2) clearly shows that the TCR interacts though its hypervariable loops with the class I molecule bound to a peptide and also with the side chains and some of the backbone of the peptide (GARCIA et al. 1996), confirming that the bound peptide is an important determinant in TCR recognition.

For prediction of T cell epitopes, most previous work has focused on binding motifs for class I molecules. However, the highest-affinity peptides for class I MHC binding may not necessarily represent the most favorable T cell epitopes, taking into account the specificity of the proteasome, TAP transport of peptides, and/or other

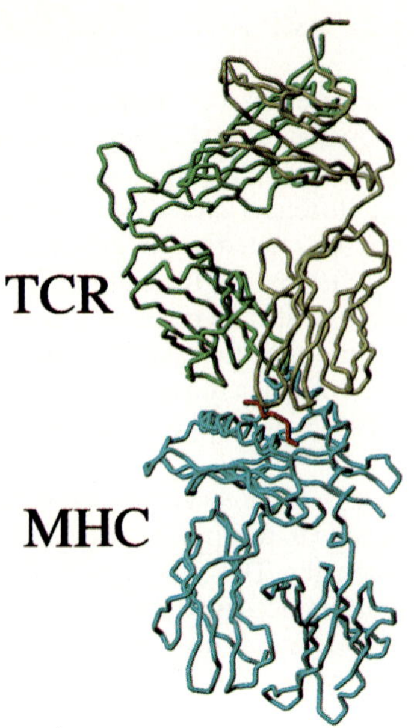

Fig. 2. Crystal structure of the oriented T cell receptor (TCR)–class I MHC–peptide complex. A backbone tube representation of a TCR (*top*, α-chain in *yellowish green* and β-chain in *green*), showing its orientation bound to a class I MHC molecule (*bottom*, *blue*; β_2-microglobulin is on the left) presenting an octamer peptide (*middle*, *red*) with the P1 residue positioned forward on the right. The TCR covers the class I binding groove, so that the Vα and Vβ complementarity-determining regions 1 and 2 are positioned over the amino-terminal region and the carboxyl-terminal region of the bound peptide, respectively. In addition, the Vα and Vβ complementarity-determining region 3 straddles the peptide between the helices around the central position of the peptide (see GARCIA et al. 1996 for details)

accessory molecules as well as the specificity of interaction with the TCR. The most relevant approach is to make use of the rapidly growing database of naturally presented peptides as a basis for an artificial neural network to predict T cell epitopes. Endogenously presented peptides have already been selected by all the specificity pressures within the cell, and this information can be captured by a neural network without the necessity of defining the exact rules of specificity at each step. Understanding the particular specificity of each particular class I allele is crucial in studying linkage of certain MHC alleles to particular diseases or autoimmunity.

5 Transport and Assembly of Class I MHC-Peptide Complexes

Intracellular transport and assembly of the class I molecules is a critical process in the antigen presentation pathway, since the overall objective is to express as many functional class I MHC-peptide complexes at the cell surface as possible. Thus assembly of class I molecules is under strict quality control, and several discrete steps have evolved to ensure that incompletely assembled class I molecules are not allowed to exit the ER (PLOEGH 1995). In vivo data demonstrate that assembly of class I molecules proceeds by the initial physical association of heavy chain-β_2-microglobulin heterodimers, followed by their subsequent interaction with TAP heterodimers

(BERGERON et al. 1994). Once heavy chain-β2-microglobulin-TAP complexes are formed, associated with additional proteins such as tapasin (see Sect. 5.2), peptides are then loaded and functional class I MHC-peptide complexes exit the ER (Fig. 3).

5.1 Endoplasmic Reticulum System of Molecular Chaperones BiP, Calnexin, and Calreticulin

Initial folding and assembly of the nascent class I heavy chain and β2-microglobulin appears to be orchestrated by the molecular chaperones of the ER, including BiP, calnexin, and calreticulin (RAJAGOPALAN 1996); Fig. 3). These chaperones prevent the formation of class I heavy chain protein aggregates by interacting with monoglucosylated N-linked oligosaccharides and/or exposed hydrophobic surfaces of the incompletely assembled class I proteins (DEGEN et al. 1992). A model for an ordered pathway for the chaperone-assisted folding of newly synthesized class I proteins has been proposed (YANG et al. 1996). Apart from being essential for the initial translocation of nascent heavy chain and β2-microglobulin into the ER, BiP, an ER resident member of the Hsp70 family, interacts in multiple ATP-controlled binding-release cycles with extended nascent class I polypeptides (HEBERT et al. 1995). During the folding cycles, BiP retains incompletely assembled class I proteins in the ER, presumably by binding to partially folded, unassembled, or misfolded class I heavy chains. However, an active role for BiP during the folding of class I molecules has not been demonstrated. Subsequently, calnexin, a type I ER membrane protein with a C-terminal RKxRRx ER retrieval sequence, interacts transiently with class I heavy chains (DEGEN et al. 1991, 1992; BERGERON et al. 1994; VASSILAKOS et al. 1996). The association of class I heavy chain and calnexin promotes binding of β2-microglobulin to the folded heavy chain and retains free class I heavy chains in the ER. However, studies in calnexin-deficient cells (SCOTT et al. 1995) indicate that there is no absolute requirement for calnexin in class I assembly. The proper interaction of heavy chain, β2-microglobulin, and peptide is further facilitated by a third chaperone, calreticulin, a soluble ER-lumenal protein with high sequence similarity to the ectodomain of calnexin and a C-terminal KDEL ER-retention/retrieval motif (SADASIVAN et al. 1996). Calnexin and calreticulin have highly similar lectin-like specificity for proteins with monoglucosylated N-linked sugars and are part of the ER quality control system. The folding of nascent proteins is thought to be accompanied by cycles of de- and reglucosylation, which drives the dissociation from and the association with calnexin and calreticulin (VAN LEEUWEN et al. 1996). Thus, once the class I molecule has achieved its fully folded conformation, it would presumably no longer be reglucosylated and would be released from calnexin and calreticulin. Recent studies show that calreticulin associates simultaneously with heavy chain/β2-microglobulin and TAP (see Sect. 5.2), whereas calnexin does not bind class I molecules coincidentally with TAP (SADASIVAN et al. 1996). Since calreticulin-associated class I glycoproteins are distinct from those associated with calnexin, but similar to those bound to TAP (VAN LEEUWEN et al. 1996), it seems likely that calreticulin serves as a different part to calnexin of the proof-reading mechanism for class I assembly and peptide loading in the ER.

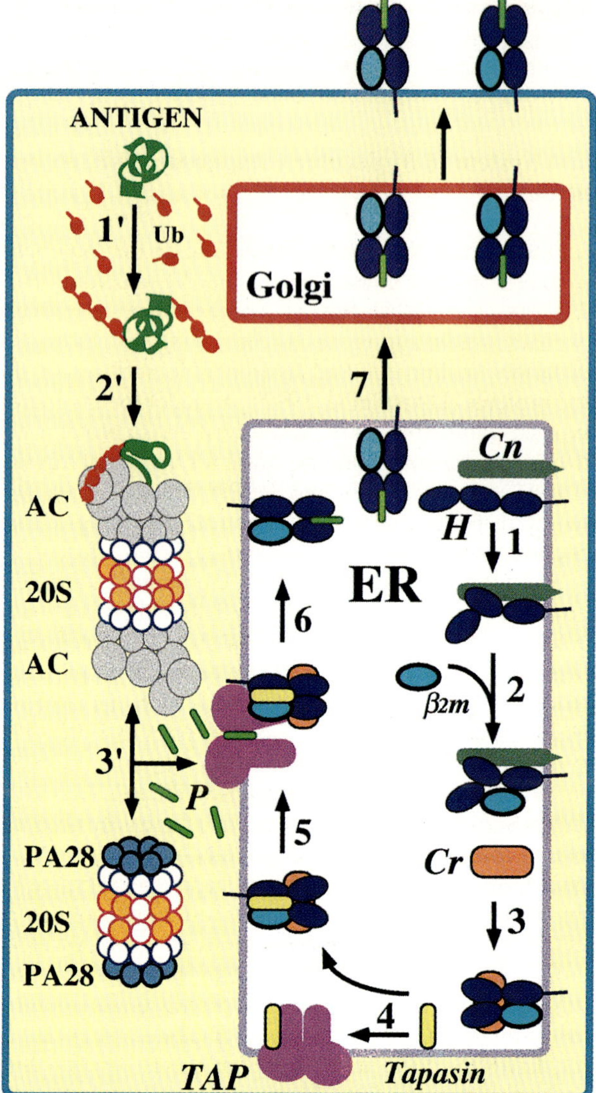

Fig. 3. Intracellular assembly of class I MHC molecules. During the process of assembly, the transmembrane class I heavy chain (*H*) initially interacts with the molecular chaperone calnexin (*Cn*) for successive folding in the endoplasmic reticulum (*ER*, *step 1*) until it has folded sufficiently to bind a soluble light-chain β_2-microglobulin ($\beta 2m$, *step 2*). The association of class I H-β_2-microglobulin heterodimer is facilitated by the chaperone calreticulin, a soluble ER-luminal homologue of calnexin (*Cr*, *step 3*). The complex of H-β_2-microglobulin-Cr is then associated with the transporter associated with antigen presentation (*TAP*) via tapasin, which binds independently to TAP and class I H-β_2-microglobulin-Cr complexes (*step 4*). Release of class I molecule from the complex requires binding of a peptide (*P*) of eight to ten amino acids, which is translocated by the TAP across the membrane of the ER (*step 5*). Only after the assembly of the trimolecular complex (*step 6*) can the class I H-β_2-microglobulin-peptide complex be expressed on the cell surface via the exocytic pathway (*step 7*) for inspection by $CD8^+$ cytotoxic T

5.2 Association of Peptide Transporters with Class I MHC Molecules

The presence of β_2-microglobulin during class I biosynthesis aids formation and maintenance of the class I disulfide bonds, which are also required for class I assembly (RIBAUDO et al. 1992). In the absence of β_2-microglobulin, class I heavy chains are labile and rapidly degraded (KOLLER et al. 1990; ZIJLSTRA et al. 1990); β_2-microglobulin is thus important for de novo folding of class I proteins. Most importantly, β_2-microglobulin association is critical for class I molecule binding to TAP (SOLHEIM et al. 1995). It has been shown that empty class I molecules are retained in the ER by a physical association with TAP, via the TAP1 subunit and β_2-microglobulin, as well as the membrane-proximal α_3-domain of the class I heavy chain (ORTMANN et al. 1994; SUH et al. 1994). The interaction of class I heavy chain-β_2-microglobulin dimers with TAP occurs via a novel MHC-encoded 48 kDa glycoprotein, tapasin (SADASIVAN et al. 1996), which binds independently to TAP and class I heavy chain-β_2-microglobulin-calreticulin complexes. In tapasin-deficient cells, class I MHC-TAP association and peptide loading appears to be defective (SADASIVAN et al. 1996). However, it is presently unclear whether class I MHC binding to TAP is a prerequisite for peptide loading. Interestingly, calreticulin is present in TAP-class I MHC complexes. Since deglucosylation of N-linked glycans is important for dissociation of class I proteins from both calreticulin and TAP (VAN LEEUWEN et al. 1996), glycan processing seems to be functionally coupled to the release of class I proteins from TAP. In addition to tapasin and calreticulin, several other proteins have also been shown to be coprecipitated with TAP (YANG et al. 1996; our unpublished data). Although the function of these coprecipitated proteins is presently unclear, it can be envisaged that peptide transport is coupled to peptide loading onto the class I molecules and that auxiliary proteins might be needed to orchestrate these events. Thus these TAP-associated proteins could be either kinases which regulate TAP activity, chaperones which are required to promote docking of empty class I complexes onto TAP, or factors which are essential for the delivery of peptides from the cytosolic peptide transport machinery to the TAP. Upon peptide addition, class I molecules associated with TAP are released, indicating that class I molecules associated with TAP require only peptides for their release from the ER.

5.3 Role of Peptides in Class I MHC Assembly and Transport

The physical properties of class I molecules are crucial factors for their presence at the cell surface (LJUNGGREN et al. 1990; JACKSON et al. 1992; SYKULEV et al. 1994),

---

cytotoxic T lymphocytes. The TAP-translocated and class I-binding peptides are derived from intracellular antigens, which are ubiquitinylated by the ubiquitin system (*Ub*, *step 1'*). The cytosolic ubiquitinylated antigens are recognized by the ATPase complex (*AC*, *step 2'*) bound to the 20S proteasome (*20S*) and degraded by the proteasomes in an ATP-dependent manner, resulting in the generation of peptides of approximately eight to ten amino acids (*step 3'*). The interferon (IFN)-inducible PA28 proteasome regulator (*PA28*), which binds the 20S proteasome, is believed to be responsible for the generation of optimal antigenic peptide as well as for delivery of peptides to the TAP transporter since stability of a class I molecule appears not only to determine whether the class I molecule is able to reach the surface, but also to determine its half-life at the cell surface. Since a high-affinity peptide is able to confer a high degree of stability to a given class I molecule, the affinity of the peptides bound to the class I molecules appears to dictate the level of class I molecules at the cell surface, consistent with the observation that in the absence of "optimal" peptides, class I molecules are short-lived at the physiological temperature of 37°C (JACKSON et al. 1992). Thus the presence of class I molecules on the cell surface is dictated by the availability of high-affinity peptides to class I molecules. Overall, the availability of a peptidic epitope and its affinity for the class I molecules govern the intracellular transport rate and the half-life, in addition to the immunogenicity (JACKSON et al. 1995), of the class I MHC-peptide complexes. As a result, incompletely assembled class I molecules (lacking bound peptides) are not transported to the cell surface, thus preventing $CD8^+$ cytotoxic T lymphocytes from killing uninfected cells during exposure to foreign antigens. It can therefore be concluded that the single most important factor deciding whether a class I molecule is functional is the antigenic peptide.

6 Regulation of Class I MHC Antigen Presentation

The MHC is tantamount to an eukaryotic operon encoding a complete kit for the processing and presentation of antigens to T lymphocytes. The MHC linkage exhibited by almost all of these MHC genes could facilitate the coordinate evolution of MHC haplotypes with compatible antigenic peptide specificities that offer an advantage in responding to the antigens of pathogenic origin (YANG et al. 1996). In addition, the expression of class I molecules on the cell surface can be increased by IFN (SONG et al. 1994). Most importantly, the MHC molecules, including the class I molecules, TAP, and proteasomal subunits LMP2 and 7, are all transcriptionally upregulated by IFN (YANG et al. 1996). Moreover, several non-MHC-encoded genes, including β_2-microglobulin, PA28, and MECL-1, that have been recruited by the immune system are also induced by IFN (SONG et al. 1994). Thus IFN inducibility provides a means of coordinately regulating several sets of genes, each of whose products are essential to the efficient processing and presentation of antigens.

IFN plays a central role in immunomodulatory functions through upregulation of the expression of a variety of genes responsible for a variety of immune responses (LEVY et al. 1988; DECKER et al. 1991; SEN et al. 1993; TANAKA et al. 1993; PINE et al. 1994). All MHC class I heavy chain genes as well as β_2-microglobulin contain upstream promoter elements which serve as recognition sites for IFN-stimulated transcription factors (PELLEGRINI et al. 1993). In contrast to heavy chain and β_2-microglobulin genes, the promoter elements of TAP and LMP do not contain the IFN-stimulated response elements (BECK et al. 1992; FRUH et al. 1992; KISHI et al. 1993). A bidirectional promoter serves both TAP1 and LMP2 (WRIGHT et al. 1995), whereas LMP7 and TAP2 each seem to have their own promoter, although the latter are not yet well characterized. The core region of the LMP2/TAP1 promotor contains two overlapping IFN consensus sequences and a γ-activation sequence as well as NFκB- and Sp1-binding sites (MIN et al. 1996). Interestingly, induction of this bidirectional promoter by IFN-regulated Stat1α results in a much more rapid induction of TAP1 and LMP2 transcripts compared to heavy chain and β_2-microglobulin induction (EPPERSON et al. 1992; MIN et al. 1996). Thus it seems that peptide supply to class I molecules is increased prior to upregulating the class I proteins, which is consistent with the assumption that peptide supply for class I molecules is limiting.

During the evolution of the immune system, it appears that the proteasome has undergone adaptations that favor antigen presentation. In addition to IFN-inducible proteasome subunits LMP and MECL-1, the other adaptation consists in the PA28 proteasome regulator. The IFN inducibility of PA28 (REALINI et al. 1994; YANG et al. 1995 AHN et al. 1995, 1996) represents an additional mechanism by which IFN can affect proteasome activity. The fact that both PA28 subunits show a cytokine response profile which corresponds exactly to the proteasomal subunits LMP2 and LMP7 (AHN et al. 1996) makes it very likely that the corresponding genes have elements similar to the TAP/LMP promoters. Indeed, an analysis of the genomic structures for PA28 reveals that the regulatory elements responsible for its induction by IFN are very similar to the ones in the TAP/LMP promoters (our unpublished data). Therefore, it is reasonable to suggest that molecules such as TAP, class I heavy chain, β_2-microglobulin, and LMP in the class I MHC antigen presentation system are upregulated by IFN in concert with PA28 proteasome regulators in order to optimize antigen presentation in an effective and synergistic manner.

Another potential mechanisn to improve antigen presentation efficiency would be to have regulatory molecules transferring peptides from the sites of generation to TAP. An efficient means to transfer peptides generated by PA28-20S proteasomal complexes to TAP would be a direct association between the 20S proteasomal complexes and the TAP on the ER membrane (YANG et al. 1996). Molecular chaperones may be involved in this process and may be required to chaperone peptides from their site of generation to TAP. Three heat shock protein genes, HSP70-1, HSP70-2, and HSP70-Hom, are encoded in the MHC, 150 kb from HLA-B-associated transcripts (MILNER et al. 1992; WALTER et al. 1994). These HSP70 genes may have evolved to perform specialized functions in antigen processing and presentation, such as the unfolding and presentation of antigens for degradation, binding and protection of antigenic peptides, or transport of peptides to the TAP in the ER membrane. It has been suggested that gp96, an ER member of the hsp90 family of chaperones, may have a role in peptide loading of class I molecules (LI et al. 1993; SRIVASTAVA et al. 1994), particularly as gp96 expression is IFN inducible (ANDERSON et al. 1994). Immunization of mice with tumor-derived gp96 was shown to protect the mice against subsequent challenge with tumor cells, and this protective mechanism involved transfer of gp96-associated antigenic peptides to class I molecules in recipient animals. Furthermore, gp96 from vesicular stomatitis virus (VSV) -infected cells contained bound peptides corresponding to the immunodominant Kb-binding peptide of VSV (NIELAND et al. 1996). From these results, an alternative model was proposed, whereby gp96 could bind peptides derived from TAP and transfer these peptides to class I molecules (SUTO et al. 1995). However, a recent study using antisense to decrease gp96 levels by 90% showed no effect on class I expression (LAMMERT et al. 1996). Thus, if gp96 is involved in peptide loading of class I molecules, it must act in addition to other mechanisms with the same function, such as direct peptide transfer from TAP.

7 Outlook

Beyond the current knowledge described in this review, we believe that the complexity of the immune system is significantly greater than is presently appreciated. Many important questions remain unanswered: Are there other molecules in the antigen presentation pathway that remain to be discovered, such as the additional proteins associated with TAP? What is the role of tapasin and/or other TAP-associated proteins in class I peptide loading? What is the mechanism of peptide transport by TAP? The regulatory mechanisms by which the proteasome and TAP activities are modulated, as well as the pathway by which peptides generated by the proteasomes interact with the peptide transport machinery, also remain to be elucidated. The most challenging task for the future is to achieve a detailed understanding of the molecular mechanism by which the class I MHC-binding peptides are generated by proteasomes. With the rapid pace of biochemical studies on proteasome-mediated antigen degradation and TAP-dependent peptide translocation, as well as a more detailed understanding of the structures and functions of the TAP and the eukaryotic proteasomes from high-resolution X-ray crystallography, the answers to these and other questions are likely to be forthcoming.

Acknowledgments. We are grateful to C. Garcia for providing Figs. 1 and 2, to P. van Endert, V. Brusic, A. DeGroot, M. Milik, C. Glass, and J. Skolnick for allowing us to cite unpublished data, to P. Sempe and C. Scott for critical reading and helpful comments, and to P.A. Peterson for encouragement.

References

Ahn JY, Tanahashi N, Akiyama K, Hisamatsu H, Noda C, Tanaka K, Chung CH, Shibmara N, Willy PJ, Mott JD, Slaughter CA, DeMartino GN (1995) Primary structures of two homologous subunits of PA28, a gamma-interferon-inducible protein activator of the 20S proteasome. FEBS Lett 366:37–42

Ahn K, Erlander M, Leturcq D, Peterson PA, Fruh K, Yang Y (1996) In vivo characterization of the proteasome regulator PA28. J Biol Chem 271:18237–18242

Aki M, Shimbara N, Takashina M, Akiyama K, Kagawa S, Tamura T, Tanahashi N, Yoshimura T, Tanaka K, Ichihara A (1994) Interferon-gamma induces different subunit organizations and functional diversity of proteasomes. J Biochem 115:257–269

Anderson SL, Shen T, Lou J, Xing L, Blachere NB, Srivastava PK, Rubin BY (1994) The endoplasmic reticular heat shock protein gp96 is transcriptionally upregulated in interferon-treated cells. J Exp Med 180:1565–1569

Androlewicz MJ, Cresswell P (1994) Human transporters associated with antigen processing possess a promicuous peptide-binding site. Immunity 1:7–14

Androlewicz MJ, Anderson KS, Cresswell P (1993) Evidence that transporters associated with antigen processing translocate a major histocompatibility complex class I-binding peptide into the endoplasmic reticulum in an ATP-dependent manner. Proc Natl Acad Sci USA 90:9130–9134

Beck S, Kelly A, Radley E, Khurshid F, Alderton RP, Trowsdale J (1992) DNA sequence analysis of 66 kb of the human MHC class II region encoding a cluster of genes for antigen processing. J Mol Biol 228:433–441

Belich MP, Glynne RJ, Senger G, Sheer D, Trowsdale J (1994) Proteasome components with reciprocal expression to that of the MHC-encoded LMP proteins. Curr Biol 4:769–776

Bergeron JJ, Brenner MB, Thomas DY, Williams DB (1994) Calnexin: a membrane-bound chaperone of the endoplasmic reticulum. Trends Biochem Sci 19:124–128

Bjorkman P, Saper M, Wiley D (1987) Structure of human class I histocompatibility antigen, HLA-A2. Nature 329:506–512

Boes B, Hengel H, Ruppert T, Multhaup G, Koszinowski UH, Kloetzel P-M (1994) Interferon γ stimulation modulates the proteolytic activity and cleavage site preference of 20S mouse proteasomes. J Exp Med 179:901–909

Brusic V, Rudy G, Harrison LC (1994) Prediction of MHC binding peptides using artificial neural networks. In: Stonier RJ, Yu XH (eds) Complex systems: mechanism of adaptation. IOS, Amsterdam, pp253–260

Brusic V, Rudy G, Kyne AP, Harrison LC (1996) MHCPEP-A database of MHC-binding peptides: update 1995. Nucleic Acids Res. 24:242–244

Campbell RD, Trowsdale J (1993) Map of the human MHC. Immunol Today 14:349–352

Chu-Ping M, Slaughter CA, DeMartino GN (1992) Identification, purification, and characterization of a protein activator (PA28) of the 20 S proteasome (Macropain). J Biol Chem 267:10515–10523

Chu-Ping M, Willy PJ, Slaughter CA, DeMartino GN (1993) PA28, an activator of the 20 S proteasome, is inactivated by proteolytic modification at its carboxyl terminus. J Biol Chem 268:22514–22519

Coux O, Tanaka K, Goldberg AL (1996) Structure and functions of the 20S and 26S proteasomes. Annu Rev Biochem 65:801–847

Davenport MP, Hill AVS (1996) Peptide associated with MHC class I and class II molecules. In: Browning M, McMicheal A (eds) HLA MHC: genes, molecules, functions. Bios, Oxford, pp277–308

Davis MM, Chien YH (1993) Topology and affinity of T-cell receptor mediated recognition of peptide-MHC complexes. Curr Opin Immunol 5:45–49

Davis MM, Chien YH (1996) T-cell antigen receptors. In: Tiley NL (ed) Transplant biology. Lippincott-Raven, Philadelphia, pp403–410

Decker T, Lew DJ, Mirkovitch J, Darnell JJE (1991) Cytoplasmic activation of GAF, an IFN-gamma-regulated DNA-binding factor. EMBO J 10:927–932

Degen E, Williams DB (1991) Participation of a novel 88-kD protein in the biogenesis of murine class I histocompatibility molecules. J Cell Biol 112:1099–1115

Degen E, Cohen-Doyle MF, Williams DB (1992) Efficient dissociation of the p88 chaperone from MHC class I molecules requires both β2-microglobulin and peptide. J Exp Med 175:1653–1661

Driscoll J, Brown MG, Finley D, Monaco JJ (1993) MHC-linked LMP gene products specifically alter peptidase activities of the proteasome. Nature 365:262–264

Engelhard VH, Appella E, Benjamin DC, Bodnar WM, Cox AL, Chen Y, Henderson RA, Huczko EL, Michel H (1993) Mass spectrometric analysis of peptides associated with the human class I MHC molecules HLA-A2.1 and HLA-B7 and identification of structural features that determine binding. Chem Immunol 57:39–62

Epperson D, Arnold D, Spies T, Cresswell P, Pober J, Johnson D (1992) Cytokines increase transporter in antigen processing-1 expression more rapidly than HLA class I expression in endothelial cells. J Immunol 149:3297–3301

Falk K, Roetzschke O, Stevanovic S, Jung G, Rammensee H-G (1991) Allele-specific motifs revealed by sequencing of self-peptides eluted from MHC molecules. Nature 351:290–296

Falk K, Rotzschke O, Takiguchi M, Grahovac B, Gnau V, Stevanovic S, Jung G, Rammensee H-G (1994) Peptide motifs of HLA-A1, -A11, -A31, and -A33 molecules. Immunogenetics 40:238–241

Falk K, Rotzschke O, Takiguchi M, Gnau V, Stevanovic S, Jung G, Rammensee H-G (1995a) Peptide motifs of HLA-B38 and B39 molecules. Immunogenetics 41:162–164

Falk K, Rotzschke O, Takiguchi M, Gnau V, Stevanovic S, Jung G, Rammensee H-G (1995b) Peptide motifs of HLA-B51, -B52 and -B78 molecules, and implications for Behcet's disease. Int Immunol 7:223–228

Falk K, Rotzschke O, Takiguchi M, Gnau V, Stevanovic S, Jung G, Rammensee H-G (1995c) Peptide motifs of HLA-B58, B60, B61, and B62 molecules. Immunogenetics 41:165–168
Fehling HJ, Swat W, Laplace C, KŸhn R, Rajewsky K, MŸller U, Von Boehmer H (1994) MHC class I expression in mice lacking the proteasome subunit LMP-7. Science 265:1234–1237
Fremont DH, Matsumura M, Stura EA, Peterson PA, Wilson IA (1992) Crystal structures of two viral peptides in complex with murine MHC class I H-2Kb. Science 257:919–927
Fruh K, Yang Y, Arnold D, Chambers J, Wu L, Waters JB, Spies T, Peterson PA (1992) Alternative exon usage and processing of the major histocompatibility complex-encoded proteasome subunits. J Biol Chem 267:22131–22140
Fruh K, Gossen M, Wang K, Bujard H, Peterson PA, Yang Y (1994) Displacement of housekeeping proteasome subunits by MHC-encoded LMPs: a newly discovered mechanism for modulating the multicatalytic proteinase complex. EMBO J 13:3236–3244
Fruh K, Karlsson L, Yang Y (1997) γ Interferon in antigen processing and presentation. In: Karupiah G (ed) γ Interferon in antiviral defense. Landes, Austin, pp 39–59
Gaczynska M, Rock KL, Goldberg AL (1993) Gamma-interferon and expression of MHC genes regulate peptide hydrolysis by proteasomes. Nature 365:264–267
Gaczynska M, Rock KL, Spies T, Goldberg AL (1994) Peptidase activities of proteasomes are differentially regulated by the major histocompatibility complex-encoded genes for LMP2 and LMP7. Proc Natl Acad Sci USA 91:9213–9217
Gaczynska M, Goldberg AL, Tanaka K, Hendil KB, Rock KL (1996) Proteasome subunits X and Y alter peptidase activities in opposite ways to the interferon-γ-induced subunits LMP2 and LMP7. J Biol Chem 271:17275–17280
Garcia KC, Degano M, Stanfield RL, Brunmark A, Jackson MR, Peterson PA, Teyton L, Wilson IA (1996) An αβ T cell receptor structure at 2.5 A and its orientation in the TCR-MHC complex. Science 274:209–219
Groettrup M, Kraft R, Kostka S, Standera S, Stohwasser R, Kloetzel P-M (1996a) A third interferon-γ-induced subunit exchange in the 20S proteasome. Eur J Immunol 26:863–869
Groettrup M, Soza A, Eggers M, Kuehn L, Dick T, Schild H, Rammensee H-G, Koszinowski U, Kloetzel P-M (1996b) A role for the proteasome regulator PA28α in antigen presentation. Nature 381:166–168
Hammer J, Takacs B, Sinigaglia F (1992) Identification of a motif for HLA-DR1 binding peptides using M13 display libraries. J Exp Med 176:1007–1013
Hebert DN, Simons JF, Peterson JR, Helenius A (1995) Calnexin, calreticulin, and Bip/Kar2p in protein folding. Cold Spring Harb Symp Quant Biol 60:405–415
Heemels MT, Ploegh HL (1995) Generation translocation, and presentation of MHC class I-restricted peptides. Annu Rev Biochem 64:463–491
Hochstrasser M (1996a) Protein degradation or regulation: Ub the judge. Cell 84:813–815
Hochstrasser M (1996b) Ubiquitin-dependent protein degradation. Annu Rev Genet 30:405–439
Hunt DF, Henderson RA, Shabanowitz J, Sakaguchi K, Michel H, Sevilir N, Cox AL, Appella E, Engelhard VH (1992) Characterization of peptides bound to the class I MHC molecule HLA-A2.1 by mass spectrometry. Science 255:1261–1263
Jackson MR, Song ES, Yang Y, Peterson PA (1992) Empty and peptide-containing class I MHC molecule conformers expressed in Drosophila Melanogaster cells. Proc Natl Acad Sci USA 89:12117–12121
Jackson MR, Fruh K, Karlsson L, Teyton L, Yang Y, Peterson PA (1995) Assembly and intracellular transport of MHC class I and class II molecules. Cold Spring Harb Symp Quant Biol 60:249–261
Jesdale BM, Deocampo G, Meisell J, Beall J, Marinello MJ, Chicz RM, de Groot AS (1997) Matrix-based prediction of MHC binding peptides: the epimatrix algorithm, reagent for HIV research. In: Brown F (ed) Vaccines 97: molecular approaches control infectious diseases. Cold Spring Harbor, New York, pp57–64
Jorgensen JL, Reay PA, Ehrich EW, Davis MM (1992) Molecular components of T-cell recognition. Annu Rev Immunol 10:835–873
Kishi F, Suminami Y, Monaco JJ (1993) Genomic organization of the mouse Lmp-2 gene and characteristic structure of its promoter. Gene 133:243–248
Kleijmer MJ, Kelly A, Geuze HJ, Slot JW, Townsend A, Trowsdale J (1992) Location of the MHC-encoded transporters in the endoplasmic reticulum and cis-golgi. Nature 357:342–344
Koller B, Marrack P, Kappler J, Smithies O (1990) Normal development of mice deficient in β2-m, MHC class I proteins, and CD8+ T cells. Science 248:1227–1230

Krishnan BR, Jamry I, Berg DE, Berg CM, Chaplin DD (1995) Construction of a genomic DNA 'feature map' by sequencing from nested deletions: application to the HLA class I region. Nucleic Acids Res 23:117–122

Lammert E, Arnold D, Rammensee H-G, Schild H (1996) Expression levels of stress protein gp96 are not limiting for major histocompatibility complex class I-restricted antigen presentation. Eur J Immunol 26:875–879

Levy D, Kessler R, Pine R, Reich N, Darnell JJ (1988) Interferon-induced nuclear factors that bind a shared promoter element correlate with positive and negative control. Genes Dev 2:383–393

Li Z, Srivastava PK (1993) Tumor rejection antigen gp96/grp94 is an ATPase: implications for protein folding and antigen presentation. EMBO J 12:3143–3151

Ljunggren HG, Stam NJ, …hlen C, Neefjes JJ, Hšglund P, Heemels MT, Bastin J, Schumacher TNM, Townsend A, KŠrre K, Ploegh HL (1990) Empty MHC class I molecules come out in the cold. Nature 346:476–480

Lšwe J, Stock D, Jap B, Zwickl P, Baumeister W, Huber R (1995) Crystal structure of the 20 S proteasome from the archaeon T. acidophilum at 3.4 A resolution. Science 268:533–539

Madden DR, Wiley DC (1992) Peptide binding to the major histocompatibility complex molecules. Curr Opin Struct Biol 2:300–304

Maryanski JL, Verdini AS, Weber PC, Salemme FR, Corradin G (1990) Competitor analogs for defined T cell antigens: peptides incorporating a putative binding motif and polyproline or polyglycine spacers. Cell 60:63–72

Matsumura M, Fremont DH, Peterson PA, Wilson IA (1992a) Emerging principles for the recognition of peptide antigens by MHC class I molecules. Science 257:927–934

Matsumura M, Saito Y, Jackson MR, Song ES, Peterson PA (1992b) In vitro peptide binding to soluble empty class I major histocompatibility complex molecules isolated from transfected Drosophila melanogaster cells. J Biol Chem 267:23589–23595

Michaelek MT, Grant EP, Gramm C, Goldberg AL, Rock KL (1993) A role for the ubiquitin dependent proteolytic pathway in MHC class I-restricted antigen presentation. Nature 363:552–554

Milner CM, Campbell RD (1992) Polymorphic analysis of the three MHC-linked HSP70 genes. Immunogenetics 36:357–362

Min W, Pober JS, Johnson DR (1996) Kinetically coordinated induction of TAP1 and HLA class I by IFN-γ. The rapid induction of TAP1 by IFN-γ is mediated by Stat1α. J Immunol 156:3174–3183

Momburg F, Neefjes JJ, HŠmmerling GJ (1994a) Peptide selection by MHC-encoded TAP transporters. Curr Opin Immunol 6:32–37

Momburg F, Roelse J, Hammerling GJ, Neefjes JJ (1994b) Peptide size selection by the major histocompatibility complex-encoded peptide transporter. J Exp Med 179:1613–1623

Momburg F, Roelse J, Howard JC, Butcher GW, Hammerling GJ, Neefjes JJ (1994c) Selectivity of MHC-encoded peptide transporters from human, mouse and rat. Nature 367:648–651

MŸller KM, Ebensperger C, Tampe R (1994) Nucleotide binding to the hydrophilic C-terminal domain of the transporter associated with antigen processing (TAP). J Biol Chem 269:14032–14037

Nieland TJF, Tan MCAA, Monnee-van Muijen M, Koning F, Kruisbeek AM, van Bleek GM (1996) Isolation of an immunodominant viral peptide that is endogenously bound to the stress protein GP96/GRP94. Proc Natl Acad Sci USA 93:6135–6139

Orlowski M (1993) The multicatalytic proteinase complex (proteasome) and intracellular protein degradation: diverse functions of an intracellular particle. J Lab Clin Med 121:187–189

Ortmann B, Androlewicz MJ, Cresswell P (1994) MHC class I/β_2-microglobulin complexes associate with TAP transporters before peptide binding. Nature 368:864–867

Parham P, Ohta T (1996) Population biology of antigen presentation by MHC class I molecules. Science 272:67–74

Parker KC, Bednarek MA, Coligan JE (1994) Scheme for handling potential HLA-A2 binding peptides based on independent binding of individual peptide side-chains. J Immunol 152:163–175

Paul WE (1993) The immune system: an introduction. In: Paul WE (ed) Fundamental immunology. Raven, New York, pp1–20

Pellegrini S, Schindler C (1993) Early events in signalling by interferons. Trends Biochem Sci 18:338–342

Peters J-M, Cejka Z, Harris JR, Kleinschmidt JA, Baumeister W (1993) Structural features of the 26 S proteasome complex. J Mol Biol 234:932–937

Pine R, Canova A, Scindler C (1994) Tyrosine phosphorylated p91 binds to a single element in the ISGF2/IRF-1 promoter to mediate induction by IFN-alpha and IFN-gamma, and is likely to aotoregulate the p91 gene. EMBO J 13:158–167

Ploegh H (1995) Biosynthesis of MHC products and its relevance for antigen presentation. In: Bell JI (ed) T cell receptor. Oxford University Press, Oxford, pp447–467

Rajagopalan S (1996) Molecular chaperones in MHC class I and class II biosynthesis and assembly. In: Urban RG (ed) MHC molecules: expression, assembly, function. Landes, Austin, pp65–71

Rammensee H-G (1995) Chemistry of peptides associated with MHC class I and class II molecules. Curr Biol 7:85–96

Rammensee H-G (1996) Antigen presentation. Recent developments. Int Arch Allergy Immunol 110:299–307

Rammensee H-G, Falk K, Roetzschke O (1993) Peptides naturally presented by MHC class I molecules. Annu Rev Immunol 11:213–244

Rammensee H-G, Friede T, Stefanovic S (1995) MHC ligands and peptide motifs: first listing. Immunogenetics 41:178–228

Realini C, Dubiel W, Pratt G, Ferrell K, Rechsteiner M (1994) Molecular cloning and expression of a gamma-interferon-inducible activator of the multicatalytic protease. J Biol Chem 269:20727–20732

Rechsteiner M, Hoffman L, Dubiel W (1993) The multicatalytic and 26 S proteases. J Biol Chem 268:6065–6068

Ribaudo R, Margulies D (1992) Independent and synergistic effects of disulfide bond formation, β2-microglobulin, and peptides on class I MHC folding and assembly in an in vitro translation system. J Immunol 149:2935–2944

Rock KL, Gramm C, Rothstein L, Clark K, Stein R, Dick L, Hwang D, Goldberg AL (1994) Inhibitors of the proteasome block the degradation of most cell proteins and the generation of peptides presented on MHC class I molecules. Cell 78:761–771

Roetzschke O, Falk K, Deres K, Schild H, Norda M, Metzger J, Jung G, Rammensee HG (1990) Isolation and analysis of naturally processed viral peptides as recognized by cytotoxic T cells. Nature 348:252–254

Ruppert J, Sidney J, Celis E, Kubo RT, Grey HM, Sette A (1993) Prominent role of secondary anchor residues in peptide binding to HLA-A2.1 molecules. Cell 74:929–937

Russ G, Esquivel F, Yewdell JW, Cresswell P, Spies T, Bennink JR (1995) Assembly, intracellular localization, and nucleotide binding properties of the human peptide transporters TAP1 and TAP2 expressed by recombinant vaccinia viruses. J Biol Chem 270:21312–21318

Sadasivan B, Lehner PJ, Ortmann B, Spies T, Cresswell P (1996) Roles for calreticulin and a novel glycoprotein, tapasin, in the interaction of MHC class I molecules with TAP. Immunity 5:103–114

Saper MA, Bjorkman PJ, Wiley DC (1991) Refined structure of the human histocompatibility antigen HLA-A2 at 2.6 Å resolution. J Mol Biol 219:277–319

Schumacher TN, Kantesaria DV, Heemels MT, Ashton-Rickardt PG, Shepherd JC, Fruh K, Yang Y, Peterson PA, Tonegawa S, Ploegh HL (1994) Peptide length and sequence specificity of the mouse TAP1/TAP2 translocator. J Exp Med 179:533–540

Scott JE, Dawson JR (1995) MHC class I expression and transport in a calnexin-deficient cell line. J Immunol 155:143–148

SeemŸller E, Lupas A, Stock D, Lowe J, Huber R, Baumeister W (1995) Proteasome from *Thermoplasma acidophilum*: a threonine protease. Science 268:579–582

SeemŸller E, Lupas A, Baumeister W (1996) Autocatalytic processing of the 20S proteasome. Nature 382:468–470

Sen GC, Ransohoff RM (1993) Interferon-induced antiviral actions and their regulation. Adv Virus Res 42:57–101

Smith JD, Solheim JC, Carreno BM, Hansen TH (1995) Characterization of class I MHC folding intermediates and their disparate interactions with peptide and β2-microglobulin. Mol Immunol 32:531–540

Solheim JC, Cook JR, Hansen TH (1995) Conformational changes induced in the MHC class I molecule by peptide and β2-microglobulin. Immunol Res 14:200–217

Song ES, Yang Y, Jackson MR, Peterson PA (1994) In vivo regulation of the assembly and intracellular transport of class I major histocompatibility complex molecules. J Biol Chem 269:7024–7029

Spies T, Bresnahan M, Bahram S, Arnold D, Blank G, Mellins E, Pious D, DeMars R (1990) A gene in the human major histocompatibility complex class II region controlling the class I antigen presentation pathway. Nature 348:744–747

Srivastava PK (1993) Peptide-binding heat shock proteins in the endoplasmic reticulum: role in immune response to cancer and in antigen presentation. Adv Cancer Res 62:153–177

Srivastava PK (1996) Immunotherapeutic stress protein–peptide complexes against cancer. PCT Int Appl, WO9610411A1

Srivastava PK, Udono H, Blachere NE, Li Z (1994) Heat shock proteins transfer peptides durine antigen processing and CTL priming. Immunogenetics 39:93–98

Suh W-K, Cohen-Doyle MF, Fruh K, Wang K, Peterson PA, Williams DB (1994) Interaction of MHC class I molecules with the transporter associated with antigen processing. Science 264:1322–1326

Suto R, Srivastava PK (1995) A mechanism for the specific immunogenicity of heat shock protein-chaperoned peptides. Science 269:1585–1588

Sykulev Y, Brunmark A, Jackson M, Cohen RJ, Peterson PA, Eisen HN (1994) Kinetics and affinity of reactions between an antigen-specific T cell receptor and peptide-MHC complexes. Immunity 1:15–22

Tanaka N, Kawakami T, Taniguchi T (1993) Recognition DNA sequences of interferon regulatory factor-1 (IRF-1) and IRF-2, regulators of cell growth and the interferon system. Mol Cell Biol 13:4531–4538

Townsend A, Bastin J, Gould K, Brownlee G, Andrew M, Coupar B, Boyle D, Chan S, Smith G (1988) Defective presentation to class I-restricted cytotoxic T lymphocytes in vaccinia-infected cell is overcome by enhanced degradation of antigen. J Exp Med 168:1211–1224

Townsend A, Ohlen C, Bastin J, Ljunggren HG, Foster L, Karre K (1989) Association of class I major histocompatibility heavy and light chains induced by viral peptides. Nature 340:443–448

Udaka K (1996) Decrypting class I MHC-bound peptides with peptide libraries. Trends Biochem Sci 21:7–11

Udaka K, Wiesmueller K-H, Kienle S, Jung G, Walden P (1995a) Decrypting the structure of major histocompatibility complex class I-restricted cytotoxic T lymphocyte epitopes with complex peptide libraries. J Exp Med 181:2097–2108

Udaka K, Wiesmueller K-H, Kienle S, Jung G, Walden P (1995b) Tolerance to amino acid variations in peptides binding to the major histocompatibility complex class I protein H-2Kb. J Biol Chem 270:24130–24134

Uebel S, Meyer TH, Kraas W, Kienle S, Jung G, WiesmŸller K-H, Tamp R (1995) Requirements for peptide binding to the human transporter associated with antigen processing revealed by peptide scans and complex peptide librairies. J Biol Chem 270:18512–18516

Van Bleek GM, Nathenson SG (1990) Isolation of an endogenously processed immunodominant viral peptide from the class I H-2Kb molecule. Nature 348:213–216

van Endert PM (1996) Peptide selection for presentation by HLA class I: a role for the human transporter associated with antigen processing? Immunol Res 15:265–279

van Endert PM, Tamp R, Meyer TH, Tish R, Bach J-F, McDevitt HO (1994) A sequential model for peptide binding and transport by the transporters associated with antigen processing. Immunity 1:491–500

Van Kaer L, Ashton-Rickard PG, Eichelberger M, Gaczynska M, Nagashima K, Rock KL, Goldberg AL, Doherty PC, Tonegawa S (1994) Altered peptidase and viral-specific T cell response in LMP2 mutant mice. Immunity 1:533–541

Van Leeuwen JEM, Kearse KP (1996) Deglucosylation of N-linked glycans is an important step in the dissociation of calreticulin-class I-TAP complexes. Proc Natl Acad Sci USA. 93:13997–14001

Vassilakos A, Cohen-Doyle MF, Peterson PA, Jackson MR, Williams DB (1996) The molecular chaperone calnexin facilitates folding and assembly of class I histocompatibility molecules. EMBO J 15:1495–1506

Vinitsky A, Cardoza C, Sepp-Lorenzino L, Michaud C, Orlowski M (1994) Inhibition of the proteolytic activity of the multicatalytic proteinase complex (proteasome) by substrate-related peptidyl aldhehydes. J Biol Chem 269:29860–29866

Walter L, Rauh F, GŸnther E (1994) Comparative analysis of the three major histocompatibility complex-linked heat shock protein 70 (hsp70) genes of the rat. Immunogenetics 40:325–330

Wang K, FrŸh K, Peterson PA, Yang Y (1994) Nucleotide binding of the C-terminal domains of the major histocompatibility complex-encoded transporter expressed in *Drosophila melanogaster* cells. FEBS Lett 350:337–341

Wright KL, White LC, Kelly A, Beck S, Trowsdale J, Ting JP (1995) Coordinate regulation of the human TAP1 and LMP2 genes from a shared bidirectional promoter. J Exp Med 181:1459–1471

Yang B, Braciale TJ (1995) Characteristics of ATP-dependant peptide transport in isolated microsomes. J Immunol 155:3889–3896

Yang Y, Fruh K, Chambers J, Waters JB, Wu L, Spies T, Peterson PA (1992a) Major histocompatibility complex (MHC)-encoded HAM2 is necessary for antigenic peptide loading onto class I MHC molecules. J Biol Chem 267:11669–11672

Yang Y, Waters JB, Fruh K, Peterson PA (1992b) Proteasomes are regulated by interferon γ: implications for antigen processing. Proc Natl Acad Sci USA 89:4928–4932

Yang Y, Fruh K, Ahn K, Peterson PA (1995) In vivo assembly of the proteasomal complexes, implications for antigen processing. J Biol Chem 270:27687–27694

Yang Y, Sempe P, Peterson PA (1996) Molecular mechanisms of class I major histocompatibility complex antigen processing and presentation. Immunol Res 15:208–233

Young ACM, Nathenson SG, Sacchettini JC (1995) Structural studies of class I major histocompatibility complex proteins: insights into antigen presentation. FASEB J 9:26–36

Zijlstra M, Bix N, Simistier N, Loring J, Raulet D, Jaenish R (1990) b2-microglobulin deficient lack CD4-8+ cytotoxic T cells. Nature 344:742–746

Peptide–Major Histocompatibility Complex Class I Complex: From the Structural and Molecular Basis to Pharmacological Principles and Therapeutic Applications

DENIS HUDRISIER[1,2] and JEAN EDOUARD GAIRIN[1]

1 Introduction

Pathologic states such as viral or parasitic infections, cancer, autoimmune diseases, or graft rejection have in common the development of an immune response that involves the recruitment of activated T cells. T cell activation depends on two

[1]Institut de Pharmacologie et de Biologie Structurale, UPR 9062 CNRS, 205 route de Narbonne, 31400 Toulouse, France

[2]*Present address:* The Ludwig Institute for Cancer Research, Lausanne Branch, 1066 Epalinges, Switzerland

fundamental events: (1) the recognition of antigenic peptides associated with major histocompatibility complex (MHC) molecules expressed at the surface of the presenting cells by antigen-specific T cell receptors (TCR) expressed by T cells and (2) the TCR-mediated signal transduction cascade triggered by the association of the TCR with its ligand, the peptide–MHC complex. Peptide–MHC interactions have been studied extensively and are now well understood. Interactions of peptides with MHC class I molecules (recognized by cytotoxic T lymphocytes, CTL) or with MHC class II molecules (recognized by helper T cells) are comparable in some aspects, but differ significantly in others. Their structural and biological principles are well documented (see, e.g., STERN and WILEY 1994; GREY et al. 1995; MADDEN 1995; RAMMENSEE 1995; JONES 1997 and references cited in these reviews). In the past few years, analysis of the molecular, structural, and biological aspects of MHC–peptide–TCR interaction has allowed the pharmacological principles of the interaction between the TCR and its ligand, the peptide–MHC complex, to be established (HUDRISIER and GAIRIN 1996). Indeed, it is now well established that TCR ligands may be endowed with agonist, partial agonist, or antagonist properties (for a review, see, e.g., JAMESON and BEVAN 1995). The structure of the antigenic peptide defines its ability to bind to the MHC molecule on the one hand and its efficiency to trigger TCR activation on the other; subsequently and/or consequently, the structure of the peptide–MHC complex dictates the functional (pharmacological) properties of the TCR ligand.

In this review, we shall focus on the peptide-MHC class I complex. The structural and molecular basis of peptide-MHC class I interactions will first be described, followed by details of the determinants of peptide binding; the consequences for interactions with T cells will then be presented, and finally the complexes will be discussed in terms of pharmacological principles and therapeutic applications.

2 Nature and Structure of the Two Components of the T Cell Receptor Ligand: The MHC Molecule and the Antigenic Peptide

Biochemical identification of peptides naturally bound to MHC and analysis by X-ray radiocrystallography of MHC molecules in complex with endogenous or viral peptides provided very precise information on the respective structures of the two components of the peptide-MHC class I complex.

2.1 MHC Class I Molecules

MHC class I molecules are encoded on chromosome 17 in mice and chromosome 6 in humans and are characterized by a very high polymorphism. They consist of a single heavy chain (44 kDa), which is noncovalently associated with an invariant light chain, β_2-microglobulin (12 kDa). The heavy chain is organized in three extracellular domains – α_1, α_2, and α_3 – of approximately 90 amino acids each, a transmembrane segment, and a short cytoplasmic tail (NATHENSON et al. 1981). Alignment of the MHC class I primary sequences indicated that the high polymorphism was mainly concentrated in the α_1- and α_2- domains (BJORKMAN 1990). This was further confirmed by radiocrystallographic analysis of the HLA-A2 molecule (BJORKMAN et al. 1987a,b). This study revealed that the α_1- and α_2- domains form one structural domain consisting of two antiparallel α-helices above a floor of eight β-pleated sheets. The unresolved material found in this groove led to the hypothesis that it might correspond to bound heterogenous peptides. Thus the domain formed by the association α_1/α_2 represents the peptide-binding groove. The α_3- and β_2-microglobulin domains are structurally related to immunoglobulin domains. The α_3-domain is the main binding site for the TCR coreceptor CD8, while β_2-microglobulin interacts primarily with the α_3-domain but also with α_1 and α_2. These structural features are shared by all the classical MHC class I molecules (Fig. 1). A direct and important consequence of this observation is the that the structure and function of MHC molecules remains the same regardless of the animal system studied.

2.2 Antigens Presented by MHC Class I Molecules

2.2.1 Nature of the Antigens

The first demonstration that antigens presented to CTL by MHC class I were of peptidic nature was made by TOWNSEND et al. (1986), who showed that cells coated with a short peptide from the sequence of an influenza virus protein were recognized by antiviral CTL as efficiently as the infected cells. Antigenic peptides associated with MHC class I molecules may be of viral, bacterial, tumoral, or endogenous (self) origin. They derive from the degradation of newly synthesized proteins and are subsequently translocated in the endoplasmic reticulum, where they bind to the MHC (HEEMELS and PLOEGH 1995). Their origins are quantitatively and qualitatively variable. Self peptides seem to come mainly from ubiquitary proteins from cytoplasmic or nuclear origin, and less frequently from membranes (HARRIS 1994). Viral peptides can be selected from proteins contributing to the virus structure or from viral enzymes (OLDSTONE et al. 1988, 1995; REHERMANN et al. 1995). Tumor antigens are derived from overexpressed proteins, differentiation antigens, or mutated proteins (BOON et al. 1994). The diversity of antigenic peptides has recently been extended to proteins translated in non-open reading frames (MALARKANNAN et al. 1995) or from introns (COULIE et al. 1995). Finally, as reviewed below (see Sect. 4.2.1), post-translationally modified peptides can also be CTL epitopes (SKIPPER et al. 1996). The diversity of peptides bound to a given MHC molecule at the cell surface has been estimated to be over several thousands, each peptide being represented by between one and approximately 1000 copies (HUNT et al. 1992). Interestingly, the set of peptides was shown to be specific for a given MHC allele (CERUNDOLO et al. 1990; FALK et al. 1990). This observation led to the basic identification of allele-specific binding motifs (FALK et al. 1991).

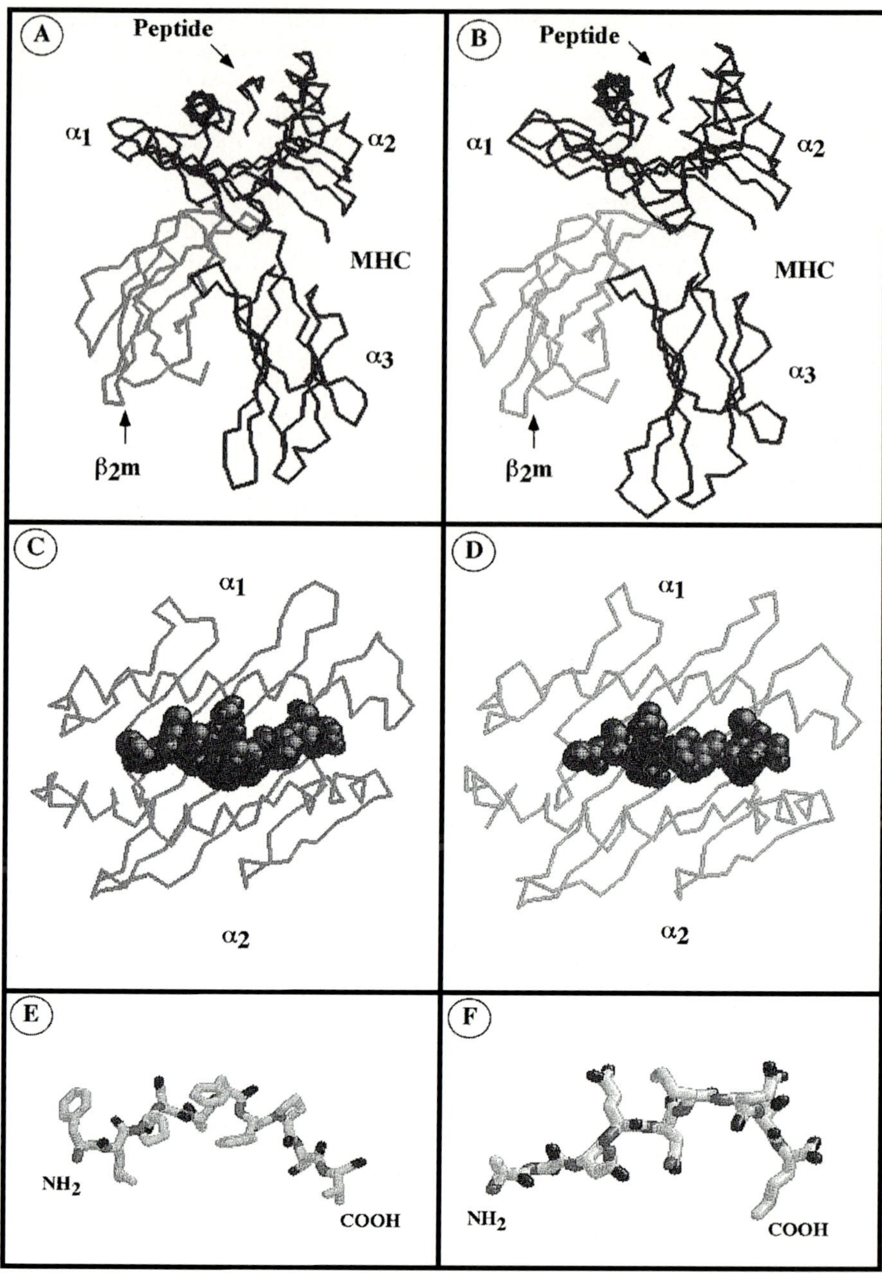
A
Peptide
α1
α2
MHC
α3
β2m
B
Peptide
α1
α2
MHC
α3
β2m
C
α1
α2
D
α1
α2
E
NH2
COOH
F
NH2
COOH

2.2.2 Allele-Specific Binding Motifs

A major contribution to our understanding of peptide-MHC class I interactions was the identification of peptides endogenously bound to MHC class I molecules. Peptide elution from MHC class I molecules in sufficient quantities allowed peptide pool-sequencing studies (FALK et al. 1991). This analysis revealed the existence of peptide-binding motifs, based on both peptide length and sequence and specific for a given MHC allele. The length of peptides bound to the MHC molecules is remarkably homogenous (eight to 11 amino acids) and depends on the MHC allele (FALK et al. 1991; RAMMENSEE et al. 1995). For example, peptides eluted from the H-2K^b molecule are octamers, whereas those eluted from H-2D^b are nonamers. Longer peptides (10- or 11-mers) are also found (JOYCE 1994). Sequencing of peptide mixtures eluted from MHC class I showed that amino acid composition at the different positions of the peptide sequence was not random. A more striking observation was that two positions, one systematically found at the C terminus and the other elsewhere in the peptide sequence, were occupied by a single amino acid (or a very limited number of amino acids), called an anchor residue (FALK et al. 1991). The C-terminal anchor is either a hydrophobic or a charged residue, and less commonly an aliphatic one. This might be a reflection of antigen processing, since (a) the enzymatic specificity of the proteasome (thought to be responsible for protein degradation and antigen maturation) generates such C-terminal extremities (GROETTRUP et al. 1996) and (b) the transport selectivity of TAP (transporter associated with antigen presentation) molecules (responsible for the translocation of antigenic peptides into the endoplasmic reticulum) is dictated in part by the peptide C termini (SCHUMACHER et al. 1994). The second anchor residue is located within the peptide sequence, most commonly at position 2 (e.g., H-2K^d, HLA-A2, or HLA-B27), but also at positions 3 (HLA-A1) or 5 (H-2K^b or H-2D^b). Peptide-binding motifs are now known for more than 40 different MHC molecules (RAMMENSEE et al. 1995). Once these motifs became known, it became a posteriori obvious that all the viral epitope sequences previously identified on the basis of CTL recognition in the context of one given allelic restriction could be aligned according to the positions of the two MHC anchor residues.

←

Fig. 1A–F. Structural comparison of **A,C,E** human and **B,D,F** murine peptide–MHC class I complexes. **A,B** The complexes are shown in association with β_2-microglobulin (β_2m). **C,D** The complexes are seen from the top of the bimolecular structure. **E,F** The extended conformation of the peptide bound to the MHC molecule is seen from the MHC α_2-domain (not shown on the figure). The molecular models were generated from crystal coordinates of a human MHC HLA-A2 molecule cocrystallized with hepatitis B virus nucleocapsid 18–27 (FLPSDFFPSV) (MADDEN et al. 1993) or a murine MHC H-2D^b molecule cocrystallized with influenza nucleoprotein 366–376 (ASNENMETM) (YOUNG et al. 1994) using Insight II (Biosym Technologies, CA)

3 Structural and Molecular Basis of Peptide–MHC Class I Interaction

3.1 Structural Basis

Major progress in our understanding of peptide-MHC class I interaction came from the resolution of the crystal structure of MHC molecules associated with single peptides. So far, seven MHC class I molecules have been crystallized with bound peptides (FREMONT et al. 1992, 1995; MADDEN et al. 1992, 1993; SILVER et al. 1992; ZHANG et al. 1992; GARBOCZI et al. 1994; YOUNG et al. 1994; REID et al. 1996). These studies allow accurate molecular modeling of the interaction between peptides and MHC class I molecules even if crystallographic data are not available (HUCZKO et al. 1993; LUESCHER et al. 1996). Peptides are bound to MHC class I molecules in an extended conformation (MATSUMURA et al. 1992). This structural feature is due to two factors: (1) the presence of bulky residues closing the ends of the binding cleft and (2) the fixation of the peptide N and C termini to the edges of the binding groove, thereby placing the peptide in the proper orientation to be held in the cleft by the hydrogen-bonding network (MATSUMURA et al. 1992; YOUNG et al. 1995). The structural constraints of the MHC class I-binding groove also explain why peptides bound to MHC molecules consist of eight to nine amino acids. However, longer peptides can be accommodated by some MHC molecules by bulging of its middle part or by extending out of the groove (COLLINS et al. 1994; JOYCE 1994). Shorter peptides that cannot reach both ends of the groove can still bind to the MHC, probably with only the C terminus fixed in the groove (HUDRISIER et al. 1995).

Radiocrystallographic studies have revealed that the NH_2- and COOH-terminal groups of the peptide interact with conserved residues located in two binding pockets of the MHC molecules named A and F (MATSUMURA et al. 1992). The peptide N-terminal residue is hydrogen bound to MHC Tyr^{7} and Tyr^{171}. The fact that its side chain is directed outside the MHC-binding groove explains the diversity of residues found at position 1. The COOH group of the peptide C-terminal residue interacts with Tyr^{84}, Thr^{143}, and Lys^{146} of the MHC F pocket, the residue side chain being buried in the groove. The peptide C terminus can be occupied by hydrophobic, aliphatic, or charged residues, depending on the nature of the residue at position 116 of the MHC. The involvement of peptide main-chain atoms in hydrogen bonds provides a general mechanism for peptide binding to all classical MHC class I molecules and contributes efficiently, in terms of energy, to the overall peptide-binding affinity (BOUVIER and WILEY 1994).

An important observation is that the peptides' backbones are highly superimposable in an allele-dependent (LEWICKI et al. 1995a) or -independent manner (MADDEN et al. 1993; YOUNG et al. 1994). Structural analysis of five peptides restricted to a given MHC allele revealed that the strong structural homology is based on the conserved bond network established between the peptide backbone and the MHC cleft and the use of identical anchor residues. Slight differences were the result of heterogeneous peptide length and/or of different peptide residue side chains directed toward the TCR (MADDEN et al. 1993). The fact that main-chain atoms of the peptide strongly interact with the MHC accounts for the diversity of peptides bound by a given MHC molecule. On the other hand, structural differences observed at the level of peptide side chains account for the diversity of TCR recognition.

Peptide-binding specificity is controlled by three parameters: (1) the overall hydrogen bond network established between the peptide main-chain atoms and polymorphic MHC side chains; (2) discrete hydrogen bonds between some peptide residue side-chain atoms and the MHC binding cleft; and (3) a series of recesses and indentations forming the pockets B–E of each binding cleft (MATSUMURA et al. 1992). Structural and physicochemical properties vary from one MHC molecule to another and are responsible for the allele-specific binding motif by accommodating peptide side chains (MATSUMURA et al. 1992). As mentionned above (see Sect. 2.2.2), the internal anchor residue located at position 2, 3, or 5 of the peptide sequence is accommodated by the pockets B, D, or C, respectively. The side chain of the anchor residue can either mediate a hydrogen bond with a residue within the specific pocket or simply fit into the pocket structure (ZHANG et al. 1992; YOUNG et al. 1994). Position 3 can still serve as an auxilliary anchor when it is not a main anchor residue (JAMESON and BEVAN 1992; SHIBATA et al. 1992; RUPPERT et al. 1993; SAITO et al. 1993; PARKER et al. 1994). Pocket E, which is poorly structured, is not essential for binding specificity (MATSUMURA et al. 1992). Finally, multiple possibilities are offered to different atoms of the peptide backbone to interact with residues of the MHC-binding cleft at short or longer distances via hydrogen bonds or by hydrophobic or Van der Walls interactions. In addition, the role of water in mediating peptide binding (FREMONT et al. 1995; SMITH et al. 1996b) and the conformational flexibility of the MHC-binding groove (SMITH et al. 1996a) are further important structural elements that must be taken into account, explaining how MHC molecules manage to bind a huge number of different peptides (YOUNG et al. 1995).

3.2 Molecular Basis

3.2.1 Kinetic and Thermodynamic Parameters

The first studies on peptide-MHC class I interaction did not clearly demonstrate selective MHC binding (BOUILLOT et al. 1989). This was due to the peptides and the methods used, the presence of peptides endogenously bound to MHC also being a serious limiting factor. In addition, since peptide-MHC interaction takes place in the endoplasmic reticulum, reconstructing the binding process in situ was difficult. Later, the use of mutant cell lines expressing empty MHC molecules (DEMARS et al. 1985; LJUNGGREN and KÄRRE 1985; SALTER and CRESSWELL 1986) allowed kinetic and thermodynamic parameters to be measured and led to the demonstration that peptide-MHC class I interaction followed the classical rules of ligand-receptor interaction (CERUNDOLO et al. 1991; CHRISTINCK et al. 1991). Two binding assays on viable cells were developed. The first one is based on the stabilization of empty MHC class I molecules at the cell surface in the presence of a specific peptide; the second one, a competition assay, was derived from classical ligand-receptor techniques and uses high-affinity, MHC-selective probes. Both measurements are, in most cases, well correlated. Stabilization assays are more stringent and might be a more accurate reflection of the in situ process (HUDRISIER et al. 1996). Competition assays allow the measurement of intrinsic binding parameters. Radioactive probes are useful tools for the direct quantitation of peptide-MHC complexes, and fluorescent ones for their direct visualization, both providing accurate measurements of kinetic and thermodynamic parameters (CHRISTINCK et al. 1991; HUDRISIER et al. 1995). In addition, fluorescent probes can be used to study antigen processing (DAY et al. 1995). Alternatively, binding assays using purified, soluble recombinant MHC class I molecules have been developed. In these assays, the experimental assessment of the peptide-MHC class I interaction is based on the refolding of stable MHC-peptide-β_2-microglobulin complexes, competition experiments with peptidic probes, or more recently plasmon resonance measurements (BOYD et al. 1992; KHILKO et al. 1993; SETTE et al. 1994b).

Association between peptide and MHC class I occurs within 1 h (CHRISTINCK et al. 1991), whereas the time necessary for the complex to dissociate, as deduced from dissociation kinetic studies, varies from minutes to hours depending on the MHC molecule and the peptide studied and the approach used (CERUNDOLO et al. 1991; CHRISTINCK et al. 1991; LUESCHER et al. 1991). Complex stability might be a critical in vivo factor in determining the immunogenicity and the hierarchial importance of high-affinity peptides (VAN DER BURG et al. 1996). Depending on the study considered, the affinity values deduced from kinetic or thermodynamic approaches are either comparable (CERUNDOLO et al. 1991; CHRISTINCK et al. 1991) or different (BOYD et al. 1992), suggesting a mechanism more complex than a simple first-order rate interaction. In addition, β_2-microglobulin plays an important role by contributing to the high affinity of complexes (TOWNSEND et al. 1989; BOYD et al. 1992). Despite the diversity of the peptide sequences bound to the MHC, the observed affinity values reflect rather strong interactions since they fall in the 1–500 n*M* range for known viral epitopes (FELTKAMP 1994; HUDRISIER et al. 1996). Interestingly enough, attempts to design synthetic peptides of higher affinity (below nanomolar values) have failed up to now. Indeed, this limit in affinity could reflect the adaptatibility of the MHC molecule to present a large number of peptides (see Sect. 3.1 above). According to the ligand-receptor rules, a higher affinity would correspond, in structural terms, to a higher stringency of binding and would result, in functional terms, in a much narrower range of peptides presentable by the MHC, a feature which is contradictory to the function of MHC.

As described below, the kinetic and thermodynamic binding parameters depend on peptide structure (GAIRIN et al. 1995; HUDRISIER et al. 1995).

3.2.2 Structural Parameters

Improved binding techniques and knowledge of peptide-binding motifs have allowed a fine delineation of peptide structural parameters such as length, primary (sequence), and secondary (conformation) structures that play a role in peptide-MHC class I interaction. As the MHC class I binding cleft is closed at each extremity, the peptide length is an important parameter, and peptides with nonoptimal lengths usually display altered binding affinity: (a) lengthening the peptide at either the N or C terminus is acceptable, but gradually decreases binding affinity, (b) deletion of the C-terminal anchor results in a dramatic loss of MHC-binding properties, and (c) N-terminal truncation can affect peptide binding to MHC dramatically or not at all, depending on the epitope sequence (CERUNDOLO et al. 1991; CHRISTINCK et al. 1991; GAIRIN et al. 1995; HUDRISIER et al. 1995).

The role of peptide primary structure on MHC binding has mainly been studied using amino acid scanning (JAMESON and BEVAN 1992; SHIBATA et al. 1992; SAITO et al. 1993; VAN DEN EYNDE et al. 1994) or peptide libraries (UDAKA et al. 1995). This has allowed the contribution of the two anchor residues to the specificity and the strength of peptide binding to be determined (SAITO et al. 1993). Replacement of the anchor residues by an alanine usually led to a dramatic alteration of the peptide-binding properties (JAMESON and BEVAN 1992; SHIBATA et al. 1992; SAITO et al. 1993; VAN DEN EYNDE et al. 1994; HUDRISIER et al. 1997). However, such alteration can be minimized if all the other positions of the peptide sequence are occupied by optimal amino acids (HUDRISIER et al. 1995). Although rare, some peptides not fitting the canonical binding motif can still bind efficiently to MHC (SHEIL et al. 1994). The C-terminal anchor can be occupied by amino acids of comparable physicochemical properties (e.g., hydrophobic or charged ones) (RAMMENSEE et al. 1995). At the other anchor position, the same observation may apply for some MHC (e.g., Asp or Glu can be found at position 3 for HLA-A1 and Tyr or Phe at position 5 for H-2K^b), but not for others for which only one amino acid is accepted (e.g., only Asn, and not Gln is accepted at position 5 of H-2D^b-restricted peptides).

Stereochemical and/or conformational changes occurring either at a main anchor (HUDRISIER 1996) or at auxilliary ones (GAIRIN and OLDSTONE 1993) result in a binding affinity decreased 1–3 logs, indicating that peptide secondary structure is also a critical factor.

3.3 Application of the MHC-Binding Rules

3.3.1 Prediction of Cytotoxic T Lymphocyte Epitopes Based on MHC-Binding Parameters

The accuracy of prediction of CTL epitopes based solely on the presence of the anchor residues in a peptide sequence is very low and has been limited to a few cases (PAMER et al. 1991; DIBRINO et al. 1993; NAYERSINA et al. 1993; CERNY et al. 1995), strongly suggesting that the allele-specific motif, although mandatory, is not sufficient to ensure MHC binding. Systematic analysis of the binding properties of peptides bearing allele-specific motifs revealed a critical role (favorable or unfavorable) for residues at non-anchor positions (RUPPERT et al. 1993; HUDRISIER et al. 1996). Amino acids that favor high-affinity binding are overrepresented in high-affinity peptides, whereas unfavorable amino acids are overrepresented in nonbinders. The function of non-anchor positions is to present residues to CTL, and a strong degeneracy of residues should thus be accepted. However, some amino acids are not tolerated at all at these positions, which therefore strongly control peptide selection by MHC class

I molecules (Ruppert et al. 1993; Hudrisier et al. 1996). The presence of multiple favorable residues at non-anchor positions does not necessarily lead to peptides of higher MHC-binding affinity (see Sect. 3.2.1 above), nor does it predict their antigenic or immunogenic properties (see Sect. 3.3.2). In contrast, the presence of a single unfavorable residue is sufficient to prevent peptide binding to MHC. Interestingly, a few non-anchor positions (usuallythree in a nonameric peptide sequence) can accept amino acids of any kind without affecting MHC binding. This structural flexibility allows a very large number of amino acid combinations (20^3, i.e., 8000) and the generation of a CTL response against a wider spectrum of peptides (Hudrisier et al. 1996). Knowledge of the anchor, auxilliary anchor, and detrimental residues for a given MHC allele leads to the concept of extended binding motifs, which greatly increase the prediction of CTL epitopes (Grey et al. 1995; Parker et al. 1994; Walden 1996).

3.3.2 Correlation Between MHC Binding, Antigenicity, and Immunogenicity

All the immunodominant viral epitopes described so far display high or intermediate binding affinity for MHC class I molecules (Feltkamp et al. 1994; Sette et al. 1994; Hudrisier et al. 1996), whereas non-immunodominant epitopes do not (Chen et al. 1994; Oukka et al. 1994), suggesting a direct link between affinity and immunogenicity of MHC class I-restricted peptides. However, only few predicted high-affinity peptides represent CTL epitopes, indicating that the correlation between affinity and immunogenicity is limited (Sette et al. 1994; van der Burg et al. 1996). Indeed, high-affinity binding results in an increased peptide-MHC complex stability, an element necessary for an efficient and productive association of the TCR (Pogue et al. 1995; Levitsky et al. 1996). In fact, stability of peptide-MHC complexes, rather than affinity, would correlate more accurately with the immunogenicity of antigenic peptides (van der Burg et al. 1996). High-affinity peptides predicted in a given foreign protein are immunogenic when injected in mice. However, not all the high-affinity viral peptides found within viral protein sequences generate a CTL response following virus infection (Oldstone et al. 1995; Hudrisier et al. 1996). Indeed, additional parameters (Barber and Parham 1994), including efficiency of antigen presentation (Oukka et al. 1994; Ossendorp et al. 1996) and appropriate T cell recognition (Oldstone et al. 1995), are required. Flanking sequences or residues within the peptide sequence can perturb protein maturation and hamper generation of the peptide by the proteasome (Einsenlhor et al. 1992; Ossendorp et al. 1996). The peptide can also be a poor substrate for the TAP transporters (Neisig et al. 1995). Absence of TCR able to recognize the peptide can be responsible for the nonrecognition of a peptide even if it is presented by the MHC at the cell surface. Within a pathogen, the immunogenicity of the different antigenic peptides is also modulated by their hierarchial importance and their respective efficiency of CTL recognition (Mylin et al. 1995; Oldstone et al. 1995). Peptides only partially conforming to the above-mentioned parameters can still be presented by MHC molecules at the cell surface and can be recognized by CTL as non-immunodominant epitopes (Oukka et al. 1994; VAN DER MOST et al. 1996). Though unclear, their role in vivo could be important in situations in which an immunodominant epitope is removed. In addition, they can be used to potentially confer immune protection against pathogens (OUKKA et al. 1996; VAN DER MOST et al. 1996).

3.3.3 Presentation of A Single Antigenic Peptide by Multiple MHC Class I Molecules

A given antigenic peptide is usually presented to CTL in the context of a given MHC molecule, a phenomenon known as "MHC restriction" (ZINKERNAGEL and DOHERTY 1974). Some antigenic peptides harbor in their sequence more than one MHC-binding motif, allowing them, in some cases, to bind efficiently to multiple MHC class I molecules. Only a very limited number of these multiple MHC-binding peptides are known to be recognized in vivo by CTL in the context of more than one MHC (OLDSTONE et al. 1992; MISSALE et al. 1993; HUDRISIER et al. 1997). Peptide binding to multiple MHC molecules involves either comparable (SIDNEY et al. 1995; LUESCHER et al. 1996) or different processes (SCHUMACHER et al. 1991; DERES et al. 1992; HUDRISIER et al. 1997) based on overlapping or nonoverlapping MHC-restricted binding motifs, respectively. In addition to the presence of several binding motifs in the sequence, the peptide structure and anchor residues must fit into the binding pockets of the different MHC molecules. This can only be made possible by a dramatic conformational change in the central part of the peptide backbone and the reorganization of the respective anchor residues and TCR contacts in the correct positions (Fig. 2). As a consequence, peptide-MHC class I complexes show quite different antigenic surfaces to TCR (HUDRISIER et al. 1997). Peptide binding to diverse MHC class I molecules has no redundancy in terms of T cell recognition; on the contrary, it is a cellular means for the infected host to increase the diversity of its T cell repertoire. In some (but not all) cases, multiple allelic antigen presentation should confer a selective advantage to the infected cells to avoid viral escape strategies (HUDRISIER et al. 1997). Indeed, such peptides are ideal candidates for vaccinal approaches and would represent a valuable alternative to the use of peptide pools (OLDSTONE et al. 1992).

4 Interaction of the Peptide–MHC Class I Complex with T Cells

4.1 T Cell Receptor Recognition of the Peptide–MHC Complex

The interaction between the TCR and its ligand occurs at the level of complementary determining regions (CDR1–3) with peptide and MHC residues (DAVIS and BJORKMAN 1988). The crystal structures of peptide-MHC complexes have shown that peptidic residues that do not interact with or are not directed toward the MHC

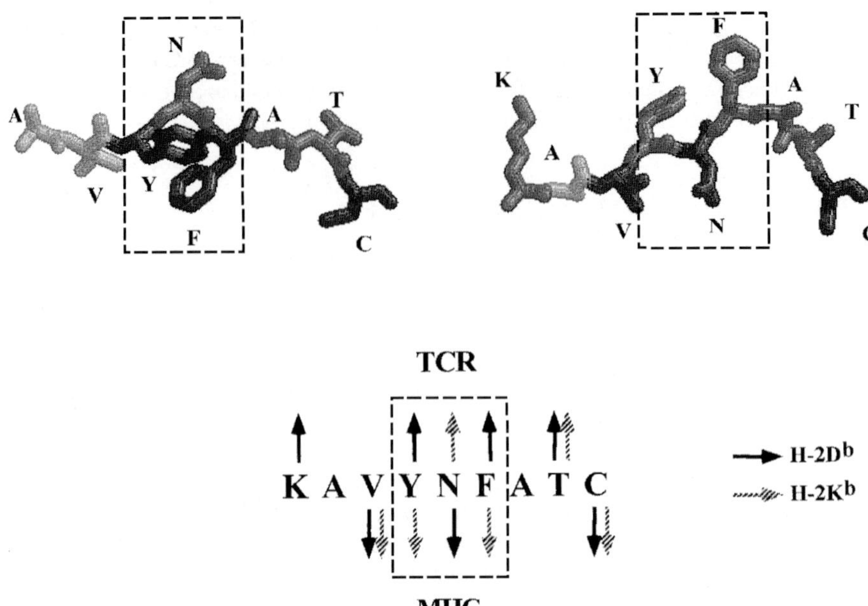

Fig. 2. Molecular modeling of the structure of an antigenic viral peptide bound to two different MHC molecules. Side view seen from the MHC α_2-domain (for the sake of clarity, the MHC structure is not shown) of the lymphocytic choriomeningitis virus glycoprotein-1 epitope (K/AVYNFATC) interacting with H-2K^b (*top left*) or with H-2D^b (*top right*). Note the dramatically different conformations adopted by the peptide core sequence YNF (*boxes*) imposed by the specific MHC binding contraints. In the peptide sequence (*bottom*), *arrows* indicate the orientation of the residues toward the MHC or the T cell receptor (*TCR*). *Solid arrows*, H-2D^b; *shaded arrows*, H-2K^b. Note that each residue of the core sequence YNF is an MHC anchor in one context and a TCR contact in the other (adapted from HUDRISIER et al. 1997)

molecule are highly accessible to the solvent and point outside the binding groove and are therefore potential TCR contacts. Study of monosubstituted analogues of known epitopes has clearly identified these TCR contacts (JAMESON and BEVAN 1992; SHIBATA et al. 1992; VAN DEN EYNDE et al. 1994). TCR-MHC interaction is less well defined, but involves residues of the α_1- and α_2-domains of the MHC (AJITKUMAR et al. 1988). It has been suggested that TCR-MHC interaction is mediated mainly through electrostatic bonds (SUN et al. 1995). More recently, studies on the first crystal structures of TCR-ligand complexes have confirmed these data and suggested a conserved orientation for the recognition of the peptide-MHC class I complex by the TCR (GARBOCZI et al. 1996; GARCIA et al. 1996). The molecular principles of ligand-TCR interactions are more difficult to establish than the peptide-MHC one, since both the TCR and its ligand, the peptide-MHC class I complex, are membrane associated. Direct measurements have been made by surface plasmon resonance using soluble molecules and by TCR photoaffinity labeling on viable cloned CTL (CORR et al. 1994; LUESCHER et al. 1994; ALRAMADI et al. 1995). These methods, together with indirect ones, have shown that the affinity of the peptide-MHC class I complex for its specific TCR is weak (FREMONT et al. 1996), but is greatly reinforced by the contribution of the coreceptor CD8 present at the surface of CTL (LUESCHER et al. 1995; GARCIA et al. 1996). Although controversial, the concept of serial engagement of several TCR by a single peptide-MHC complex is interesting, since it reconciles the efficiency of CTL activation with the fact than only a few peptide-MHC complexes are present at the surface of a target cell and interact weakly with the TCR (VALITUTTI et al. 1995).

4.2 Structural Modifications of the Antigen and Their Consequences

From synthesis to degradation, proteins can be subject to several alterations, including post-translational, metabolic, or chemical modifications or mutations that lead to regulation or dysregulation of their function.

4.2.1 Post-translational, Metabolic, and Chemical Modifications

The main post-translational, metabolic, and chemical modifications include phosphorylation (Ser, Thr, Tyr), lipidation, bond rearrangement (Asn, Gln), oxidation (Met, Cys), and hydroxylation (Lys, Pro). They take place in virtually all subcellular compartments, including those in which antigens are processed. Among all the possible post-translational modifications, attention has been focused on the glycosylation process (HAURUM et al. 1994, 1995; FERRIS et al. 1996; SKIPPER et al. 1996). Glycosylation and antigen presentation take place in the same subcellular locations, i.e., the cytosol and the endoplasmic reticulum. The first evidence that such modifications could play an important role came from in vitro experiments based on the use of model peptides. HAURUM et al. (1994) have shown that O- or N-glycosylation could influence antigen selection by the MHC class I molecules and CTL recognition. These authors have also shown that the TCR could recognize a glycosylated peptide and that the glycosylated residue could confer specificity to TCR recognition. A similar observation has recently been made for phosphorylated peptides (HAURUM 1997). In a very elegant manner, CHEN et al. (1996) showed that chemically altered peptides (in this case, altered by bond rearrangement) are also recognized by CTL. All these data indicate that the TCR repertoire covers a wider range of products than unmodified peptides.

However, the in vivo importance of such events remains unclear. For example, it is interesting to note that all but one of the known T cell epitopes bears a glycosylation motif. This could reflect either a preferential use of unmodified peptide as epitope or alternatively a technical bias, post-translationally modified peptides not being specifically sought. For instance, it is not known whether the processing of glycoproteins occurs before or after completion of post-translational or metabolic modifications (FERRIS et al. 1996). One possibility is that proteins are deglycosylated or dephosphorylated before entering the antigen-processing pathway. This hypothesis is supported by the observation that the unique post-translational epitope characterized so far is a deglycosylated peptide (SKIPPER et al. 1996).

Therefore, the pathophysiological impact of post-translational modification of antigens remains to be studied. In addition to its role in diversity of the immune response, it would be interesting to consider the possibility that such altered antigens could lead to an unwanted autoimmune response or alternatively that they may form neoantigens in transformed or virally infected cells in an environment-dependent context (e.g., stress, tissue- or cell-specific metabolic pathway).

4.2.2 Mutations of the Antigen Sequence and Their Impact on Antigen Recognition and T Cell Activation

Pathogens use numerous strategies to avoid the immune response (KOUP 1994). These include the ability to downregulate MHC expression or the blocking of antigen processing. Another important strategy consists in generating mutations in the sequence of CTL epitopes. The relevance in vivo of antigenic variants for escape strategies is still controversial (MOSKOPHIDIS and ZINKERNAGEL 1995; OLDSTONE et al. 1995). Many pathogens escape CTL recognition by a single mutation at the level of an epitope residue (FRANCO et al. 1995). Mutations can affect an anchor or an auxilliary anchor residue, thus preventing peptide selection by the MHC molecule (DECAMPOS-LIMA et al. 1993; COUILLIN et al. 1994; HUDRISIER et al. 1997). Mutations can also affect CTL recognition by inducing a conformation change in the TCR aggretope (PIRCHER et al. 1990). Conversely, the mutation can affect a TCR contact of the antigenic peptide (LEWICKI et al. 1995a,b; HUDRISIER et al. 1997). By affecting a crucial TCR contact, a mutation can prevent TCR-ligand interaction and consequently CTL activation even in the context of a polyclonal epitope-specific immune response.

A much more sophisticated mechanism, TCR antagonism, used by naturally occurring viral variants to evade the CTL response has been evidenced recently (BERTOLETTI et al. 1994; KLENERMAN 1994; REHERMANN et al. 1995; REID 1996). This mechanism, which is presented below (see Sect. 4.3.1), might have important implications not only for pathogen escape strategies, but also for normal development or manipulation of the cellular immune response (JAMESON and BEVAN 1995).

4.3 Pharmacological Principles and Therapeutic Applications

4.3.1 Peptide-MHC Class I Complexes Can Act as T Cell Receptor Agonists or Antagonists

The concept of TCR antagonism implies altered CTL functions as a consequence of TCR interactions with structurally altered peptide-MHC complexes. First demontrated for peptides-MHC class II complexes, it was rapidly extended to peptide-MHC class I complexes. Indeed, while the process of MHC-peptide-TCR recognition is antigen specific, it is now well known that altered synthetic or natural variants of MHC class I-restricted peptides can still be recognized by $CD8^+$ T cells (JAMESON and BEVAN 1993; SETTE et al. 1994; KESSLER et al. 1996; MARTIN et al. 1996). CTL functions comprise several activities, such as cytotoxicity, proliferation, cytokine production, and memory, which are fully triggered by a so-called agonist. In the manner of many other ligand-receptor interactions, peptide modification can alter ligand-TCR interactions and subsequently CTL functions. Depending on the structural modification, such peptides can lead to either full or partial T cell activation or antagonism. Partial activation means that lymphocyte function is affected in a qualitative way (e.g., cytotoxicity without cytokine production), whereas antagonism represents a nonfunctional ligand-TCR interaction. Therefore, such cross-reactive peptides are thought to play a role in physiopathological situations such as thymic or peripheral T cell selection, anergy, memory, viral escape, or autoimmune disorders (JAMESON and BEVAN 1995). How these mutations act to alter CTL activation is unknown. Comparison of the crystal structures of an altered peptide endowed with antagonist properties versus an agonist bound to HLA-B8 suggested that antagonism might result from a conformational change in the TCR ligand (REID et al. 1996), but this hypothesis remains controversial. Interestingly, CD8-dependent CTL clones seem more prone to antagonism following epitope modification (KESSLER et al. 1996).

Taken together, these data imply that a single TCR can functionally interact with several different ligands; these ligands can be classified as agonists, partial agonists, or antagonists. A strong sequence homology between these ligands is not required, since poorly related or minimally homologous peptides can stimulate the same TCR (EVAVOLD et al. 1995; KERSH and ALLEN 1996). In addition, $CD8^+$ T cells can be partially activated by self-derived peptides (CAO et al. 1995). These findings have at least two important consequences: (1) the concept of T cell epitope molecular mimicry cannot be limited to sequence homology, but must be extended to antigenic surface similarity (QUARATINO et al. 1995), and (2) multiple MHC class I-peptide complexes that are ligands of a single TCR likely exist in the endogenous repertoire.

Recent evidence that antagonist peptides can play a role in the process of positive or negative selection of T lymphocytes suggests their importance in pathophysiological conditions. Their participation in such events, although speculative, might help us to understand the role of peptides in shaping the lymphocyte repertoire and in the maintenance of T cell memory or activation (JAMESON and BEVAN 1995). This might also have important consequences in terms of therapeutic applications, since knowledge of the multiple TCR contacts in the antigen sequence, of their differential use by various CTL clones (COLE et al. 1995), and of their hierarchial role in T cell activation could lead to a better design of peptidic or nonpeptidic immunomodulatory molecules.

4.3.2 Therapeutic Applications

Analysis of peptide-MHC interaction has resulted in various strategies to identify antigens of therapeutic interest. Numerous viral or tumor antigens have now been characterized in humans. A prime example is the identification of melanoma-specific antigens of the melanoma antigen (MAGE) family (BOON et al. 1994). Recently, a new technique based on the use of multimeric peptide-MHC complexes has been developed to detect antigen-specific CTL (ALTMAN et al. 1996). Improvement of this technique could help in the future to monitor the benefits of a peptide-based immunotherapy.

The fact that exogenously added peptides efficiently bind to cell-surface MHC molecules opens up the possibility of using synthetic peptides to enhance or inhibit an immune response. The use of synthetic peptides to modulate the CTL response has already been investigated. In order to inhibit CTL response, high-affinity, MHC class I-specific synthetic peptides have been used in vitro to inhibit the $CD8^+$ CTL response (GAIRIN and OLDSTONE 1992). In this strategy, peptides act as MHC blockers by preventing the access of viral epitopes to MHC molecules. The limitation of this approach is the requirement of constantly high amounts of synthetic peptides in cell culture or biological fluids (ISHIOKA et al. 1994). A promising alternative approach migh be the use of antagonist peptides, which, since they interact with the TCR, are active even at low doses and are able to modulate lymphocyte recognition by affecting cytokine response (KUCHROO et al. 1994). In order to enhance the immune response, synthetic peptides corresponding to the sequence of natural epitopes can be used in antitumor or antiviral vaccine strategies. Alternatively, covalent peptide-MHC complexes can be very useful tools to induce a CTL response (LUESCHER et al. 1992; ANJUERE et al. 1995) and might represent potent immunomodulatory agents. However, in order to fulfill this objective, the vaccination protocols and the route of vaccine administration need to be carefully controlled. Indeed, depending on the route of injection, the same peptide can either trigger CTL activation or induce tolerance (AICHELE et al. 1995). Peptide vaccination has been also shown to promote the rejection of a tumor (KAST et al. 1989) or its propagation (TOES et al. 1996). Another limiting factor in the use of peptides is their possible degradation in biological media and the short half-life of complexes in vivo. Two recently successful approaches have been developed in order to increase the stability of antigenic peptides presented by MHC class I by introducing retro-inverso or reduced bonds (GUICHARD et al. 1995, 1996) or non-natural amino acids (AYYOUB et al. 1997) into the peptide sequence. Finally, exotic peptides or peptides bearing nonpeptidic portions in their sequences have also been successfully designed (ROGNAN et al. 1995; WEISS et al. 1995; BOUVIER and WILEY 1996). Such structurally modified peptidic or nonpeptidic MHC ligands might ultimately be of therapeutic interest.

5 Conclusion

The past decade of study on MHC class I-peptide-TCR interactions has revolutionized our understanding of the CTL response to pathogens or cancer. The first half of the decade saw the establishment of the general structural and molecular principles of the MHC class I-peptide complex. In the past few years, these principles have been more precisely defined. Further, and perhaps more importantly, the last 2 or 3 years have seen spectacular progress in our knowledge of MHC class I-peptide-TCR interactions and the emergence of their pharmacological principles. Nevertheless, their role in the development of the immune repertoire, T cell selection, activation, or anergy has yet to be fully elucidated. Establishment of the structural and pharmacological principles of MHC class I-peptide-TCR interactions is a key step in our future understanding of the mechanisms involved in T cell ontogenesis, the evasion of the immune system by pathogens, and their manipulation for therapeutic purposes. In particular, the design of pharmacologically active TCR ligands (such as synthetic peptide or nonpeptide vaccines, MHC blockers, TCR antagonists) that modulate recognition of target cells by CTL and their use in immunotherapeutic approaches against viral infection, cancer, or autoimmune disease represent a major challenge for the next decade in scientific and clinical research.

References

Aichele P, Brduschariem K, Zinkernagel RM, Hengartner H, Pircher H (1995) T cell priming versus T cell tolerance induced by synthetic peptides. J Exp Med 182:261–266

Ajitkumar P, Geier SS, Kesari KV, Borriello F, Nakagawa M, Bluestone JA, Saper MA, Wiley DC, Nathenson SG (1988) Evidence that multiple residues on both the alpha-helices of the class I MHC molecule are simultaneously recognized by the T cell receptor. Cell 54:47–56

Alramadi BK, Jelonek MT, Boyd LF, Margulies DH, Bothwell ALM (1995) Lack of strict correlation of functional sensitization with the apparent affinity of MHC/peptide complexes for the TCR. J Immunol 155:662–673

Altman JD, Moss PAH, Goulder PJR, Barouch DH, McHeyzer-Williams MG, Bell JI, McMichael AJ, Davis MM (1996) Phenotypic analysis of antigen-specific T lymphocytes. Science 274:94–96

Anjuere F, Horvath C, Cerottini JC, Luescher IF (1995) Induction of CTL in vivo by major histocompatibility complex class I-peptide complexes covalently associated on the cell surface. Eur J Immunol 25:1535–1540

Ayyoub M, Mazarguil H, Monsarrat B, van den Eynde B, Gairin JE (1997) A structurally modified, peptidase-resistant analogue of human tumor antigen MZ2-E is a potent activator of anti-melanoma cytotoxic T lymphocytes. Proceedings of the 13th European Immunology Meeting, 22–25 June, Amsterdam, p 5.10

Barber LD, Parham P (1994) The essence of epitopes. J Exp Med 180:1191–1194

Bertoletti A, Sette A, Chisari FV, Penna A, Levrero M, De Carli M, Fiaccadori F, Ferrari C (1994) Natural variants of cytotoxic epitopes are T-cell receptor antagonists for antiviral cytotoxic T cells. Nature 369:407–410

Bjórkman PJ (1990) Structure, function, and diversity of class I major histocompatibility complex molecules. Annu Rev Immunol 59:253–288

Bjørkman PJ, Saper MA, Samraoui B, Bennet WS, Strominger JL, Wiley DC (1987a) The foreign antigen binding site and T cell recognition region of class I histocompatibility antigens. Nature 329:513–518

Bjorkman PJ, Saper MA, Samraoui B, Bennet WS, Strominger JL, Wiley DC (1987b) Structure of the human class I histocompatibility antigen, HLA-A2. Nature 329:506–512

Boon T, Cerottini JC, Van den Eynde B, Vanderbruggen P, Vanpel A (1994) Tumor antigens recognized by T lymphocytes. Annu Rev Immunol 12:337–365

Bouillot M, Choppin J, Cornille F, Martinon F, Papo T, Gomard E, Fournié-Zaluski MC, Levy J-P (1989) Physical association between MHC class I molecules and immunogenic peptides. Nature 339:473–475

Bouvier M, Wiley DC (1994) Importance of peptide amino and carboxyl termini to the stability of MHC class I molecules. Science 265:398–402

Bouvier M, Wiley DC (1996) Antigenic peptides containing large PEG loops designed to extend out of the HLA-A2 binding site form stable complexes with class I major histocompatibility complex molecules. Proc Natl Acad Sci USA 93:4583–4588

Boyd LF, Kozlowski S, Margulies DH (1992) Solution binding of an antigenic peptide to a major histocompatibility complex class I molecule and the role of β_2-microglobulin. Proc Natl Acad Sci USA 89:2242–2246

Cao W, Tykodi SS, Esser MT, Braciale VL, Braciale TJ (1995) Partial activation of CD8+ T cells by a self-derived peptide. Nature 378:295–298

Cerny A, Mchutchison JG, Pasquinelli C, Brown ME, Brothers MA, Grabscheid B, Fowler P, Houghton M, Chisari FV (1995) Cytotoxic T lymphocyte response to hepatitis C virus-derived peptides containing the HLA A2.1 binding motif. J Clin Invest 95:521–530

Cerundolo V, Alexander J, Anderson K, Lamb C, Cresswell P, McMichael A, Gotch F, Townsend A (1990) Presentation of viral antigen controlled by a gene in the major histocompatibility complex. Nature 345:449–452

Cerundolo V, Elliott T, Elvin J, Bastin J, Rammensee H-G, Townsend A (1991) The binding affinity and dissociation rate of peptides for class I major histocompatibility complex molecules. Eur J Immunol 21:2069–2075

Chen W, Khilko S, Fecondo J, Margulies DH, McCluskey J (1994) Determinant selection of major histocompatibility complex class I-restricted antigenic peptides is explained by class I-peptide affinity and is strongly influenced by nondominant anchor residues. J Exp Med 180:1471–1483

Chen W, Ede NJ, Jackson DC, McCluskey J Purcell AW (1996) CTL recognition of an altered peptide associated with asparagine bond rearrangement. J Immunol 157:1000–1005

Christinck ER, Luscher MA, Barber BH, Williams DB (1991) Peptide binding to class I MHC on living cells and quantitation of complexes required for CTL lysis. Nature 352:67–69

Cole GA, Hogg TL, Woodland DL (1995) T cell recognition of the immunodominant Sendai virus NP324-332/K^b epitope is focused on the center of the peptide. J Immunol 155:2841–2848

Collins EJ, Garboczi DN, Wiley DC (1994) Three-dimensional structure of a peptide extending from one end of a class I MHC binding site. Nature 371:626–629

Corr M, Slanetz AE, Boyd LF, Jelonek MT, Khilko S, Al-Ramadi BK, Kim YS, Maher SE, Bothwell ALM, Margulies DH (1994) T cell receptor-MHC class I peptide interactions: affinity, kinetics, and specificity. Science 265:946–9949

Couillin I, Culmann-Penciolelli B, Gomard E, Choppin J, Levy J-P, Guillet J-G, Saragosti S (1994) Impaired cytotoxic T lymphocyte recognition due to genetic variations in the main immunogenic region of the human immunodeficiency virus 1 NEF protein. J Exp Med 180:1129–1134

Coulie PG, Lehmann F, Lethe B, Lurquin C, Andrawiss M, Boon T (1995) A mutated intron sequence codes for an antigenic peptide recognized by cytolytic T lymphocytes on a human melanoma. Proc Natl Acad Sci USA 92:7976–7980

Davis MM, Bjorkman PJ (1988) T-cell receptor genes and T-cell recognition. Nature 334:395–402

Day PM, Esquivel F, Lukszo J, Bennink JR, Yewdell JW (1995) Effect of TAP on the generation and intracellular trafficking of peptide-receptive major histocompatibility complex class I molecules. Immunity 2:137–147

DeCampos-Lima P-O, Gavioli R, Zhang Q-J, Wallace LE, Dolcetti R, Rowe M, Rickinson AB, Masucci MG (1993) HLA-A11 epitopes loss isolates of Epstein-Barr virus from a highly A11+ population. Science 260:98–100

DeMars R, Rudersdorf R, Chang C, Petersen J, Standtmann J, Korn N, Sidwell B, Orr HT (1985) Mutations that impair post-transcriptional step in expression of HLA-A and -B antigens. Proc Natl Acad Sci USA 82:8183–8187

Deres K, Schumacher TNM, Wiesmuller K-H, Stevanovic S, Greiner G, Jung G, Ploegh HL (1992) Preferred size of peptides that bind to H-2Kb is sequence dependant. Eur J Immunol 22:1603–1608

Dibrino M, Tsuchida T, Turner RV, Parker KC, Coligan JE, Biddison WE (1993) HLA-A1 and HLA-A3 T-Cell epitopes derived from influenza virus proteins predicted from peptide binding motifs. J Immunol 151:5930–5935

Einsenlhor LC, Yewdell JW, Bennink JR (1992) Flanking sequences influence the presentation of an endogenously synthesized peptide to cytotoxic T lymphocytes. J Exp Med 175:481–487

Evavold BD, Sloanlancaster J, Wilson KJ, Rothbard JB, Allen PM (1995) Specific T cell recognition of minimally homologous peptides: evidence for multiple endogenous ligands. Immunity 2:655–663

Falk K, Rotzschke O, Rammensee H-G (1990) Cellular peptide composition governed by major histocompatibility complex class I molecules. Nature 348:248–251

Falk K, Rotzschke O, Stevanovic S, Jung G, Rammensee HG (1991) Allele-specific motifs revealed by sequencing of self-peptides eluted from MHC molecules. Nature 351:290–296

Feltkamp MCW, Vierboom, MPM, Kast WM, Melief CJM (1994) Efficient MHC class I-peptide binding is required but does not ensure MHC class I-restricted immunogenicity. Mol Immunol 31:1391–1399

Ferris RL, Buck C, Hammond SA, Woods AS, Cotter RJ, Takiguchi M, Igarashi Y, Ichikawa Y, Siliciano RF (1996) Class I-restricted presentation of an HIV-1 gp41 epitope containing an N-linked glycosylation site. J Immunol 156:834–840

Franco A, Ferrari C, Sette A, Chisari FV (1995) Viral mutations, TCR antagonism and escape from the immune response. Curr Opin Immunol 7:524–531

Fremont DH, Matsumura M, Stura EA, Peterson PA, Wilson IA (1992) Crystal structure of two viral peptides in complex with murine MHC class I H-2K^b. Science 257:919–927

Fremont DH, Stura EA, Matsumura M, Peterson PA, Wilson IA (1995) Crystal structure of an H-2K^b-ovalbumin peptide complex reveals the interplay of primary and secondary anchor positions in the major histocompatibility complex binding groove. Proc Natl Acad Sci USA 92:2479–2483

Fremont DH, Rees WA, Kozono H (1996) Biophysical studies of T-cell receptors and their ligands. Curr Opin Immunol 8:93–100

Gairin JE, Oldstone MBA (1992) Design of high-affinity major histocompatibility complex-specific antagonist peptides that inhibit cytotoxic T lymphocyte activity: implications for control of viral disease. J Virol 66:6755–6762

Gairin JE, Oldstone MBA (1993) Virus and cytotoxic T lymphocytes: crucial role of viral peptide secondary structure in major histocompatibility complex class I interactions. J Virol 67:2903–2907

Gairin JE, Mazarguil H, Hudrisier D, Oldstone MBA (1995) Optimal lymphocytic choriomeningitis virus sequences restricted by H-2D^b major histocompatibility complex class I molecules and presented to cytotoxic T lymphocytes. J Virol 69:2297–2305

Garboczi DN, Madden DR, Wiley DC (1994) Five viral peptide-HLA-A2 co-crystals – simultaneous space group determination and X-ray data collection. J Mol Biol 239:581–587

Garboczi DN, Ghosh P, Utz U, Fan QR, Biddison WE, Wiley DC (1996) Structure of the complex between human T-cell receptor, viral peptide and HLA-A2. Nature 384:134–141

Garçia KC, Degano M, Stanfield RL, Brunmark A, Jackson MR, Peterson PA, Teyton L, Wilson IA (1996a) An $\alpha\beta$ T cell receptor structure at 2.5 Å and its orientation in the TCR-MHC complex. Science 274:209–219

Garcia KC, Scott CA, Brunmark A, Carbone FR, Peterson PA, Wilson IA, Teyton L (1996b) CD8 enhances formation of stable T-cell receptor/MHC class I molecule complexes. Nature 384:577–581

Grey HM, Ruppert J, Vitiello A, Sidney J, Kast WM, Kubo RT, Sette A (1995) Class I MHC-peptide interactions: structural requirements and functional implications. Cancer Surv 22:37–49

Groettrup M, Soza A, Kuckelhorn U, Kloetzel P-M (1996) Peptide antigen production by the proteasome: complexity provides efficiency. Immunol Today 17:429–435

Guichard G, Calbo S, Muller S, Kourilsky P, Briand J-P, Abastado J-P (1995) Efficient binding of reduced peptide bond pseudopeptides to major histocompatibility complex class I molecule. J Biol Chem 270:26057–26059

Guichard G, Connan F, Graff R, Ostankovitch M, Muller S, Guillet JG, Choppin J, Briand JP (1996) Partially modified retro-inverso pseudopeptides as non-natural ligands for the human class I histocompatibility molecule HLA-A2. J Med Chem 39:2030–2039

Harris PE (1994) Self-peptides bound to HLA molecules. Crit Rev Immunol 14:61–87

Haurum JS (1997) Class I MHC-restricted presentation of phosphopeptides and recognition by cytotoxic T lymphocytes. Proceedings of the Keystone Symposium on Cell Immunology and Immunotherapy of Cancer, 1–7 February, Copper Mountain, Colo. USA, p 25

Haurum JS, Arsequell G, Lellouch AC, Wong SYC, Dwek RA, McMichael AJ, Elliott T (1994) Recognition of carbohydrate by major histocompatibility complex class I-restricted, glycopeptide-specific cytotoxic T lymphocytes. J Exp Med 180:739–744

Haurum JS, Tan L, Arsequell G, Frodsham P, Lellouch AC, Moss PAH, Dwek RA, McMichael AJ, Elliott T (1995) Peptide anchor residue glycosylation: effect on class I major histocompatibility complex binding and cytotoxic T lymphocyte recognition. Eur J Immunol 25:3270–3276

Heemels MT, Ploegh H (1995) Generation, translocation, and presentation of MHC class I-restricted peptides. Annu Rev Biochem 64:463–491

Huczko EL, Bodnar WM, Benjamin D, Sakaguchi K, Zhu NZ, Shabanowitz J, Henderson RA, Appella E, Hunt DF, Engelhard VH (1993) Characteristics of endogenous peptides eluted from the class-I MHC molecule HLA-B7 determined by mass spectrometry and computer modeling. J Immunol 151:2572–2587

Hudrisier D (1996) Structure, function and pharmacology of antigens presented by the MHC class I molecules to cytotoxic T lymphocytes. PhD Thesis, University of Toulouse, no 2383

Hudrisier D, Gairin JE (1996) The pharmacology of the T cell recptor and its ligands. Med Sci 12:1198–1208

Hudrisier D, Mazarguil H, Oldstone MBA, Gairin JE (1995) Relative implication of peptide residues in binding to major histocompatibility complex class I H-2D^b: application to the design of high-affinity, allele-specific peptides. Mol Immunol 32:895–907

Hudrisier D, Mazarguil H, Laval F, Oldstone MBA, Gairin JE (1996) Binding of viral antigens by Major Histocompatibility Complex H-2D^b molecules are restricted by dominant elements at peptide nonanchor residues: implications for selection and presentation of T cell epitopes. J Biol Chem 271:17829–17836

Hudrisier D, Oldstone MBA, Gairin JE (1997) The signal sequence of lymphocytic choriomeningitis virus contains an immunodominant cytotoxic T cell epitope that is restricted by both H-2D^b and H-2K^b molecules. Virology 234:62–73

Hunt DF, Henderson RA, Shabanowitz J, Sakaguchi K, Michel H, Sevilir N, Cox AL, Appella E, Engelhard VH (1992) Characterisation of peptides bound to the class I MHC molecule HLA-A2.1 by mass spectometry. Science 255:1261–1263

Ishioka GY, Adorini L, Guery JC, Gaeta FCA, Lafond R, Alexander J, Powell MF, Sette A, Grey HM (1994) Failure to demonstrate long-lived MHC saturation both in vitro and in vivo: implications for therapeutic potential of MHC-blocking peptides. J Immunol 152:4310–4319

Jameson SC, Bevan M (1992) Dissection of major histocompatibility complex (MHC) and T cell receptor contact residues in a K^b-restricted ovalbumin peptide and an assessment of the predictive power of MHC-binding motifs. Eur J Immunol 22:2663–2667

Jameson SC, Bevan M (1995) T cell receptor antagonists and partial agonists. Immunity 2:1–11

Jameson SC, Carbone FR, and Bevan MJ (1993) Clone-specific T cell receptor antagonists of major histocompatibility complex class I-restricted cytotoxic T cells. J Exp Med 177:1541–1550

Jones EY (1997) MHC class I and class II structures. Curr Opin Immunol 9:75–79

Joyce S, Kuzushima K, Kepecs G, Angeletti RH, Nathenson SG (1994) Characterisation of an incompletely assembled major histocompatibility class I molecule (H-2K^b) associated with unusually long peptides: implications for antigen processing and presentation. Proc Natl Acad Sci USA 91:4145–4149

Kast WM, Offringa R, Peters PJ, Woordouw AC, Meloen RH, Vandereb AJ, Melief CJM (1989) Eradication of adenovirus E1-induced tumors by E1A-specific cytotoxic T lymphocytes. Cell 59:603–614

Kersh GJ, Allen PM (1996) Essential flexibility in the T-cell recognition of antigen. Nature 380:495–498

Kessler BM, Bassanini P, Horvath C, Cerottini JC, Luescher IF (1996) Effects of epitope modification on TCR-ligand binding and antigen recognition by seven H-2K^d-restricted CTL clones specific for a photoreactive peptide derivative. J Exp Med 185:629–640

Khilko SN, Corr M, Boyd LF, Lees A, Inman JK, Margulies DH (1993) Direct detection of major histocompatibility complex Class-I binding to antigenic peptides using surface plasmon resonance – peptide immobilization and characterization of binding specificity. J Biol Chem 268:15425–15434

Klenerman P, Rowland-Jones S, McAdam S, Edwards J, Daenke S, Lalloo D, Koppe B, Rosenberg W, Boyd D, Edwards A, Giangrande P, Phillips RE, McMichael AJ (1994) Cytotoxic T-cell activity antagonized by naturally occuring HIV-1 Gag variants. Nature 369:403–407

Koup RA (1994) Virus escape from CTL recognition. J Exp Med 180:779–782

Kuchroo VK, Greer JM, Kaul D, Ishioka G, Franco A, Sette A, Sobel RA, Lees MB (1994) A single TCR antagonist peptide inhibits experimental allergic encephlomyelitis mediated by a diverse T cell repertoire. J Immunol 153:3326–3336

Levitsky V, Zhang QJ, Levitskaya J, Masucci MG (1996) The life span of major histocompatibility complex peptide complexes influences the efficiency of presentation and immunogenicity of two class I restricted cytotoxic T lymphocyte epitopes in the Epstein Barr virus nuclear antigen 4. J Exp Med 183:915–926

Lewicki H, Tishon A, Borrow P, Evans CF, Gairin JE, Hahn KM, Jewell DA, Wilson IA, Oldstone MBA (1995a) CTL escape viral variants. I. Generation and molecular characterization. Virology 210:29–40

Lewicki H, von Herrath M, Evans CF, Whitton JL, Oldstone MBA (1995b) CTL escape viral variants. II. Biologic activity in vivo. Virology 211:443–450

Ljunggren HG, Kärre K (1985) Host resistance directed selectively against H-2 deficient lymphoma variants. J Exp Med 162:1745–1759

Luescher IF, Romero P, Cerottini JC, Maryanski JL (1991) Specific binding of antigenic peptides to cell-associated MHC class I molecules. Nature 351:72–74

Luescher IF, Lopez JA, Malissen B, Cerrotini JC (1992) Interaction of antigenic peptides with MHC class I molecules on living cells studied by photoaffinity labeling. J Immunol 148:1003–1011

Luescher IF, Cerottini JC, Romero P (1994) Photoaffinity labeling of the T cell receptor on cloned cytotoxic T lymphocytes by covalent photoreactive ligand. J Biol Chem 269:5574–5582

Luescher IF, Vivier E, Layer A, Mahiou J, Godeau F, Malissen B, Romero P (1995) CD8 modulation of T-cell antigen receptor-ligand interactions on living cytotoxic T lymphocytes. Nature 373:353–356

Luescher IF, Romero P, Kuznetsov D, Rimoldi D, Coulie P, Cerottini JC, Jongeneel CV (1996) HLA photoaffinity labeling reveals overlapping binding homologous melanoma-associated gene peptides by HLA-A1, HLA-A29 and HLA-B44. J Biol Chem 271:12463–12471

Madden DR (1995) The three-dimensional structure of peptide-MHC complexes. Annu Rev Immunol 13:587–622

Madden D, Gorga J, Strominger J, Wiley DC (1992) The three-dimensional structure of HLA B27 at 2.1 Å resolution suggests a general mechanism for tight peptide binding to MHC. Cell 70:1035–1048

Madden DR, Garboczi DN, Wiley DC (1993) The antigenic identity of peptide-MHC complexes – a comparison of the conformations of five viral peptides presented by HLA-A2. Cell 75:693–708

Malarkannan S, Afkarian M, Shastri N (1995) A rare cryptic translation product is presented by K^b major histocompatibility complex class I molecules to alloreactive T cells. J Exp Med 182:1739–1750

Martin S, Kohler H, Weltzien HU, Leipner C (1996) Selective activation of CD8 T cell effector functions by epitope variants of lymphocytic choriomeningitis virus glycoprotein. J Immunol 157:2358–2365

Matsumura M, Fremont DH, Peterson PA, Wilson IA (1992) Emerging principles for the recognition of peptide antigens by MHC class I molecules. Science 257:927–934

Missale G, Redeker A, Person J, Fowler P, Guilhot HS, Schlicht J, Ferrari C, Chisari FV (1993) HLA-A31 and HLA-w68-restricted cytotoxic T cell responses to a single hepatitis B virus nucleocapsid epitope during acute viral hepatitis. J Exp Med 177:751–762

Moskophidis D, Zinkernagel RM (1995) Immunobiology of cytotoxic T-cell escape mutants of lymphocytic choriomeningitis virus. J Virol 69:2187–2193

Mylin LM, Bonneau RH, Lippolis JD, Tevethia SS (1995) Hierarchy among multiple $H\text{-}2^b$-restricted cytotoxic T-lymphocyte epitopes within simian virus 40 T antigen. J Virol 69:6665–6677

Nathenson SG, Uehara H, Ewenstein BM (1981) Primary structural analysis of the transplantation antigens of the murine H-2 major histocompatibility complex. Annu Rev Biochem 50:1025–1052

Nayersina R, Fowler P, Guilhot S, Missale G, Cerny A, Schlicht H-G, Vitiello A, Chesnut R, Person JL, Redeker AG, Chisari FV (1993) HLA A2 restricted cytotoxic T lymphocyte responses to multiple hepatitis B surface antigen epitopes during hepatitis B virus infection. J Immunol 150:4659–4671

Neisig A, Roelse J, Sijts AJA, Ossendorp F, Feltkamp MCW, Kast WM, Melief CJM, Neefjes JJ (1995) Major differences in transporter associated with antigen presentation (TAP)-dependent translocation of MHC class I-presentable peptides and the effect of flanking sequences. J Immunol 154:1273–1279

Oldstone MBA, Whitton JL, Lewicki H, Tishon A (1988) Fine dissection of a nine amino acid glycoprotein epitope, a major determinant recognized by lymphocytic choriomeningitis virus class I restricted $H\text{-}2D^b$ cytotoxic T lymphocytes. J Exp Med 168:559–570

Oldstone MBA, Tishon A, Geckeler R, Lewicki H, Whitton JL (1992) A common antiviral cytotoxic T-lymphocyte epitope for diverse major histocompatibility complex haplotypes: implication for vaccination. Proc Natl Acad Sci USA 89:2752–2755

Oldstone MBA, Lewicki H, Borrow P, Hudrisier D, Gairin JE (1995) Discriminated selection among viral peptides with the appropriate anchor residues: implications for the size of the CTL repertoire and control of viral infection. J Virol 69:7423–7429

Ossendorp F, Eggers M, Neisig A, Ruppert T, Groettrup M, Sijts A, Mengede E, Kloetzel PM, Neefjes J, Koszinowski U, Melief, CJM (1996) A single residue exchange within a viral CTL epitope alters proteasome-mediated degradation resulting in lack of antigen presentation. Immunity 5:115–124

Oukka M, Riche N, Kosmatopoulos K (1994) A nonimmunodominant nucleoprotein-derived peptide is presented by influenza a virus-infected $H\text{-}2^b$ Cells. J Immunol 152:4843–4851

Oukka M, Manuguerra JC, Livaditis N, Tourdot S, Riche N, Vergnon I, Cordopatis P, Kosmatopoulos K (1996) Protection against lethal viral infection by vaccination with nonimmunodominant peptides. J Immunol 157:3039–3047

Pamer EG, Harty JT, Bevan MJ (1991) Precise prediction of a dominant class I MHC-restricted epitope of Listeria monocytogenes. Nature 353:852–855

Parker KC, Bednarek MA, Coligan JE (1994) Scheme for ranking potential HLA-A2 binding peptides based on independent binding of individual peptide side-chains. J Immunol 152:163–175

Pircher H, Moskophidis D, Rohrer U, Burki K, Hengartner H, Zinkernagel RM (1990) Viral escape by selection of cytotoxic T cell-resistant virus variants in vivo. Nature 346:629–633

Pogue RR, Eron J, Frelinger JA, Matsui M (1995) Amino terminal alteration of the HLA A0201 restricted human immunodeficiency virus pol peptide increases complex stability and in vitro immunogenicity. Proc Natl Acad Sci 92:8166–8170

Quaratino S, Thorpe CJ, Travers PJ, Londei M (1995) Similar antigenic surfaces, rather than sequence homology, dictate T cell epitope molecular mimicry. Proc Natl Acad Sci USA 92:10398–10402

Rammensee HG (1995) Chemistry of peptides associated with MHC class I and class II molecules. Curr Opin Immunol 7:85–96

Rammensee HG, Friede T, Stevanovic S (1995) MHC ligands and peptide motifs: first listing. Immunogenetics 41:178–228

Rehermann B, Fowler P, Sidney J, Person J, Redeker A, Brown M, Moss B, Sette A, Chisari FV (1995) The cytotoxic T lymphocyte response to multiple hepatitis B virus polymerase epitopes during and after acute viral hepatitis. J Exp Med 181:1047–1058

Reid SW, McAdam S, Smith KJ, Klenerman P, O'Callaghan CA, Harlos K, Jakobsen BK, McMichael AJ, Bell JI, Stuart DI, Jones EY (1996) Antagonist HIV-1 Gag peptides induce structural changes in HLA B8. J Exp Med 184:2279–2286

Rognan D, Scapozza L, Folkers G, Daser A (1995) Rational design of nonnatural peptides as high-affinity ligands for the HLA-B*2705 human leukocyte antigen. Proc Natl Acad Sci USA 92:753–757

Ruppert J, Sidney J, Celis E, Kubo RT, Grey HM, Sette A (1993) Prominent role of secondary anchor residues in peptide binding to HLA-A2.1 molecules. Cell 74:929–937

Saito Y, Peterson PA, Matsumura M (1993) Quantitation of peptide anchor residue contributions to class-I major histocompatibility complex molecule binding. J Biol Chem 268:21309–21317

Salter RD, Cresswell P (1986) Impaired assembly and transport of HLA-A and -B antigens in a mutant TxB cell hybrid. EMBO J 5:943–949

Schumacher TNM, DeBruijn MLH, Bernie LN, Kast WM, Melief CJM, Neefjes JJ, Ploegh HL (1991) Peptide selection by MHC class I molecules. Nature 350:703–706

Schumacher TNM, Kantesaria DV, Heemels MT, Ashtonrickardt PG, Shepherd JC, Fruh K, Yang Y, Peterson PA, Tonegawa S, Ploegh HL (1994) Peptide length and sequence specificity of the mouse Tap1/Tap2 translocator. J Exp Med 179:533–540

Sette A, Alexander J, Ruppert J, Snoke K, Franco A, Ishioka G, Grey HM (1994a) Antigen analogs/MHC complexes as specific T cell receptor antagonists. Annu Rev Immunol 12:413–431

Sette A, Sidney J, DelGuercio M-F, Southwood S, Ruppert J, Dalhberg C, Grey HM, Kubo RT (1994b) Peptide binding to the most frequent HLA-A class I alleles measured by quantitative molecular binding assays. Mol Immunol 31:813–822

Sette A, Vitiello A, Reherman B, Fowler P, Nayersina R, Kast,WM, Melief CJM, Oseroff C, Yuan L, Ruppert J, Sidney J, Delguercio MF, Southwood S, Kubo RT, Chesnut RW, Grey HM, Chisari FV (1994c) The relationship between class I binding affinity and immunogenicity of potential cytotoxic T cell epitopes. J Immunol 153:5586–5592

Sheil JM, Schell TD, Shepherd SE, Klimo GF, Kioschos JM, Paterson Y (1994) Presentation of a horse cytochrome c peptide by multiple H-2^b class I major histocompatibility complex (MHC) molecules to C57BL/6- and bm1-derived cytotoxic T lymphocytes: presence of a single MHC anchor residue may confer efficient peptide-specific CTL recognition. Eur J Immunol 24:2141–2149

Shibata KI, Imarai M, VanBleek GM, Joyce S, Nathenson SG (1992) Vesicular somatitis virus antigenic octapeptide N52-59 is anchored onto the groove of the H-$2K^b$ molecule by the side chains of three amino acids and the main-chain atoms of the amino-terminus. Proc Natl Acad Sci USA 89:3135–3139

Sidney J, Delguercio MF, Southwood S, Engelhard VH, Appella E, Rammensee HG, Falk K, Rotzschke O, Takiguchi M, Kubo RT, Grey HM, Sette A (1995) Several HLA alleles share overlapping peptide specificities. J Immunol 154:247–259

Silver ML, Guo H-C, Strominger JL, Wiley DC (1992) Atomic structure of a human MHC molecule presenting an influenza virus peptide. Nature 360:367–369

Skipper JCA, Hendrickson RC, Gulden PH, Brichard V, VanPel A, Chen Y, Shabanowitz J, Wolfel T, Slinguff CLJ, Boon T, Hunt DF, Engelhard VH (1996) An HLA-A2-restricted tyrosinase antigen on melanoma cells results from posttranslational modification and suggests a novel pathway for processing of membrane proteins. J Exp Med 183:527–534

Smith KJ, Reid SW, Harlos K, McMichael AJ, Stuart DI, Bell JI, Jones EY (1996a) Bound water structure and polymorphic amino acids act together to allow the binding of different peptides to MHC class I HLA-B53. Immunity 4:215–228

Smith KJ, Reid SW, Stuart DI, McMichael AJ, Jones EY, Bell JI (1996b) An altered position of the α2 helix of MHC class I is revealed by the crystal structure of HLA-B*3501. Immunity 4:203–213

Stern LJ, Wiley DC (1994) Antigenic peptide binding by class I and class II histocompatibility proteins. Structure 2:245–251

Sun R, Sheperd SE, Geier SS, Thomson CT, Sheil JM, Nathenson SG (1995) Evidence that the antigen receptors of cytotoxic T lymphocytes interact with a common pattern on the H-2K^b molecule. Cell 3:573–582

Toes RE, Offringa R, Blom RJ, Melief CJ, Kast WM (1996) Peptide vaccination can lead to enhanced tumor growth through specific T-cell tolerance induction. Proc Natl Acad Sci USA 93:7855–7860

Townsend AR, Rothbard J, Gotch FM, Bahadur G, Wraith D, McMichael AJ (1986) The epitopes of influenza nucleoprotein recognized by cytotoxic T lymphocytes can be defined with short synthetic peptides. Cell 44:959–968

Townsend A, Claes O, Bastin J, Ljunggren HG, Foster L, Kärre K (1989) Association of class I major histocompatibility heavy and light chains induced by viral peptides. Nature 340:443–448

Udaka K, Wiesmuller K-H, Kienle S, Jung G, Walden P (1995) Tolerance to amino acid variations in peptide binding to the major histocompatibility complex class I protein H-2K^b. J Biol Chem 270:24130–24134

Valitutti S, Muller S, Cella M, Padovan E, Lanzavecchia A (1995) Serial triggering of many T-cell receptors by a few peptide-MHC complexes. Nature 375:148–151

van den Eynde B, Mazarguil H, Lethe B, Laval F, Gairin JE (1994) Localization of two cytotoxic T lymphocyte epitopes and three anchoring residues on a single nonameric-peptide that binds to H-2L^d and is recognized by cytotoxic T lymphocytes against mouse tumor P815. Eur J Immunol 24:2740–2745

van der Burg SH, Visseren MJW, Brandt RMP, Kast WM, Melief CJM (1996) Immunogenicity of peptides bound to MHC class I molecules depends on the MHC-peptide complex stability. J Immunol 156:3308–3314

van der Most RG, Sette A, Oseroff C, Alexander J, Murali-Krishna K, Lau LL, Southwood S, Sidney J, Chesnut RW, Matloubian M, Ahmed R (1996) Analysis of cytotoxic T cell responses to dominant and subdominant epitopes during acute and chronic lymphocytic choriomeningitis virus infection. J Immunol 157:5543–5554

Walden P (1996) T cell epitope determination. Curr Opin Immunol 8:68–74

Weiss GA, Collins EJ, Garbczi DN, Wiley DC, Schreiber SL (1995) A tricyclic ring system replaces the variable regions of peptides presented by three alleles of human MHC class I molecules. Chem Biol 2:401–407

Young ACM, Zhang WG, Sacchettini JC, Nathenson SG (1994) The 3-dimensional structure of H-2D^b at 2.4 angstrom resolution: implications for antigen-determinant selection. Cell 76:39–50

Young ACM, Nathenson SG, Sacchettini JC (1995) Structural studies of class I major histocompatibility complex proteins: insights into antigen presentation. FASEB J 9:26–36

Zhang W, Young ACM, Imarai M, Nathenson SG (1992) Crystal structure of the major histocompatibility complex class I H-2K^b molecule containing a single viral peptide: implications for peptide binding and T-cell receptor recognition. Proc Natl Acad Sci USA 89:8403–8407

Zinkernagel RM, Doherty PC (1974) Restriction of in vitro T cell-mediated cytotoxicity in lymphocytic choriomeningitis within asyngeneic or semiallogeneic system. Nature 248:701–702

Immune Response of β_2-Microglobulin-Deficient Mice to Pathogens

JEFFREY A. FRELINGER[1] and JONATHAN SERODY[2]

1 Introduction

The ability of hosts to respond to pathogens is one of the driving forces in the selection of the immune system. The immune response to both bacteria and viruses involves the ability of T cell to recognize and to respond to pathogens. Both $CD4^+$ and $CD8^+$ T cells are important in the overall immune response. In the most basic paradigm, $CD4^+$ T cells respond to exogenous antigen, which is internalized, degraded, and presented on the surface of antigen-presenting cells by major histocompatibility complex (MHC)-encoded class II molecules. This arm of the response includes not only helping B cells produce antibodies, but also the production of proinflamatory cytokines, including interferon (INF)-γ, tumor necrosis factor (TNF)-α, and interleukin (IL)-2. It is often stated that these responses are important in inhibiting the spread of virus from cell to cell by aiding the production of antibodies which neutralize virus and creating a cytokine milieu which does not support virus replication. In contrast, $CD8^+$ T cells respond primarily to peptides derived from endogenously synthesized proteins. The T cells are able to recognize and kill the cells

[1]Department of Microbiology and Immunology and [2]Department of Medicine, University of North Carolina, Chapel Hill, NC 27599, USA

expressing these peptides by at least two major pathways, one utilizing perforin exocytosis and the other utilizing fas-fas ligand interactions. This response has the net effect of removing cells which are producing viruses or other pathogens which have access to the cytoplasm of infected cells. This has given rise to the simple idea that CD4 responses (and antibodies) block virus spread, while CD8 responses remove the source of infection, resulting in clearance of pathogens which hide intracellularly.

Recently, ZINKERNAGEL (1996) has proposed a modification of this idea, suggesting that CD8 responses are effective against virus infections which are nonlytic and result in persistent infections in the absence of a CD8 response, while CD4 responses (and antibody) are effective against viruses which are themselves directly destructive to cells.

Genetic null alleles have been used for many years to characterize the function of genes and the pathways in which the gene products function. Such experiments of nature were critical in the development of our understanding of the immune system. For example, DiGeorge's syndrome was crucial in establishing the role of the thymus. These experiments have provided a mirror image of physiologists' surgical ablation experiments with the advantage that genetic ablation occurs before birth, from the earliest time of development. Recent advances in molecular genetics, particularly the production of transgenic mice and the ability to perform homologous recombination in embryonic stem cells, has allowed the reverse engineering of new mutations of genes already cloned. This has resulted in a cottage industry for immunologists studying the effects of deletions of important molecules and their subsequent effects on the immune response in a variety of systems. This has been aided by the relatively rapid availability of these new mouse strains.

This chapter will examine one of the first immunologically relevant knockout mice to become available, i.e., the insertional mutation of the mouse β_2-microglobulin gene. This mutation was produced almost simultaneously in two different laboratories (ZIJLSTRA et al. 1989; KOLLER et al. 1990). While the mutants were constructed slightly differently, there is no evidence that they differ in phenotype, and as a result we will discuss all of the results here without regard to the mutant stock used in the experiments. There is an additional genetic complication. The knockout mice were constructed in 129/Sv embryonic stem cells. The chimeric mice were mated in the first generation with B6 mice to simplify identification of the progeny derived from embryonic stem cell-derived germ cells. These heterozygous animals were then intercrossed, and homozygous mutant mice were selected and used to establish the cell line. It is immediately clear that these animals have a mixed 129/C57BL6 background, and each separate line is a presumptive and distinct recombinant inbred strain. Thus there can be substantial genetic differences between mice used in different laboratories after only a few generations. This is now better controlled by backcrossing the mutants onto a homogeneously inbred background. Several different inbred backgrounds are currently commercially available, and in the future this should be less of a problem.

The initial characterization of the mutants was of profound importance. Several theories had suggested that nonclassical MHC molecules might be important in development, and the fact that these mice were largely normal disproved these hypotheses. Of more immunologic interest was the composition of the peripheral T cell compartment. As expected, the mice lacked expression of functional class I molecules on the cell surface. This had been expected from in vitro expression studies (PARNES et al. 1986). Even more interesting was the lack of CD8 T cell expresssion in the periphery despite normal CD8 expression in the thymus. These mice were profoundly deficient in functional $CD4^-CD8^+$ T cells in the periphery. This result itself showed the importance of MHC class I expression and positive selection in the developing T cell compartment in the thymus. The lack of $CD8^+$ cells provided a model system for testing the role of $CD8^+$ cells in infections in a way that is much more effective than either in vivo ablations with antibodies or adoptive transfer experiments. Both these methods always have the limitation that removal of the target cell population is incomplete, and most workers believed that the β2-microglobulin-deficient mice would be a more effective system for examining the role of $CD8^+$ T cells.

An interesting question soon arose concerning the ability of β2-microglobulin-deficient mice to produce functional natural killer (NK) cells. Since NK cells use class I molecules as markers of the downregulation of killing, it was possible that NK activity would not correctly develop in β2-microglobulin-deficient mice (HOGLUND et al. 1997). An early report suggested that β2-microglobulin-deficient mice had lowered NK activity (LIAO et al. 1991). Later experiments showed that, while they were less able to respond to the NK inducer, poly I:C, they were able to produce NK cells with other inducers (SU et al. 1994; TAY et al. 1995; ZAJAC et al. 1995). The specificity of these NK cells was interesting. While NK cells from normal mice were able to kill blast targets from β2-microglobulin-deficient mice, NK cells from β2-microglobulin-deficient mice did not recognize and kill syngeneic targets, but could kill conventional targets and function to reject hematopoietic stem cells (BIX et al. 1991; OHLEN et al. 1995; ZAJAC et al. 1995). This specificity was unexpected, since the NK cells were not known to have a learning process to recognize self as T cells do. However, it is clear that either the NK class I receptors can distinguish among levels of class I expression or there is another, independent mechanism of discrimination which has not been previously appreciated.

Early on, several investigators appreciated that some $CD8^+$ cells were produced in their studies. Indeed, early papers showed that some $CD8^+$ cells were produced by alloimmunization (BIX and RAULET 1992; GLAS et al. 1992). Similarly, two groups demonstrated that immunization with allogeneic tumor cells could stimulate the production of αβ, $CD8^+$ cytotoxic T lymphocytes (CTL) in β2-microglobulin-deficient mice (APASOV and SITKOVSKY 1993, 1994; LAMOUSE-SMITH et al. 1993). It was also appreciated that the development of γδ-T cells was essentially normal in these mice (RAULET et al. 1991). Thus it is important to remember that β2-microglobulin-deficient mice are not completely deficient in $CD8^+$ cells and that, under some circumstances, potent CD8 cell-mediated responses can be elicited. Nonetheless, the mice are profoundly deficient in $CD8^+$ cells, which allows the examination of the role of $CD8^+$ T cells in a variety of systems. The best studies reported have taken pains to examine the possible involvement of $CD8^+$ cells in the responses studied and have ruled out significant contributions of the $CD8^+$ cells in the particular system under study.

2 Responses to Viruses

The existence of CD8-deficient mice should, in principle, allow the rigorous determination of the role of $CD8^+$ T cells in resistance to and recovery from virus infection. Not surprisingly, β_2-microglobulin-deficient mice have now been tested in many different virus systems to attempt to determine the role of $CD8^+$ cells in the resistance and clearance of virus. In this section, we will review in detail the work on the two best-studied systems, lymphocytic choriomeningitis virus (LCMV) and influenza. Both are illustrative of the complexities which arise from an innate lack of $CD8^+$ T cells.

2.1 Lymphocytic Choriomeningitis Virus

The immune response to LCMV has been studied extensively in many laboratories (ASANO and AHMED 1995). This virus has become the prototypical case of virus-induced, immunomediated pathology. In brief, when normal mice are infected with LCMV intraperitoneally, they mount an efficient immune response, including both antibodies and cytotoxic T cells. This response results in the clearance of the virus from the mouse and resistance to reinfection by any route of infection. In contrast, when mice are infected by intracranial injection, the mice regularly develop acute encephalitis and die after about 1 week. Mice previously infected by the intraperitoneal route are resistant to lethal disease following intracranial challenge. Adoptive transfer experiments and ablation experiments showed that both protection from reinfection and development of lethal disease completely depend on $CD8^+$ T cells. Depletion of $CD8^+$ cells either by specific antibody treatment or by nonspecific methods, such as sublethal irradiation, results in survival following intracranial virus infection. Interestingly, while these animal survive, they become persistently viremic.

LCMV provides an ideal model to test the role of $CD8^+$ cells in β_2-microglobulin-deficient mice. There are clear predictions as to the outcomes of infection. The mice should not die following intracranial challenge and should not be able to clear the virus. The results of the first experiments were surprising. When β_2-microglobulin-deficient mice were infected intracranially with LCMV, many of the infected animals died, but survival was longer (MULLER et al. 1992). Other investigators found a similar situation with variable survival, which seemed to depend on both the exact virus stock used and the genetic makeup of the β_2-microglobulin-deficient mice (DOHERTY et al. 1993; LEHMANN-GRUBE et al. 1993; QUINN et al. 1995). A widespread finding was that β_2-microglobulin-deficient mice produced $CD4^+$ T cells which were capable of lysing LCMV infected but uninfected target cells, providing that the target cells expressed class II molecules (MULLER et al. 1992; DOHERTY et al. 1993; QUINN et al. 1993). These studies have all shown that the wasting disease in these mice is caused by the $CD4^+$ cells based on adoptive transfer of disease.

A great surprise is the lack of CNS pathology. Although intracranial infection resulted in the most severe disease, there was little evidence of encephalitis in any of the published experiments, although the virus used in all the experiments was able to provoke lethal encephalitis in normal mice (MULLER et al. 1992; DOHERTY et al. 1993; LEHMANN-GRUBE et al. 1993; QUINN et al. 1993; CHRISTENSEN et al. 1994). Thus the disease must be mediated by a different mechanism than the disease mediated by $CD8^+$ cells in wild-type mice. Recent experiments have shown that, while $CD4^+$ CTL are able to kill target cells using a fas-dependent mechanism, wasting disease is not dependent on the ability to carry out fas-dependent killing. Mice which are fas-deficient are equally susceptible to wasting disease as those whose fas expression is normal (ZAJAC et al. 1996). A major possible mediator of this wasting disease is TNF-α. β_2-Microglobulin-deficient mice produce TNF-α after infection, and weight loss is a well-documented feature of high levels of TNF expression. In contrast, fas-deficient, β_2-microglobulin-deficient mice are resistant to lethal disease mediated by transfer of $CD4^+$ T cells. Thus, while a dose of LCMV-immune $CD4^+$ T cells transferred to fas^+, β_2-microglobulin-deficient mice results in lethal disease, the same dose transferred to $fas^-$$\beta_2$-microglobulin-deficient mice results in survival (QUINN et al. 1993; ZAJAC et al. 1996). It therefore appears that there are two pathologic processes occurring in LCMV-infected, β_2-microglobulin-deficient mice. One is mediated by direct damage to an unknown target organ and is dependent on fas-mediated lysis of cells. This is most apparent in the lethal disease seen on adoptive transfer. The second is independent of fas-mediated lysis, but is probably dependent on cytokines, with major candidates being TNF-α and INF-γ. Consistent with a role for cytokines, male and female animals show differences in their susceptibility to disease (MULLER et al. 1995).

Problems remain in how to interpret the above studies. As noted above, there are some $CD8^+$ cells in β_2-microglobulin-deficient mice, but it is unclear whether they play a role in these responses. There is evidence for the production of $CD8^+$ T cells in response to both virus and peptide (LEHMANN-GRUBE et al. 1994; COOK et al. 1995). This was expected to some extent, since it was known that, in cell lines, a small amount of both correctly folded L^d and D^b could be expressed and was associated with peptide. Our own recent experiments show that $CD8^+$ L^d-restricted CTL can be produced from $H2^d$ β_2-microglobulin-deficient mice by infection with LCMV and that these CTL have a peptide specificity identical to that of wild-type $CD8^+$ CTL from normal $H2^d$ mice (QUINN et al. 1997). Thus the role of $CD8^+$ T cells in response to infections is unclear. It is clear that, while β_2-microglobulin-deficient mice are profoundly depleted of $CD8^+$ T cells, these cells are not entirely absent. It is important to note that this depletion also has effects on the $CD4^+$ cells present . Thus a disease mediated by $CD4^+$ T cells is readily apparent in β_2-microglobulin-deficient mice, but is not seen even on adoptive transfer in normal mice. Thus the $CD4^+$ population in a β_2-microglobulin-deficient mouse is not simply the normal mouse complement of $CD4^+$ cells without $CD8^+$ cells, but must contain cells with a different spectrum of specificity as well.

2.2 Other Viruses

Much less is known about the response of β2-microglobulin-deficient mice to other viruses, and much of the work involved has been of a more superficial nature, with mice being infected with the virus in question and the outcome observed without much detailed study of the underlying mechanisms.

A protective immune response to influenza virus has been reported to depend on both $CD4^+$ and $CD8^+$ T cells. The prospect of using β2-microglobulin-deficient mice to elucidate the relative contribution of each compartment was enticing. In retrospect, the results are not surprising. Influenza-infected β2-microglobulin-deficient mice produced $CD4^+$ CTL that were class II restricted, as did LCMV-infected mice (TAYLOR and BENDER 1995). The consequences of influenza infection were also different in β2-microglobulin-deficient mice. The virus was cleared from the lungs more slowly than from wild-type mice (EICHELBERGER et al. 1991; BENDER et al. 1992). In addition, priming for resistance to subsequent infection was effective (BENDER et al. 1994). In a similar manner, β2-microglobulin-deficient mice could be protected by vaccinia recombinants (EPSTEIN et al. 1993). In one report, the incidence of fatal disease was increased (BENDER et al. 1992). Thus, consistent with the earlier reports using $CD4^+$ and $CD8^+$ cells in adoptive transfer experiments, both $CD4^+$ and $CD8^+$ cells play a role in the resistance to and/or recovery from influenza virus infection (MCDERMOTT et al. 1987; GRAHAM et al. 1994).

Sendai virus infection has also been investigated. There is delayed clearance of Sendai virus, and there are some subtle alterations in IgA production (HOU et al. 1992; HYLAND et al. 1994). In other viruses, such as reovirus, where Ig is the most important mechanism of clearance, β2-microglobulin-deficient mice clear the infection normally (BARKON et al. 1996). In a similar manner, rotavirus clearance also occurred, although it was slightly delayed, and β2-microglobulin-deficient mice were resistant to rechallenge (FRANCO and GREENBERG 1995). Neither mouse cytomegalovirus nor herpes simplex virus was sensitive to the loss of class I molecules and $CD8^+$ T cells (MANICKAN and ROUSE 1995; POLIC et al. 1996). As mentioned above, it was possible to document the production of $CD4^+$ cells with CTL activity (NIEMIALTOWSKI et al. 1994). In contrast to these viruses, gamma herpesvirus 68 shows a pattern similar to other viruses where clearance is delayed (WECK et al. 1996). Vaccinia virus infection also results in survival, but subtle alterations of the antibody response occur (SPRIGGS et al. 1992).

The response of β2-microglobulin-deficient mice to Theiler's virus is interesting. Theiler's virus causes a demyelinating disease that is in some ways similar to multiple sclerosis. Genetic evidence based on resistance to disease mapping to the D^b locus has suggested that $CD8^+$ T cells might play an important role in the disease. As in many other systems, β2-microglobulin-deficient mice failed to clear Theiler's virus following infection (FIETTE et al. 1993; RODRIGUEZ et al. 1993). It is of particular interest that, although these mice regularly developed areas of demyelination, they did not die or develop other neurologic symptoms. Indeed, β2-microglobulin-deficient mice showed areas of remyelination, which normal mice do not (MILLER et al. 1995). Together, these observations suggest that $CD8^+$ cells do play an important role in the production of disease and can inhibit remyelination, but are not required for demyelination itself. This is similar to the results using the coronavirus mouse hepatitis virus (MHV) A59. In this system, too, virus is only cleared slowly, but the mice are more sensitive to acute disease (LAVI and WANG 1995). Interestingly, even at low virus doses, demyelination can occur, suggesting that $CD8^+$ cells are not required for demyelination (GOMBOLD et al. 1995).

β_2-Microglobulin-deficient mice have also been used to investigate other potentially autoimmune diseases, including diabetes in NOD mice. This disease was reported to require CD4 responses because of the requirement for the unusual IA allele expressed by NOD mice (SLATTERY and MILLER 1996). Surprisingly, when the β_2-microglobulin mutation is crossed onto the NOD background, mice are resistant to diabetes (SERREZE et al. 1994; SUMIDA et al. 1994; WICKER et al. 1994; WANG et al. 1996). This provides a compelling argument for a role for class I molecules and $CD8^+$ T cells in the production of diabetes and insulitis in NOD mice.

3 Response to Nonviral Pathogens

The β_2-microglobulin-deficient mouse has been employed by several investigators as a model system to determine the role of $CD8^+$ T lymphocytes in the immune response to nonviral pathogens. These studies must be interpreted cautiously for several reasons. We have shown that wild-type $CD8^+$ T lymphocytes are capable of mediating protection against a lethal dose of *Listeria monocytogenes* (LM) in β_2-microglobulin-deficient mice (SERODY et al. 1996). Despite the lack of H-$2K^b$ molecules on the surface of cells in β_2-microglobulin-deficient mice, as shown by flow cytometry, we have shown that K^b-restricted $CD8^+$ lymphocytes are effective in the adoptive transfer of protection.

The class I-restricted response has been best demonstrated in the $H2^d$ β_2-microglobulin-deficient mouse. Using peptide pulsed target cells, COOK et al. (1995) showed that L^d-restricted $CD8^+$ T lymphocytes could be produced in the β_2-microglobulin-deficient mouse. Therefore, changes in the response of the β_2-microglobulin-deficient mouse to nonviral pathogens cannot only be attributed to the lack of class I-restricted $CD8^+$ T lymphocytes. Additionally, β_2-microglobulin-deficient mice have an exaggerated compensatory $CD4^+$ response to pathogens (MARUSIC-GALESIC et al. 1993). This enhanced response may be capable of compensating for the lack of cytolytic $CD8^+$ T cells in these mice. Thus an equivalent response in these mice does not indicate that $CD8^+$ T cells have no role in the immune response to different nonviral pathogens.

3.1 *Listeria monocytogenes*

LM is a facultative intracellular bacterium that causes meningoencephalitis in immunocompromised hosts (MIELKE et al. 1993). LM is actively taken up from the blood by macrophages in the reticuloendothelial (RE) system after infection. Organisms are engulfed into the phagolysosome of the macrophage. The secretion of the protein listeriolysin O allows for the organism to escape the phagolysosome and replicate unrestricted in the cytoplasm. LM is capable of using the cellular actin machinery in concert with the product of the ACT A gene to spread from cell to cell without entering an extracellular phase (KOCKS et al. 1992).

The murine model of listeroisis is the best-characterized immune response to a bacterial pathogen. In the 1960s, several groups demonstrated that T cell activation of macrophages was critical for the eradication of LM after murine infection (MACKANESS 1962; SIMON and SHEAGREN 1972). These groups also showed that the hallmark of T cell activity in the response to LM was the formation of granulomata.

Further studies using severe combined immunodeficient (SCID) and nude mice challenged the hypothesis that T cell activation of macrophages was the critical effect in the clearance of LM (EMMERLING et al. 1975). Current work suggests that multiple arms of the immune response are critical in this clearance. By day 5 after infection, detectable numbers of listeria-specific $CD8^{+}$T cells can be found in the liver and spleen of infected animals (MIELKE et al. 1993). The expansion and activation of these T cells appears to be responsible for complete clearance of the bacterium. In addition, $\gamma\delta$ T lymphocytes precede the appearance of $\alpha\beta$ T lymphocytes after infection with LM, and depletion of this population of lymphocytes resulted in enhanced numbers of LM 5 days after an intraperitoneal infection.

ROBERTS et al. (1993) used β_2-microglobulin-deficient mice to evaluate the specific role of $CD8^{+}$ $\alpha\beta$ T lymphocytes during a primary infection. Mice were infected with LM intravenously and the course of infection was followed by plating splenic and liver homogenates. These investigators found a statistically significant increase in the number of LM at days 8 and 16 after infection in β_2-microglobulin-deficient mice as compared to C57BL/6 mice. No changes were seen at days 1, 3, and 5 after infection. Using monoclonal antibodies, $\gamma\delta$ or NK cells were depleted in β_2-microglobulin-deficient mice on days 0 and 2 after infection. Depletion of $\gamma\delta$ T lymphocytes led to a statistically significant (p.<001) 1- to 2-log increase in the numbers of LM in the spleen and liver after infection as compared to untreated control mice. Treatment with antiasialo-GM1 to deplete NK cells had no effect.

From this work, the authors concluded that $CD8^{+}$ T lymphocytes were important in mediating late clearance in murine listeriosis. In the absence of $CD8^{+}$ $\alpha\beta$ T cells, $\gamma\delta$ T lymphocytes were critically important in the early immune response. However, there were several factors that rendered definite conclusions from this work difficult to interpret. β_2-Microglobulin-deficient mice initially had a mixed background of 129/J and C57BL/6 genes. Therefore, the C57BL/6 mouse, which the authors used, is not a true control animal when comparing the in vivo response of the β_2-microglobulin-deficient mouse. In addition, the clearance of LM in the β_2-microglobulin-deficient mice in these experiments is exaggerated at the later time points compared to the clearance observed by our group (SERODY et al. 1996) and by LADEL et al. (1994).

The latter authors extended the observations made by ROBERTS et al. (1993) by comparing the response of β_2-microglobulin-deficient mice with $A\beta^{-/-}$ class II knockout mice. $A\beta^{-/-}$ mice are devoid of $CD4^{+}$ T lymphocytes as a result of the inability to select this subset from the $CD4^{+}/CD8^{+}$ double-positive subset in the thymus (LADEL et al. 1994). In agreement with the results obtained by ROBERTS et al. (1994), β_2-microglobulin-deficient mice had a delayed clearance of LM. However, LADEL et al. (1994) used heterozygous littermate control animals in the comparison. β_2-microglobulin-deficient mice showed a delay in clearance of LM as compared to $A\beta^{-/-}$ mice. In both model systems, treatment with a anti-$\gamma\delta$ T cell monoclonal antibody exacerbated the infection. The LD_{50} for LM of the β_2-microglobulin-deficient mice was fourfold less than that of C57BL/6 mice.

LADEL et al. (1994) also investigated the role of $CD4^+$ and $CD8^+$ cells in the protective memory response against LM. Mice were infected with a sublethal dose of LM and then challenged with a dose 20 times the LD_{50} for β_2-microglobulin-deficient mice. Both β_2-microglobulin-deficient and $A\beta^{-/-}$ mice were able to respond to a lethal infection with LM after a sublethal inoculum. However, again, the mutant mice did not clear the second infection as quickly as control mice. Finally, the same group of investigators evaluated the role of $CD4^+$ and $CD8^+$ lymphocytes in the formation of granulomata. In both of the knockout mice, granulomata formation was markedly impaired as compared to wild-type mice.

The natural route of infection of LM is through the gastrointestinal tract. EMOTO et al. 1996 evaluated the response of β_2-microglobulin-deficient mice to oral infection with LM. They focused on the role of $CD8_{\alpha\alpha}$ and $CD8_{\alpha\beta}$ T lymphocytes during an intestinal infection with LM. $CD8_{\alpha\alpha}$ T cells are selected independent of the presence of β_2-microglobulin; conversely, $CD8_{\alpha\beta}$ cells require the presence of β_2-microglobulin. Thus, by using β_2-microglobulin-deficient mice, they were able to focus on the role of different intestinal intraepithelial cells (IEL).

This group found that the number of IFN-γ-producing cells after T cell receptor (TCR) ligation from IEL was less in β_2-microglobulin-deficient mice than in heterozygous littermates ($p.<01$). Interestingly, they showed that the population of $CD8_{\alpha\alpha}$ lymphocytes was markedly reduced in β_2-microglobulin-deficient mice. There was a compensatory increase in the number of $\gamma\delta$ IEL in β_2-microglobulin-deficient mice. Finally, IFN-γ production and cytolytic activity were less in β_2-microglobulin-deficient mice than in heterozygote littermates. However, this decrease was not as great as the decrement in cell numbers, suggesting that the residual IEL present in β_2-microglobulin-deficient mice are potent effectors.

We used β_2-microglobulin-deficient mice to evaluate the specific role of class I and class II-restricted CTL in the adoptive transfer of protection against a lethal infection with LM. We showed that class II-restricted CTL are easily produced in β_2-microglobulin-deficient mice. These lymphocytes were capable of mediating protection if given before an infection. However, we showed that early in infection, class II-negative parenchymal cells are infected and limit the ability of class II-restricted lymphocytes to respond to an ongoing infection.

In accordance with previous work, we showed that $H\text{-}2K^b$-restricted lymphocytes were effective in protecting β_2-microglobulin-deficient mice despite the lack of K^b complexes on the surface of the cells as determined by flow cytometry. Finally, we demonstrated that the production of IFN-γ was not a requirement for activity of class II-restricted effector cells.

3.2 Mycobacteria and Fungi

Similar to the initial work in LM, multiple investigators have shown that the $CD4^+$ T lymphocyte is critical for the immune response to *Mycobacterium tuberculosis* (ORME 1994). Early in the first 3 weeks after infection, $CD4^+$ T lymphocytes are potent producers of IFN-γ. This cytokine activates macrophages that have engulfed the bacillus and is pivotal in the eradication of the organism. A specific role for $CD8^+$ lymphocytes in the immune response to *Mycobacterium tuberculosis* has been more difficult to identify.

Previous investigators had shown that mycobacterial-specific $CD8^+$ T lymphocytes could be produced in mice (DELIBERO et al. 1988). The in vivo role of mycobacteria-specific $CD8^+$ T lymphocytes in the immune response to this organism has been investigated by FLYNN et al. (1992) using β_2-microglobulin-deficient mice. β_2-Microglobulin-deficient mice backcrossed onto a C57BL/6 background were infected with acid-fast bacillus (AFB) and compared to C57BL/6 mice. Seventy percent of the β_2-microglobulin-deficient mice died after infection with 10^6 AFB. All of the wild-type mice survived this infection. In histologic sections, β_2-microglobulin-deficient mice consistently had 1–3 logs greater AFB than wild-type mice. Granulomata formation was similar in both types of mice.

The authors compared the response of β_2-microglobulin-deficient and C57BL/6 mice after infection with bacille Calmette-Guérin (BCG). No differences were seen in mortality or colony-forming units (CFU) from spleen, liver, or lungs after infection. Vaccination with BCG increased the time to death after infection with AFB, but did not alter survival in β_2-microglobulin-deficient mice. After mycobacterial infection, β_2-microglobulin-deficient mice produced less IFN-γ than C57BL/6 mice, but this was not statistically significant.

The conclusion drawn from this work was that $CD8^+$ T lymphocytes were important in the control of pulmonary infection with *Mycobacterium tuberculosis*. This may be due to infection of class I-expressing parenchymal cells of the lung by AFB. The particular role of class I-restricted CTL or lymphocytes that recognize class Ib molecules in the immune response to AFB cannot be ascertained from this work.

Histoplasma capsulatum is a fungus that causes a clinical illness closely resembling tuberculosis. $CD4^+$ lymphocytes and macrophages are pivotal to the immune response to *H. capsulatum*. DEEPE (1994) evaluated the role of $CD8^+$ lymphocytes in the immune response to *H. capsulatum* in β_2-microglobulin-deficient mice. Fungal CFU were followed in β_2-microglobulin-deficient mice and heterozygote littermates after infection. While no difference was seen 1 week after infection, statistically significant differences were noted in CFU from spleen and liver in the knockout mice at weeks 2 and 3 after infection (p.<01). While the magnitude of the infection was greater in β_2-microglobulin-deficient mice, these mice were able to control infection with *H. capsulatum*.

3.3 Parasitic Infections

3.3.1 Leishmania

Murine infection with the amastigote form of *Leishmania major* has provided critical insights into the role of $CD4^+$ subsets in the immune response. Infection with *L. major* in BALB/c mice produces a Th2 response and a progressive course that eventually leads to death (HEINZEL et al. 1989). C57Bl/6 mice produce a Th1 response and survive. Blockage of the Th2 response in BALB/c mice by the administration of anti-IL-4 monoclonal antibodies allows BALB/c mice to control the infection (SADICK et al. 1990). In contrast to the crucial role of $CD4^+$ T cells, direct evidence of a role for $CD8^+$ lymphocytes has not been easily demonstrated.

WANG et al. (1996) infected β_2-microglobulin-deficient mice with *L. major* to assess the role of $CD8^+$ lymphocytes in the immune response. β_2-Microglobulin-deficient mice controlled the infection as well as control mice. This work suggested that $CD8^+$ lymphocytes are not critical in the clearance of *L. major*.

3.3.2 Plasmodia

The immune response to malaria is complex. Immunity to *Plasmodia* species is dependent on persistent exposure to the organism. Presentation of plasmodial antigens by hepatic cells can initiate a T cell response that may be critical in the maintenance of immunity. $CD8^+$ CTL that recognize circumsporozoite proteins (CSP) or sporozoite surface proteins (SPZ) can be demonstrated in humans (MALIK et al. 1991). WHITE et al. (1996) used β_2-microglobulin-deficient mice to investigate the immune response to attenuated *P. berghei* sporozoites. Spleen cells from β_2-microglobulin-deficient mice generated proliferative activities to SPZ antigens similar to those of wild-type mice. However, unlike wild-type mice, which show a decrement in proliferative activity after boosting, β_2-microglobulin-deficient mice showed an increase in activity. This work suggested that $CD8^+$ lymphocytes might downregulate the immune response to SPZ after initial exposure.

WHITE et al. (1996) also evaluated whether SPZ from attenuated *P. berghei* were protective in β_2-microglobulin-deficient mice. After SPZ priming and two boost immunizations, wild-type and heterozygote littermates were protected against a 10 000-dose challenge with *P. berghei* SPZ. All β_2-microglobulin-deficient mice were not protected by attenuated SPZ and became parasitemic at day 9 after infection. This lack of protection in β_2-microglobulin-deficient mice was independent of the dose of SPZ used. Unlike the response in wild-type recipients, wild-type splenocytes from SPZ immunized animals did not protect β_2-microglobulin-deficient mice despite the presence of antibody specific for CSP in these mice. Splenocytes from β_2-microglobulin-deficient mice were incapable of protecting wild-type animals.

VAN DER HEYDE et al. (1993) evaluated the response of β_2-microglobulin-deficient mice to infection with the erythrocytic stage of infection of *Plasmodia chabaudi adami*, *yoelii yoelii*, and *chabaudi chabaudi*. No differences were seen in C57BL/6 or β_2-microglobulin-deficient mice after infection with any of the species of plasmodia used. This work suggests that $CD8^+$ T lymphocytes are important in the immune response to the exoerythrocytic stage of infection with *P. berghei*. As class I-restricted CTL should not be capable of lysing infected erythrocytes, the lack of difference between the clearance of several species of plasmodia in β2-microglobulin-deficient mice is not surprising.

3.3.3 Toxoplasma

The immune response to the intracellular protozoan *Toxoplasma gondii* is dependent on the production of IFN-γ (SUZUKI et al. 1988). The transfer of T lymphocytes from immunized animals to naive recipients mediates adoptive transfer of protection. $CD8^+$ lymphocytes mediate this process more efficiently than $CD4^+$ T lymphocytes (GAZZINELLI et al. 1991). DENKERS et al. (1993) evaluated the immune response to *T. gondii* in β2-microglobulin-deficient recipients. Unexpectedly, both β2-microglobulin-deficient and heterozygote littermate controls after vaccination with tachyzoites were protected when challenged with virulent *T. gondii*. Resistance in both sets of mice was dependent on the production of IFN-γ. There was a striking increase in $NK1.1^+$ cells in β2-microglobulin-deficient mice after vaccination, which was the source of IFN-γ in these mice, replacing the lost $CD8^+$ T cells.

3.3.4 Chlamydia

T lymphocytes are important mediators of the immune response to *Chlamydia* sp. β2-Microglobulin-deficient mice showed higher levels of organisms, but cleared the infection with similar kinetics to wild-type mice (MAGEE et al. 1995).

MORRISON et al. (1995) evaluated the immune response to genital infection with *C. trachomatis*. C57BL/6 and β2-microglobulin-deficient mice were no longer colonized by *Chlamydia* by 4–5 weeks after infection. In contrast, class II-deficient mice shed organisms over a 10-week period and showed a delay in resolution of the primary infection. Histopathologic sections from the genitourinary tract of infected animals confirmed the previous findings and demonstrated persistent inflammation only in the class II-deficient mice.

4 Concluding Remarks

It is clear from the assembled data that β2-microglobulin-deficient mice are not simply the same as normal mice with $CD8^+$ T cells removed. There are substantive effects of maturation of the immune system without class I expression which go beyond positive selection of $CD8^+$ T cells. The role of $CD8^+$ T cells includes both positive and negative effects on subsequent immune responses. While the responses to both bacteria and viruses are complex, β2-microglobulin-deficient mice are able to mount substantial and protective immune responses. It must be stressed that, even if a class I-deficient mouse can respond, this does not suggest that $CD8^+$ T cells do not have a role in those immune responses, but rather that $CD4^+$ cells can compensate for the lack of $CD8^+$ T cells. Thus the responses are complex and interrelated. β_2-Microglobulin-deficient mice have revealed more about the potential of $CD4^+$ cells than the depletion of $CD8^+$ cells during development.

References

Asano MS, Ahmed R (1995) Immune conflicts in lymphocytic choriomeningitis virus. Springer Semin Immunopathol 17:247–259

Bix M, Liao NS, Zijlstra M, Loring J, Jaenisch R, Raulet D (1991) Rejection of class I MHC-deficient haemopoietic cells by irradiated MHC-matched mice. Nature 349:329–331

Apasov S, Sitkovsky M (1993) Highly lytic CD8+, alpha beta T-cell receptor cytotoxic T cells with major histocompatibility complex (MHC) class I antigen-directed cytotoxicity in beta 2-microglobulin, MHC class I-deficient mice. Proc Natl Acad Sci USA 90:2837–2841

Apasov SG, Sitkovsky MV (1994) Development and antigen specificity of CD8+ cytotoxic T lymphocytes in beta 2-microglobulin-negative, MHC class I-deficient mice in response to immunization with tumor cells. J Immunol 152:2087–2097

Barkon ML, Haller BL, Virgin HWT (1996) Circulating immunoglobulin G can play a critical role in clearance of intestinal reovirus infection. J Virol 70:1109–1116

Bender BS, Croghan T, Zhang L, Small PA Jr (1992) Transgenic mice lacking class I major histocompatibility complex-restricted T cells have delayed viral clearance and increased mortality after influenza virus challenge. J Exp Med 175:1143–1145

Bender BS, Bell WE, Taylor S, Small PA Jr (1994) Class I major histocompatibility complex-restricted cytotoxic T lymphocytes are not necessary for heterotypic immunity to influenza. J Infect Dis 170:1195–1200

Bix M, Raulet D (1992) Functionally conformed free class I heavy chains exist on the surface of beta 2 microglobulin negative cells. J Exp Med 176:829–834

Christensen JP, Marker O, Thomsen AR (1994) The role of CD4+ T cells in cell-mediated immunity to LCMV: studies in MHC class I and class II deficient mice. Scand J Immunol 40:373–382

Cook, JR, Solheim JC, Connolly JM, Hansen TH (1995) Induction of peptide-specific CD8+ CTL clones in beta 2-microglobulin-deficient mice. J Immunol 154:47–57

Deepe GS Jr (1994) Role of CD8+ T cells in host resistance to systemic infection with histoplasma capsulatum in mice. J Immunol 152:3491–3500

DeLibero G, Flesch I, Kaufmann SHE (1988) Mycobacteria-reactive Lyt-2+ T cell lines. Eur J Immunol 18:59

Denkers EY, Gazzinelli RT, Martin D, Sher A (1993) Emergence of NK1.1+ cells as effectors of IFN-gamma dependent immunity to toxoplasma gondii in MHC class I-deficient mice. J Exp Med 178:1465–1472

Doherty PC, Hou S, Southern PJ (1993) Lymphocytic choriomeningitis virus induces a chronic wasting disease in mice lacking class I major histocompatibility complex glycoproteins. J Neuroimmunol 46:11–17

Eichelberger M, Allan W, Zijlstra M, Jaenisch R, Doherty PC (1991) Clearance of influenza virus respiratory infection in mice lacking class I major histocompatibility complex-restricted CD8+ T cells. J Exp Med 174:875–880

Emmerling P, Finger H, Bockemuhl J (1975) Listeria monocytogenes infection in nude mice. Infect Immun 12:437–439

Emoto M, Neuhaus O, Emoto Y, Kaufmann SH (1996) Influence of β_2-microglobulin expression on γ-interferon secretion and target cell lysis by intraepithelial lymphocytes during intestinal Listeria monocytogenes infection. Infect Immun 64:569–575

Epstein SL, Misplon JA, Lawson CM, Subbarao EK, Connors M, Murphy BR (1993) Beta 2-microglobulin-deficient mice can be protected against influenza A infection by vaccination with vaccinia-influenza recombinants expressing hemagglutinin and neuraminidase. J Immunol 150:5484–5493

Fiette L, Aubert C, Brahic M, Rossi CP (1993) Theiler's virus infection of beta 2-microglobulin-deficient mice. J Virol 67:589–592

Flynn JL, Goldstein MM, Triebold KJ, Koller B, Bloom BR (1992) Major histocompatibility complex class I-restricted T cells are required for resistance to Mycobacterium tuberculosis infection. Proc Natl Acad Sci USA 89:12013–12017

Franco MA, Greenberg HB (1995) Role of B cells and cytotoxic T lymphocytes in clearance of and immunity to rotavirus infection in mice. J Virol 69:7800–7806

Gazzinelli RT, Hakim FT, Hieny S, Shearer GM, Sher A (1991) Synergistic role of CD4+ and CD8+ T lymphocytes in IFN-γ production and protective immunity induced by an attenuated *Toxoplasma gondii* vaccine. J Immunol 146:286

Glas R, Franksson L, Ohlen C, Hoglund P, Koller B, Ljunggren HG, Karre K (1992) Major histocompatibility complex class I-specific and -restricted killing of beta 2-microglobulin-deficient cells by CD8+ cytotoxic T lymphocytes. Proc Natl Acad Sci USA 89:11381–11385

Gombold JL, Sutherland RM, Lavi E, Paterson Y, Weiss SR (1995) Mouse hepatitis virus A59-induced demyelination can occur in the absence of CD8+ T cells. Microbial Pathogen 18:211–221

Graham MB, Braciale VL, Braciale TJ (1994) Influenza virus-specific CD4+ T helper type 2 T lymphocytes do not promote recovery from experimental virus infection. J Exp Med 180:1273–1282

Heinzel FP, Sadick MD, Holaday BJ, Coffman RL, Locksley RM (1989) Reciprocal expression of interferon-γ or interleukin-4 during the resolution or progression of murine leishmaniasis. Evidence for expansion of distinct helper T cell subsets. J Exp Med 169:59–72

Hoglund P, Sundback J, Olsson-Alheim M, Johansson M, Salcedo M, Ohlen C, Ljunggren H, Sentman C, Karre K (1997) Host MHC class I gene control of NK-cell specificity in the mouse. Immunol Rev 155:11–28

Hou S, Doherty PC, Zijlstra M, Jaenisc R, Katz JM (1992) Delayed clearance of Sendai virus in mice lacking class I MHC-restricted CD8+ T cells. J Immunol 149:1319–1325

Hyland L, Hou S, Coleclough C, Takimoto T, Doherty PC (1994) Mice lacking CD8+ T cells develop greater numbers of IgA-producing cells in response to a respiratory virus infection. Virology 204:234–241

Kocks C, Gouin E, Tabouret M, Berche P, Ohayon H, Cossart P (1992) L. monocytogenes-induced actin assembly requires the *actA* gene product, a surface protein. Cell 68:521–531

Koller B, Marrack P, Kappler J, Smithies O (1990) Normal development of mice deficient in β2-microglobulin, MHC class I proteins and CD8+ T cells. Science 248:1227–1230

Ladel CH, Flesch IEA, Arnoldi J, Kaufmann SHE (1994) Studies with MHC-deficient knock-out mice reveal impact of both MHC I- and MHC II-dependent T cell responses on *Listeria monocytogenes* infection. J Immunol 153:3116–3122

Lamouse-Smith E, Clements VK, Ostrand-Rosenberg S (1993) Beta 2M-/- knockout mice contain low levels of CD8+ cytotoxic T lymphocyte that mediate specific tumor rejection. J Immunol 151:6283–6290

Lavi E, Wang Q (1995) The protective role of cytotoxic T cells and interferon against coronavirus invasion of the brain. Adv Exp Med Biol 380:145–149

Lehmann-Grube F, Lohler J, Utermohlen O, Gegin C (1993) Antiviral immune responses of lymphocytic choriomeningitis virus-infected mice lacking CD8+ T lymphocytes because of disruption of the beta 2-microglobulin gene. J Virol 67:332–339

Lehmann-Grube F, Dralle H, Utermohlen O, Lohler J (1994) MHC class I molecule-restricted presentation of viral antigen in beta 2-microglobulin-deficient mice. J Immunol 153:595–603

Liao NS, Bix M, Zijlstra M, Jaenisch R, Raulet D (1991) MHC class I deficiency: susceptibility to natural killer (NK) cells and impaired NK activity. Science 253:199–202

Mackaness GB (1962) Cellular resistance to infection. J Exp Med 118:381–406

Magee DM, Williams DM, Smith JG, Bleicker CA, Grubbs BG, Schachter J, Rank RG (1995) Role of CD8 T cells in primary chlamydia infection. Infect Immun 63:516–521

Malik A, Egan JE, Houghten RA, Sadoff JC, Hoffman SL (1991) Human cytotoxic T lymphocytes against the *Plasmodium falciparum* circumsporozoite protein. Proc Natl Acad Sci USA 88:3300–3305

Manickan E, Rouse BT (1995) Roles of different T-cell subsets in control of herpes simplex virus infection determined by using T-cell-deficient mouse-models. J Virol 69:8178–8179

Marusic-Galesic S, Udaka K, Walden P (1993) Increased number of cytotoxic T cells within CD4+8- T cells in beta 2-microglobulin, major histocompatibility complex class I-deficient mice. Eur J Immunol 23:3115–3119

McDermott MR, Lukacher AE, Braciale VL, Braciale TJ, Bienenstock J (1987) Characterization and in vivo distribution of influenza-virus-specific T-lymphocytes in the murine respiratory tract. Am Rev Respir Dis 135:245–249

Mielke MEA, Ehlers S, Hahn H (1993) The role of cytokines in experimental listeriosis. Immunobiol 189:285–315

Miller DJ, Rivera-Quinones C, Njenga MK, Leibowitz J, Rodriguez M (1995) Spontaneous CNS remyelination in beta 2 microglobulin-deficient mice following virus-induced demyelination. J Neurosci 15:8345–8352

Morrison RP, Feilzer K, Tumas DB (1995) Gene knockout mice establish a primary protective role for major histocompatibility complex class II-restricted responses in *Chlamydia trachomatis* genital infection. Infect Immun 63:4661–4668

Muller D, Koller BH, Whitton JL, LaPan KE, Brigman KK, Frelinger JA (1992) LCMV-specific, class II-restricted cytotoxic T cells in beta 2-microglobulin-deficient mice. Science 255:1576–1578

Muller D, Chen M, Vikingsson A, Hildeman D, Pederson K (1995) Oestrogen influences CD4+ T-lymphocyte activity in vivo and in vitro in beta 2-microglobulin-deficient mice. Immunology 86:162–167

Niemialtowski MG, Godfrey VL, Rouse BT (1994) Quantitative studies on CD4+ and CD8+ cytotoxic T lymphocyte responses against herpes simplex virus type 1 in normal and beta 2-m deficient mice. Immunobiol 190:183–194

Ohlen C, Hoglund P, Sentman CL, Carbone E, Ljunggren HG, Koller B, Karre K (1995) Inhibition of natural killer cell-mediated bone marrow graft rejection by allogeneic major histocompatibility complex class I, but not class II molecules. Eur J Immunol 25:1286–1291

Orme I (1994) Protective and memory immunity in mice infected with Mycobacterium tuberculosis. Immunobiol 191:503–508

Parnes JR, Sizer KC, Seidman JG, Stallings V, Hyman R (1986) A mutational hot-spot within an intron of the mouse beta 2-microglobulin gene. EMBO J 5:103–111

Polic B, Jonjic S, Pavic I, Crnkovic I, Zorica I, Hengel H, Lucin P, Koszinowski UH (1996) Lack of MHC class I complex expression has no effect on spread and control of cytomegalovirus infection in vivo. J Gen Virol 77:217–225

Quinn DG, Zajac AJ, Frelinger JA, Muller D (1993) Transfer of lymphocytic choriomeningitis disease in beta 2-microglobulin-deficient mice by CD4+ T cells. Int Immunol 5:1193–1198

Quinn D, Zajac A, Frelinger J (1995) Immune responses to LCMV in b2-microglobulin-deficient mice. Immunol Rev 148:151–169

Quinn DG, Zajac AJ, Hioe C, Frelinger J (1997) Virus-specific, $CD8^+$ Majorhistocompatibility complex class I restricted cytotoxic T lymphocytes in LCMV infected β_2-microglobulin-deficient mice. J Virol 71:8329–8396

Raulet DH, Spencer DM, Hsiang YH, Goldman JP, Bix M, Liao NS, Zijstra M, Jaenisch R, Correa I (1991) Control of gamma delta T-cell development. Immunol Rev 120:185–204

Roberts AD, Ordway DJ, Orme IM (1993) Listeria monocytogenes infection in beta 2 microglobulin-deficient mice. Infect Immun 61:1113–1116

Rodriguez M, Dunkel AJ, Thiemann RL, Leibowit J, Zijlstra M, Jaenisch R (1993) Abrogation of resistance to Theiler's virus-induced demyelination in H-2b mice deficient in beta 2-microglobulin. J Immunol 151:266–276

Sadick MD, Heinzel FP, Holaday BJ, Pu RT, Dawkins RS, Locksley RM (1990) Cure of murine leishmaniasis with anti-interleukin 4 monoclonal antibody. Evidence for a T cell-dependent, interferon-γ-independent mechanism. J Exp Med 171:115–127

Serody JS, Poston RM, Weinstock D, Kurlander RJ, Frelinger JA (1996) CD4+ cytolytic effectors are inefficient in the clearance of *Listeria monocytogenes*. Immunology 88:544–550

Serreze DV, Leiter EH, Christianson GJ, Greiner D, Roopenian DC (1994) Major histocompatibility complex class I-deficient NOD-B2m null mice are diabetes and insulitis resistant. Diabetes 43:505–509

Simon HB, Sheagren JN (1972) Enhancement of macrophages bactericidal capacity by antigenically stimulated immune lymphocytes. Cell Immunol 4:163–174

Slattery RM, Miller JF (1996) Influence of T lymphocytes and major histocompatibility complex class II genes on diabetes susceptibility in the NOD mouse. Curr Top Microbiol Immunol 206:51–66

Spriggs MK, Koller BH, Sato T, Morrissey PJ, Fanslow WC, Smithies O, Voice RF, Widmer MB, Maliszewski CR (1992) Beta 2-microglobulin-, CD8+ T-cell-deficient mice survive inoculation with high doses of vaccinia virus and exhibit altered IgG responses. Proc Natl Acad Sci USA 89:6070–6074

Su HC, Orange JS, Fast LD, Chan AT, Simpson SJ, Terhorst C, Biron CA (1994) IL-2-dependent NK cell responses discovered in virus-infected beta 2-microglobulin-deficient mice. J Immunol 153:5674–5681

Sumida T, Furukawa M, Sakamoto A, Namekawa T, Maeda T, Zijlstra M, Iwamoto I, Koike T, Yoshida S, Tomioka H et al. (1994) Prevention of insulitis and diabetes in beta 2-microglobulin-deficient non-obese diabetic mice. Int Immunol 6:1445–1449

Suzuki Y, Orellana MA, Schrieber RD, Remington JS (1988) Interferon-γ: the major mediator of resistance against *Toxoplasma gondii*. Science 516:240–242

Tay CH, Welsh RM, Brutkiewicz RR (1995) NK cell response to viral infections in beta 2-microglobulin-deficient mice. J Immunol 154:780–789

Taylor SF, Bender BS (1995) Beta 2-microglobulin-deficient mice demonstrate class II MHC restricted anti-viral CD4+ but not CD8+ CTL against influenza-sensitized autologous splenocytes. Immunol Lett 46:67–73

Van der Heyde H, Manning DD, Roopenian DC, Weidanz WP (1993) Resolution of blood-stage malarial infections in CD8+ cell-deficient beta$_2$-microglobulin deficient mice. J Immunol 151:3187–3191

Wang B, Gonzalez A, Benoist C, Mathis D (1996) The role of CD8+ T cells in the initiation of insulin-dependent diabetes mellitus. Eur J Immunol 26:1762–1769

Weck KE, Barkon ML, Yoo LI, Speck SH, Virgin HI (1996) Mature B cells are required for acute splenic infection, but not for establishment of latency, by murine gammaherpesvirus 68. J Virol 70:6775–6780

White KL, Snyder HL, Krzych U (1996) MHC class I-dependent presentation of exoerythrocytic antigens to CD8+ T lymphocytes is required for protective immunity against Plasmodium berghei. J Immunol 156:3374–3381

Wicker LS, Leiter EH, Todd JA, Renjilian RJ, Peterson E, Fischer PA, Podolin PL, Zijlstra M, Jaenisch R, Peterson LB (1994) Beta 2-microglobulin-deficient NOD mice do not develop insulitis or diabetes. Diabetes 43:500–504

Zajac AJ, Muller D, Pederson K, Frelinger JA, Quinn DG (1995) Natural killer cell activity in lymphocytic choriomeningitis virus-infected beta 2-microglobulin-deficient mice. Int Immunol 7:1545–1556

Zajac AJ, Quinn DG, Cohen PL, Frelinger JA (1996) Fas-dependent CD4+ cytotoxic T-cell-mediated pathogenesis during virus infection Proc Natl Acad Sci USA 93:14730–14735

Zijlstra M, Bix M, Simister N, Loring J, Raulet D, Jaenisch R (1989) β2 Microglobulin deficient mice lack CD4-CD8+ cytolytic T cells. Nature 344:742–745

Zinkernagel RM (1996) Immunology taught by viruses. Science 271:173–178

CNS Neurons: The Basis and Benefits of Low Class I Major Histocompatibility Complex Expression

GLENN F. RALL

1 Introduction: Host Cells That Restrict "Counterproductive" MHC Recognition

The host immune response is generally thought to consist of cells with "professional" immunologic functions, such as B and T cells, macrophages, and natural killer (NK) cells. However, differentiated cells which do not normally participate in immune surveillance may be recruited to serve an integral function in the immune-mediated elimination of foreign intracellular pathogens such as viruses. As discussed elsewhere in this volume, most cells have the ability to present immunogenic, "non-self" peptides (called epitopes) in association with "self" class I major histocompatibility complex (MHC) molecules. This cell surface complex is engaged by the T cell

The Fox Chase Cancer Center, Department of Basic Science, Divisions of Virology and Immunology, 7701 Burholme Avenue, Philadelphia, PA 19111, USA receptor (TCR) of cytotoxic T lymphocytes (CTL). Appropriate MHC-epitope-TCR interaction leads to the CTL-mediated lysis of the epitope-expressing target cell via the perforation of the plasma membrane, introduction of CTL-derived proteolytic enzymes (granzymes) into the target cell cytosol, and eventual cell death, presumably via apoptosis.

For the majority of virally infected tissues, this strategy is ultimately beneficial: the net gain (reduction in the number of virus-producing cells) outweighs the consequence of lysis of an infected differentiated cell. This is true because, for most tissues, immune-mediated cell loss is accompanied by increased cell proliferation, which restores the damaged tissue. For example, hepatitis B virus (HBV) infects between 5% and 40% of all hepatocytes within the livers of chronically infected humans (BIANCHI and GUDAT 1980). An estimated 7% of HBV-infected hepatocytes are lost each day (NOWAK et al. 1996), implying that between 0.3% and 3% of all hepatocytes must be replenished daily to maintain a stable liver cell mass. Thus, for many virally infected tissues, a balance exists between cell loss and cell renewal.

The presentation of immunogenic epitopes by virally infected cells can be seen as an altruistic event. Because the expression of these antigenic molecules often precedes production of infectious viral progeny, immune-mediated lysis of the infected cell destroys a "factory" of virus production and therefore decreases the total viral burden. Thus these unfortunate few die for the benefit of the common good: reduced viral load and recruitment of T cells to the site of a productive infection minimizes the opportunity of the virus to infect a neighboring cell.

However, altruistic antigen presentation may not always benefit an infected host. While expression of MHC molecules is highest in cells such as those of the lymphoid system, the gut, and the kidney, MHC class I is expressed to a low to undetectable degree in other cells, including those of the brain, sperm cells at certain stages of differentiation, and trophoblasts within the placenta (DAVID-WATINE et.al. 1990). In this chapter, we discuss recent findings concerning the molecular basis and potential benefits of low MHC expression, predominately focusing on class I MHC presentation in central nervous system (CNS) neurons and trophoblasts within the placenta. We also take a broader perspective concerning the lack of CTL recognition in the CNS and describe work which suggests that neurons may employ several mechanisms, low MHC expression being one of them, to avoid immune cell recognition.

1.1 Obligatory Haplotype Mismatch: The Maternal Fetal Interface

Pregnancy poses an unique challenge to the maternal immune response, since the developing fetus can be viewed as a natural graft. The placenta contains fetal and paternally derived histocompatibility antigens that are potentially immunogenic for the maternal immune system (VOLAND et.al. 1994); the activation of the maternal host defenses against what it perceives as "foreign" would obviously be detrimental to the developing fetus. One factor contributing to the lack of a maternal anti-fetus response is that the placental trophoblasts, which are in direct contact with maternal blood and tissues, express little or no classical MHC antigens (FAULK and TEMPLE 1976; HUNT and ORR 1992). The deleterious consequences of trophoblast MHC expression on fetal survival were confirmed in transgenic mouse studies in which a mouse class I MHC gene (D^d) was expressed in murine trophoblasts under the transcriptional control of the c-fos promoter, known to be highly expressed in trophoblasts during development (DESCHAMPS et.al. 1985). Expression of this MHC molecule in vivo resulted in a higher incidence of spontaneous abortion, measured as a function of reduced transgene transmission to surviving pups. Thus, if the MHC expression had no effect on survival of the pups, one would expect approximately 50% transmission of the transgene in heterozygous (D^d transgenic×normal) crosses. However, in the litters born to these parents, transmission was reduced to 20%–30%, indicating in utero death of 20%–30% of the transgenic fetuses. Breeding of the c-fos-MHC male transgenic with β_2-microglobulin knockout females (who lack expression of the classical class I MHC and therefore fail to positively select $CD8^+$ T cells) restored transmission of the transgene to 50% of the progeny. Thus, while the expression of MHC molecules in the trophoblasts does not inexorably lead to spontaneous abortion (20%–30% of surviving neonates of (D^dnormal) crosses were transgenic) there does exist a correlation between transgene transcription and MHC expression on placental tissues (VOLAND et al. 1994).

The absence of MHC may not be the only way in which the fetus is protected from the maternal immune response. In humans, the highly polymorphic "classical" MHC class I molecules are encoded in the HLA-A, -B, and -C gene families. At least three additional class I genes, HLA-E, -F, and -G have been identified. These nonclassical (class Ib) genes are highly homologous to the classical HLA genes, and all associate with β_2-microglobulin. However, in contrast to the classical HLA genes, HLA-E, -F, and -G are nonpolymorphic: for example, HLA-G is derived from alternative splicing and a premature stop codon (reviewed in WOOD 1994), resulting in a protein that lacks a full cytoplasmic tail (GERAGHTY et.al. 1987). HLA-G is predominately expressed on trophoblasts in fetal placental tissues at the materno-fetal interface, where the classical class I and II genes are absent (CAROSELLA et al. 1996), implying that it plays a potentially important role in maternal tolerance to the developing fetus.

Whether this nonclassical MHC expression on trophoblasts can present peptides to the maternal immune response remains unresolved (SANDERS et.al. 1991; VINCE and JOHNSON 1995). However, a link between HLA-G expression and resistance to NK cell-mediated lysis has been demonstrated, suggesting that HLA-G acts as a "surrogate" cell surface MHC molecule to prevent NK-mediated lysis (LIAO et.al. 1991). HLA-G is capable of inhibiting the NK activity of decidual large granular leukocytes against the trophoblasts (FERRY et.al. 1991). Blocking the HLA-G antigen with antibody restored NK cytolytic activity (CHUMBLEY et al. 1994). Since HLA-G polymorphism is likely to be low, it is possible that HLA-G can serve as the public ligand for the NK cell receptor, protecting fetal trophoblasts from NK killing and conferring immunological tolerance to fetal tissue.

1.2 Essential and Nonrenewable: CNS Neurons

For many viral infections, lysis of infected cells by the class I MHC-mediated CTL response does not often result in long-term deficits since, following infection, many of these tissues can recover by upregulation of cell division. However, neurons of the CNS are an essential, but nondividing population; therefore, loss of these cells, either by virus-mediated cytolysis or CTL-mediated killing, would have obvious deleterious consequences to a host. Thus the immune response to a virus may be more damaging than the viral infection itself. Perhaps to provide protection against such an immune response, normal uninfected CNS neurons lack surface expression of class I MHC molecules. Whether neurons can be induced to express MHC is still debated, although recent reports have convincingly shown that MHC expression on CNS neurons can be detected under certain pathological circumstances. The debate centers around the "inability" versus "reluctance" of CNS neuron class I MHC expression. The inability of some neuronal culture systems to present epitopes has been attributed to an intrinsic defect in the synthesis of the antigen-presenting machinery. Alternatively, reluctant expression may depend on stimulating factors such as cytokines that are normally absent within the parenchyma; the presence of these inducers may allow for upregulation of MHC on neurons.

2 Constraints on Immune Cell-Target Cell Interaction in the Brain

The low level of class I MHC on all brain parenchymal cells implies that immune responses in the CNS differ significantly from those in the periphery. In addition to the low levels of MHC expression, may of additional features unique to the CNS contribute to the hypothesis that immune responses are suppressed within the brain. These features, including the presence of the blood-brain barrier and absence of lymphatic drainage, have been extensively reviewed elsewhere (WEKERLE et al. 1986; SEDGWICK and DORRIES 1991; CSERR and KNOPF 1992; RALL and OLDSTONE 1995). Here, we briefly discuss recent contributions to understanding the role of the blood-brain barrier, lymphatic drainage, and brain-derived gangliosides in restricting immune recognition in the brain and describe how the neuronal environment may restrict viral infections, indirectly altering the generation of immunogenic peptides.

2.1 The First Obstacle: The Blood-Brain Barrier Restricts Trafficking of Immune Mediators into the CNS

The interface between the circulation and the brain parenchyma, known as the blood-brain barrier, is comprised of endothelial cells, basal membranes, and astrocyte endfeet (RAPOPORT 1976; BRIGHTMAN et al. 1983). This complex barrier controls the exchange of cells and metabolites between the blood and the brain, predominately due to tight junctions within the capillary endothelium. Consequently, many circulating lymphocytes are restricted from crossing the endothelial barrier and patrolling the brain parenchyma. It is this high degree of surveillance imposed by the tightly packed endothelial cells which has contributed to the belief that the CNS is immune privileged.

Continuous tight junctions are not the only feature that makes the capillaries of the CNS distinctive. Due to the need for blood-borne cells and metabolites to pass between endothelial cells to enter the brain, a decisive factor in determining how easily a molecule enters the CNS is lipid solubility (GOLDSTEIN and BETZ 1986). Lipid-soluble molecules, such as nicotine, ethanol, and heroin, can readily cross the barrier, while water-soluble molecules, including many of the immune-mediating proteins such as antibodies, cytokines, and complement components, cannot gain access to the parenchyma. For non-lipid-soluble molecules to enter the CNS parenchyma, specific transporter molecules present on the endothelial cells are required.

From an immunological perspective, the blood-brain barrier restricts, but does not prohibit, immune responses from occurring within the CNS. For example, lymphocytes are not completely blocked from gaining access to the CNS (WEKERLE et al. 1986; HICKEY and KIMURA 1987; TYOR et al. 1989; HICKEY et al. 1991). Hickey and coworkers have shown that activated T cells enter the CNS in a random manner (HICKEY et al. 1991), regardless of antigen specificity or MHC compatibility. Thus it appears that any T cell clone which is activated in the periphery will ultimately gain access to the CNS parenchyma. This condition, as Hickey points out, is supported experimentally by grafting studies in which an allograft, normally tolerated if grafted into the CNS, will be rejected rapidly when the host is exposed to alloantigens in the periphery (MASON et al. 1986). Furthermore, the barrier changes in response to viral infections and inflammatory responses by altering the profile of adhesion molecules expressed on the endothelial surface and by possible transient increases in blood-brain barrier permeability (LASSMANN et al. 1991; WEKERLE et al. 1991; RALL et al. 1995). Thus CNS "immune privilege" may be a constantly changing state, influenced both by intra- and extraparenchymal conditions.

2.2 Absence of Conventional Lymphatic Drainage

While the blood-brain barrier was thought to restrict the efferent movement of peripheral cells into the CNS, the absence of a conventional lymphatic system within the CNS was believed to interfere with the afferent arm of immunity, alerting the peripheral immune response to antigens within the CNS. However, despite the absence of typical lymphatic channels in the brain, a connection between the CNS and the lymphatics was demonstrated when a large fraction of radioiodinated albumin injected into the brain was isolated from cervical lymphatics (YAMADA et al. 1991). These and other results (CSERR et al. 1992; WELLER et al. 1996) show that cerebral extracellular fluids drain into the blood from the CNS along cranial nerves, primarily the olfactory nerve, and along spinal nerve root ganglia. Thus the outflow of cerebrospinal fluid (CSF) along cranial nerves and spinal nerve roots is the conduit for cross-talk between the brain parenchyma, the brain vasculature, and the lymphatic system (YOFFEY and COURTICE 1970; CSERR and KNOPF 1992).

2.3 Influence of an Immunosuppressive Environment

The paucity of MHC expression on resident brain cells, the lack of lymphatic drainage, and the presence of the highly selective blood-brain barrier collectively suggest that immunological nonresponsiveness within the CNS is due to an absence of lymphocyte access to the brain or an absence of recognition of infected cells. However, the isolation and characterization of brain-enriched molecules with potent immunosuppressive qualities suggests that the brain may also actively restrict immune responses.

One group of molecules with recently described immunomodulatory effects on neurotropic viral infections are the sialic acid-containing glycosphingolipids known as gangliosides. Gangliosides are ubiquitous cell membrane components that can also be shed into the extracellular environment. Their multifunctional roles in cell metabolism include regulation of cell-cell interaction, differentiation, signal transduction, and growth regulation (reviewed in BERGELSON 1995).

The influence of soluble gangliosides on the immune response has been studied for over a quarter of a century. In general, gangliosides have been shown to have potent and apparently generalized immunosuppressive qualities, including the suppression of CTL, T helper cell, and NK cell proliferation (reviewed in BERGELSON 1995). Interestingly, one form of ganglioside (GM3) can activate cells, but these lymphocytes have a suppressor phenotype (DYATLOVITSKAYA et al. 1991). While the precise mechanisms used by gangliosides to downregulate the immune response are not fully understood, their effects appear to be diverse, influencing cell surface expression by insertion into the lipid bilayer and altering soluble mediator function by binding to such proteins as interleukin-2 (IL-2), interferon gamma (IFN-γ), and tumor necrosis factor (TNF) (DYATLOVITSKAYA and BERGELSON 1987).

Gangliosides are enriched within the CNS (WIEGANDT 1971; IRANI et al. 1996), especially within neurons. Recently, gangliosides were shown to selectively block the in vitro production of Th1-associated cytokines, such as IL-2 and IFN-γ, perhaps by blocking NF-κB activation (IRANI et al. 1996). Furthermore, gangliosides inhibited T cell proliferation by preventing their entry into the cell cycle (IRANI et al. 1996). The immunosuppressive role of gangliosides in the resolution of CNS viral infections was substantiated in mice when intraparenchymal T cells, recruited to the brain parenchyma in response to the neuronal infection caused by Sindbis virus, were arrested in the cell cycle, despite the presence of activation markers (IRANI et al. 1997). Thus, while activated cells routinely patrol the CNS (HICKEY et al. 1991), molecules within the CNS microenvironment may suppress or impair the effector mechanisms of intraparenchymal lymphocytes.

2.4 Restricted Replication of Viruses

Because of these unique aspects of the CNS environment, the degree of immunological surveillance and the vigor of the immune response is significantly less in the brain than in peripheral organs. Taking these facts into consideration, it is not unexpected that neurons have evolved unique capacities to prevent their destruction by viral infections. Indeed, probably because CNS neurons are a harbor from the immune response, a large number of DNA and RNA viruses are neurotropic.

Many of these viruses convert to a persistent phenotype upon infection of neurons. In such cases, viral nucleic acids and proteins are readily detected in the absence of direct cell death or induction of the immune response, and often in the absence of production of extracellular infectious progeny. While the detailed mechanisms by which viruses establish and maintain long-term persistent infections are largely unknown, both viral genes and host cellular genes have been shown to cooperatively promote viral persistence.

In the persistent infection caused by the murine virus lymphocytic choriomeningitis virus (LCMV), expression of nucleoprotein (NP) antigen and short (S) RNA (RODRIGUEZ et al. 1983; FAZAKERLEY et al. 1991) is readily detected in the CNS, but electron microscopy examination has failed to reveal mature viral particles budding from infected neurons within the brain parenchyma, and almost no extracellular infectious virus can be detected. This was also observed in recent studies using the PC12 neuronal cell culture system (DE LA TORRE et al. 1993), in which differentiation of the chromaffin-like PC12 cells to a neuronal phenotype was accompanied by a 3- to 4-log decrease in the production of infectious virus.

How neurons block the production of infectious virus despite viral gene expression remains unresolved and is relevant to the human infections caused by measles, rabies, influenza, poliovirus, and mumps and the animal infections caused by Borna disease virus (BDV), the coronaviruses, and the arenaviruses, among others. Since each of these viruses can establish productive infections in other differentiated cells, the dramatic reduction in infectious virus production in neurons suggests that the neuronal environment is in some way inhospitable for viral propagation.

While the cellular mechanisms that restrict infectious virus production in neurons are not known, a number of observations indicate that the intracellular level of cyclic nucleotides in neurons may be involved. For example, the shift from acute to persistent measles virus infection in mouse neuroblastoma cultures depends on functions that affect endogenous cyclic adenosine monophosphate (cAMP) and cyclic guanosine monophosphate (cGMP) levels (MILLER and CARRIGAN 1982). In human fibroblasts, the addition of cAMP-enhancing compounds significantly decreased the yield of herpes simplex virus (HSV) I (STANWICK et al. 1977).

Neural cells contain abundant cAMP, with levels being closely correlated with terminal differentiation, cessation of cell division, and induction of neuron-specific functions (GREENGARD 1978). Since cAMP activates cAMP-dependent kinases and cAMP-gated ion channels and can also interact with other second messenger signals, it is reasonable to speculate that changes in the intracellular levels of cAMP or its downstream effectors may be able to alter the efficiency of viral replication in CNS neurons.

How might decreased viral synthesis affect antigen presentation and resultant immune recognition in the CNS? Reduced viral production, especially of extracellular virus, would limit the availability of viral proteins to enter the circulation via the lymphatics and activate or recruit the peripheral immune response. Furthermore, reduced levels of replication also impact on the number of endogenously synthesized viral peptides which can serve as epitopes for class I MHC recognition. Finally, many viral infections have been shown to enhance the transcription of the antigen-presenting machinery; perhaps slowly replicating viruses would not trigger this activation, resulting in a lower level of class I MHC-presented molecules on infected neurons.

3 MHC Expression in Neurons

3.1 Expression of Class I MHC in Peripheral Nervous System Neurons

The hypothesis that has been proposed as to why CNS neurons do not express class I MHC molecules relates to the consequences of CTL-mediated lysis of these nondividing cells: depletion of a critical and nonrenewable population would have dramatic effects on the host. Therefore, if MHC expression and terminal differentiation are linked, one might expect that peripheral neurons, which can regenerate, might express class I MHC under certain circumstances. Indeed, both in vivo and in vitro studies have confirmed that peripheral neurons can express class I MHC RNA. However, whether these cells can present endogenously generated peptides on the cell surface, and whether peripheral nervous system (PNS) neurons serve as targets for CTL recognition and attack, remains unresolved.

In an effort to assay whether brain-derived cells could serve as targets for alloantigen-specific CTL lysis, primary PNS neurons from the superior cervical ganglia were cultured and used as targets in standard chromium release assays (KEANE et al. 1992). PNS neurons were lysed both by allospecific CTL and by purified granules obtained from lymphokine-activated killer (LAK) cells, whereas CNS neurons remained refractory to CTL killing, even after 3 weeks in culture. Interestingly, no expression of MHC class I was detected on the surface of the PNS neurons by immunohistochemistry; however, the authors argue that the degree of killing seen in these cultures (as much as 60% at effector to target ratios of 40:1) and the purity of their CTL population strongly suggests that neuronal death is mediated through interaction of T cells with target cells via T cell receptor-class I MHC association.

In vivo, the ability of peripheral neurons to express MHC and to be lysed by CTL is less clear. Following a peripheral nerve lesion of rat facial and sciatic motor nerves, an induction of class I MHC antigens was detected. The MHC expression was transient for some lesions (e.g., MHC expression decreased 21 days after nerve crush, coincident with nerve regeneration), although MHC antigen remained detectable in nonregenerating lesions, such as a cut nerve (MAEHLEN et al. 1988, 1989; O'MALLEY and MACLEISH 1993). While there is no evidence that upregulated MHC expression results in the CTL-mediated lysis of these neurons, MAEHLEN et al. (1989) speculate that MHC expression may serve to signal cell death by a nonimmunological process, a hypothesis supported by later studies with "electrically silent" CNS neurons (NEUMANN et al. 1995).

Increased neuronal MHC expression could be due to the local inflammation, edema, and neighboring cell dysregulation that is induced by physical damage to neurons. Therefore, the effect of a neuron-specific, relatively noninflammatory lesion, such as a viral infection, may help to understand the link between intra- and extraneuronal stimuli. In an in vivo model of acute HSV infection of peripheral nerves (SIMMONS 1989; SIMMONS and TSCHARKE 1992), class I MHC protein expression in the PNS was upregulated transiently in satellite and Schwann cells (both of glial origin), but not in neurons within the HSV-infected spinal ganglia (PEREIRA et al. 1994). The authors report that, while no protein was expressed on HSV-infected neurons, class I MHC mRNA was detected, leading to two possible interpretations. In one case, these mRNAs correspond to classical MHC molecules, implying that the block in cell surface protein expression is post-transcriptionally controlled. This block could restrict translation of the MHC RNA themselves or could abrogate the function of accessory antigen-presenting molecules required to bring the MHC antigens to the cell surface. Alternatively, these transcripts could correspond to nonclassical MHC antigens, which would not be detected as proteins with the antibodies used in this study.

Thus the variability in PNS neuron MHC, RNA and protein expression is governed in part by the system (in vivo or in vitro) and the stimulating event (viral infection, nerve crush, axotomy). In general, however, PNS neurons appear to respond to deleterious stimuli; the role of upregulated MHC "classical or nonclassical" on the neuronal surface remains to be elucidated.

3.2 Expression of Class I MHC in CNS Neurons In Vitro

Most studies that evaluate neuronal antigen-presenting capacity in vitro find that MHC expression is low or absent in quiescent neurons, but can be induced by such stimuli as cytokines, viral infections, or direct injury. These reports collectively argue that the normal, low to negligible expression of class I molecules reflects a regulatory control rather than a genetic lesion in CNS neurons (LAMPSON and FISHER 1984). Two reports, however, suggest that MHC proteins are either constitutively absent or constitutively expressed on cells of neuronal origin. One study used the neuroblastoma line N2A to convincingly demonstrate constitutive expression of MHC, equaling that of a control ependymoblastoma line (TING et al. 1987). These cells were also efficient targets for CTL, but not for NK cells, implying that "classical" TCR-MHC interactions between CTL and these neuroblastoma cells occurred. These results differed from previous reports using the same cell line that were not able to detect MHC expression (LAMPSON et al. 1983; WONG et al. 1984; MAIN et al. 1985). TING et al. (1987) noted that detection strategies that depend on antibodies are exquisitely sensitive to antibody concentration and suggested that this may explain the discrepancy between their findings and previous reports.

In 1993, MASSA et al. (1993) evaluated the ability of IFN-γ to induce MHC expression in primary embryonic neuronal cultures and found that, while IFN-γ could activate MHC transcription in cultured astrocytes, primary CNS neurons remained refractory to the cytokine stimulus. As discussed later in this chapter, the block was attributed to an absence of induction of a nuclear factor which binds to the IFN-responsive consensus sequence.

These disparate findings may not be as controversial as they initially appear; TING et al. (1987) correctly notes the inherent problem of comparing a neuroblastoma cell line with neurons in vivo: it is possible that only dividing neurons, e.g., neuroblastoma, normally express class I MHC, linking MHC expression with mitotic activity and differentiation status. Consistent with this hypothesis, the primary neurons used by MASSA et al. (1993) were evaluated 8 days post-culture, a time when these cells are no longer mitotically active.

Cell division may not be the only parameter that determines the inducibility of MHC genes in neurons. Recently, NEUMANN et al. (1995, 1997) used the combined approaches of electrophysiology and reverse transcriptase-polymerase chain reaction (RT-PCR) to monitor MHC transcription in the presence or absence of cytokine stimulation. While other reports had evaluated the effect of cytokine addition to neurons in culture (LAMPSON and FISHER 1984; JOLY et al. 1991; DREW et al. 1993), the added approach of patch clamping allowed correlations to be drawn between MHC expression and functional activity in individual neurons. In this recent report, transcription of MHC class I genes was very rare in neurons with spontaneous action potentials, as determined by whole-cell patch clamping of cultured hippocampal neurons. However, in electrically silent neurons, defined as those without spontaneous action potentials, class I MHC transcription was detected; class I MHC protein expression could be seen only in electrically silent neurons treated with IFN-γ. Thus neurons may possess the basic requirements to interact with $CD8^+$ cells, but expression of these molecules is tightly regulated. In cases of overt neural damage and loss of neural activity, MHC upregulation might serve to target a damaged cell population. In a follow-up study, NEUMANN et al (1997) confirmed that basal neuronal expression of MHC class I genes differed significantly from gene expression in other cells; while transcription of the heavy chain appeared intact in most primary neurons, there was a failure to transcribe either β_2-microglobulin (the light chain of the class I MHC heterodimer) or the transporter proteins TAP-1 and -2 (NEUMANN et al. 1997). Consequently, unstimulated primary neurons failed to express MHC on the cell surface. The addition of IFN-γ induced β_2-microglobulin and TAP transcription in some, but not all cultured neurons. To understand how the expressing and nonexpressing cells differed, reagents which suppressed electrical activity were added to the cultures and MHC surface expression was monitored. Suppression of the neuronal electric activity by the addition of a sodium channel blocker led to induction of class I expression on virtually all neurons, reinforcing the connection between MHC expression and a lack of electrical activity reported earlier (NEUMANN et al. 1995).

In one of the first papers to appear on the issue of neuronal MHC expression (WONG et al. 1984), IFN-γ was found to induce a dramatic expression of H-2 antigens on virtually all oligodendrocytes, astrocytes, and microglia, while MHC expression was found on at least "some" neurons. While later reports noted the potential problem of non-neuronal contamination in these studies (BARTLETT et al. 1989), the disparity in MHC expression observed in the original report, when viewed in the context of recent findings, may reflect replicative or functional differences between neuron populations.

Viral infection has also been shown to increase MHC expression on cultured neuronal cell lines. In one study, persistent infection of neuroblastoma cells (C1300) with measles virus (Edmonston strain) was shown to induce MHC protein expression on the cell surface (GOPAS et al. 1992). The motivation for this work was not to correlate the relevance of their neuroblastoma line to in vivo neuronal MHC expression, but rather to parallel neuroblastomas of children. The authors argue that viral infection of malignant neuroblasts may serve both to upregulate MHC expression and to provide unique immunogenic (viral) peptides for CTL recognition and lysis (GOPAS et al. 1992). In support of this hypothesis, persistently infected C1300 cells not only expressed higher levels of cell surface MHC, but also could be recognized and lysed by measles virus-elicited CTL.

A second example of virus-induced MHC expression in vitro evaluated the susceptibility of BDV-infected neurons to CTL recognition and lysis (PLANZ et al. 1993). In the wild, infection of horses and sheep with BDV often leads to an encephalomyelitis accompanied by severe neurological disorders (LUDWIG et al. 1988); the inflammation has led to the characterization of Borna disease as an immunopathological condition (NARAYAN et al. 1983). Both $CD4^+$ T helper cells and $CD8^+$ CTL are found in CNS lesions. Effector lymphocytes isolated from brains of persistently infected rats showed haplotype-specific cytotoxic activity when incubated with BDV-infected primary neurons, confirming that these neurons can be targets for virus-specific CTL (PLANZ et al. 1993).

3.3 Expression of Class I MHC in CNS Neurons In Vivo

While in vitro studies of neuroblastomas and cultured primary neurons have told us much about the constraints on neuronal MHC induction and the stimuli that can overcome such constraints, these systems may not reflect the MHC status of neurons in vivo. For example, immunosuppressive molecules normally present in the CNS, such as gangliosides, would not be expected to function in vitro. Furthermore, the process of culturing primary neurons often involves trypsinization and incubation with serum-containing medium; these conditions could provide essential "background" stimulation for MHC expression on cultured cells. Consequently, to establish whether certain pathological conditions are accompanied by enhanced MHC expression, an immunohistochemical study (LAMPSON and HICKEY 1986) correlated MHC expression in human brain biopsies evaluated as "histologically normal" with those containing a range of neuropathological lesions, including glial tumors. HLA expression could be found adjacent to vessel walls in all brains, regardless of their condition. However, MHC expression did not overlap with any parenchymal cell, including neurons, oligodendrocytes, microglia, or astrocytes (LAMPSON and HICKEY 1986). It should be noted that none of the specimens examined in this communication showed obvious lymphocytic inflammation. Given the suggestion that activated lymphocytes and their products may induce neuronal MHC expression (FONTANA et al. 1984; HICKEY et al. 1985), the absence of MHC, even in glial tumor-bearing individuals, may not be surprising. However, as these authors point out, exposure to circulating lymphocytes alone is not sufficient to induce MHC; class I expression is also absent from parenchymal cells in areas of the brain that lack a blood-brain barrier (WHELAN and LAMPSON 1985).

A more extensive study of 40 human brain specimens by SOBEL and AMES (1988) was able to detect class I expression in parenchymal cells, but notably no expression was seen in neurons (as determined by their location in the brain, not by neuronal antibody double labeling). The expression patterns, similar to the conclusions drawn by Lampson and Hickey, did not change as a function of age, sex, duration between death and tissue preservation, systemic illnesses at the time of death, or CNS lesions. Importantly, the 40 brains surveyed included a number of neuropathologic diseases expected to induce immune cell infiltrates, including atherosclerosis, hydrocephalus, and metastatic carcinoma.

Neither of these studies, however, included patients with neuronal viral infections which had been shown in vitro to upregulate MHC expression. In subacute sclerosing panencephalitis (SSPE), the rare but lethal neurodegenerative disease of the CNS following acute measles virus infection, MHC induction on neurons was found in five out of six autopsy specimens (GOGATE et al. 1996). However, when MHC and neuronal markers were used to colocalize the signals in these sections, MHC expression was rarely found on infected neurons. The disparity between measles virus infection and MHC expression could be explained by two hypotheses. In one scenario, enhanced neuronal MHC expression may not be a direct consequence of viral infection, but may be controlled by exogenous factors induced by viral replication in the CNS. Alternatively, MHC-expressing, measles virus-infected neurons may serve as efficient targets for antiviral CTL, with infected cells cleared by the cytolytic immune response prior to acquisition of the brain tissue (GOGATE et al. 1996).

Because analysis of autopsy specimens does not afford the opportunity to evaluate the progression of the disease and the factors that govern MHC expression, animal model systems of neurotropic viral infections have proven informative. In experimental viral infections of mice and rats and in one case of spongiform encephalopathy caused by the scrapie agent, infection of neurons correlated with MHC induction (DUGUID and TRZEPACZ 1993; BILZER and STITZ 1994; PEARCE et al. 1994).

Infection of mice with a highly neurotropic variant of mouse hepatitis virus (MHV) establishes a limited infection within the brain and has a low mortality rate (PEARCE et al. 1994). Clearance of virus from the brain is associated with infiltration of $CD8^+$ lymphocytes, suggesting that CTL-mediated lysis may contribute to viral clearance. Infection of astrocytes by MHV had previously been shown to induce MHC expression on these infected cells (GOMBOLD and WEISS 1992); in the study by Pearce and colleagues, upregulation of class I antigens in the nerve fiber layer was also seen, suggesting that some neurons may be capable of MHC antigen presentation. A similar correlation between CD8 T cell infiltration and MHC upregulation on neurons was found for BDV infection (BILZER and STITZ 1994). In this paper, treatment of BDV-infected rats with the anti-CD8 monoclonal antibody OX-8 inhibited the immunopathologic reaction and reduced MHC class I antigen expression on neurons.

Collectively, it has been demonstrated that all neurons, peripheral or central, in vitro or in vivo, can be induced to express class I MHC. The stimuli that result in induction may be viral (e.g., measles), cellular (e.g., $CD8^+$ T cells), or soluble (e.g., IFN-γ), although not all members of these groups can induce class I antigen expression. For example, persistent infection with the arenavirus LCMV does not induce neuronal class I MHC (Mucke and Oldstone 1992). Discovering the aspects of viral infection that govern MHC regulation remains a major effort in virology and immunology research, and the list of viral proteins that interfere with the antigen-presenting pathway continues to grow (reviewed in Spriggs 1996). It is important to note that, despite evidence of MHC expression on neurons, no reports exist of neuronal lysis in vivo by CTL. Consequently, despite transitory changes in blood-brain barrier permeability, movement of CNS antigens to the periphery via a lymphoid route, and presentation of foreign peptides by class I antigens, the "immune privilege" of the mammalian CNS is apparently preserved.

4 Basis of Neuronal Block in MHC Expression

4.1 Reduced Activity of Transcription Factors

If neurons can be induced to present antigens via the class I pathway, but normally do not, what intracellular mechanisms are responsible for suppression and how does an inducing signal override this blockade?

Amplification of the N-*myc* gene is correlated with an increased growth rate and enhanced metastatic ability of human neuroblastomas (Bernards et al. 1986). While the issue remains controversial (Feltner et al. 1989), there appears to be an inverse correlation between expression of the N-*myc* gene and MHC class I expression: as the neuroblastoma progresses, the level of N-*myc* increases while the level of class I antigens decreases. The effect of N-*myc* on MHC expression was reversible with IFN treatment. Interestingly, this association between N-*myc* expression and class I synthesis was true only for neuroblastoma; expression of N-*myc* in fibroblasts was not correlated with levels of class I antigen expression (Bernards et al. 1986).

It was later found that two distinct elements in the MHC promoter render the class I genes susceptible to *myc*-induced suppression (Lenardo et al. 1989). N-*myc* reduced the binding of a transcription factor specific for one of these elements, the MHC class I enhancer. Consequently, in the presence of high concentrations of intracellular *myc*, the activity of this enhancer is compromised and cellular transcription of MHC genes is constitutively low. The identity of this nuclear transcription factor was shown to be the p50 subunit of the inducible transcription factor NF-κB (vant Veer et al. 1993). Introduction of the NF-κB p50 subunit into neuroblastoma cells restored expression of class I molecules.

Why do neuroblastomas express such high levels of N-*myc*? To test the idea that N-*myc* expression results from the inactivation or loss of some N-*myc* repressor, Versteeg et al. (1990) fused neuroblastomas overexpressing N-*myc* with lines that do not express N-*myc*. N-*myc* expression was turned off in the resulting hybrids, confirming the hypothesis that the absence of the N-*myc* suppressor can account for the unusually high levels in neuroblastomas. As predicted, the level of class I MHC transcription in these hybrids was restored, verifying the causal link between N-*myc* and class I expression.

Defects in MHC class I transcription are not the only obstacles to antigen presentation in neuronal cell lines. JOLY and OLDSTONE (1992) found that the OBL-21 cell line, previously shown to express low levels of the class I heavy chain (JOLY et al. 1991), also expressed negligible levels of the TAP-1 and -2 transporter proteins (previously called HAM molecules), which are responsible for shuttling proteolyzed peptide fragments into the endoplasmic reticulum. While the levels of N-*myc* in this cell line have not been established, nor is it known whether N-*myc* overexpression is the cause of decreased transcription of the TAP genes, it is generally appreciated that many antigen-presenting genes share common regulatory motifs, leaving open the possibility that N-*myc* can sequester nuclear transcription factors that regulate multiple genes involved in antigen presentation.

The above studies employed transformed cell lines as model systems, again raising the relevance of neuroblastoma lines to primary neurons or neurons in vivo. MASSA and colleagues (1993) evaluated the ability of class I antigens to be upregulated by IFN-γ in primary astrocytes, neurons, and oligodendrocytes. They found that expression in oligodendrocytes and astrocytes could be enhanced by the addition of the stimulatory cytokine, but that neurons remained refractory to the effects of IFN-γ. The differences in inducibility among these cell types was due to differences in the expression of nuclear factors specific for the NF-κB-like region I enhancer within the class I promoter. Thus, similar to neuroblastomas, the absence of class I MHC in neurons appears to be linked to a decreased abundance or activity of DNA-binding transcription factors. A critical difference between primary neurons and neuroblastomas is their response to cytokine stimulation. Addition of IFN-γ or TNF-α induces MHC expression in most neuroblastoma cells (DREW et al. 1993), whereas primary neurons are generally refractory to cytokine stimulation.

The expression of the IFN-β gene is regulated by enhancers similar to those found in the MHC genes (BURKE and OZATO 1989). One might therefore predict that the IFN-β gene would be similarly downregulated in primary neurons, although surprisingly the data indicate that it is highly expressed in neurons following treatment with IFN-β-inducing stimuli (WARD and MASSA 1995). The authors discuss the potential benefit of this differential regulation: neurons successfully escape MHC class I-restricted lysis by downmodulating MHC cell surface expression, while retaining the beneficial properties of IFN-β production, which include restriction of viral infection within a cell population (SCHIJNS et al. 1991).

Finally, while transcription factors do play a pivotal role in MHC regulation, *cis*-acting signals appear to also regulate cell type-specific MHC expression (MURPHY et al. 1996). Cell-specific expression of class I antigens is achieved in part by a series of negative and positive regulatory elements located within the extended class I MHC promoter (WEISSMAN and SINGER 1991). The constitutive ability of this promoter to function in a given cell type is dependent on the activity of the activating and silencing domains within the promoter. Thus it is possible that the class I promoter in neurons is constitutively functional, but is actively repressed by upstream elements. MURPHY and coworkers (1996) demonstrated that the removal of four elements upstream of the transcriptional start site of the class I promoter of the non-MHC-expressing human neuroblastoma line CHP-126 restored constitutive MHC expression. Two of these "silencer" elements display differing functions in CHP-126 and HeLa cells. Silencer B, located between nucleotides 402 and 503 within the class I MHC promoter, suppresses MHC expression in neuroblastomas, but enhances expression in HeLa cells; likewise, silencer A, which suppresses expression in the CHP-126 cells, has no detectable function in HeLa cells.

The authors acknowledge that NF-κB plays a critical role in cytokine-induced expression. However, while NF-κB may be necessary for cytokine-induced expression, it is not necessary for constitutive expression, since an upstream enhancer is able to activate the promoter in the absence of NF-κB. Thus suppression of constitutive MHC synthesis appears to reside, at least in part, at the level of *cis*-acting elements within the MHC promoter, in addition to the altered function of transcriptional regulators (MURPHY et al. 1996).

4.2 Reduced Susceptibility to Cytotoxic T Lymphocyte-Mediated Lysis and the Potential Role of Cytokines

An underlying assumption of this chapter is that increased neuronal MHC expression confers a higher susceptibility to CTL-mediated recognition and lysis. Three studies suggest that this may not be an accurate assumption and that neurons may be refractory to CTL lysis at a level beyond TCR-MHC recognition.

As described earlier in this chapter, cytokine treatment of neuroblastoma cell lines can upregulate cell surface MHC expression. However, despite increased levels of MHC, MAIN and colleagues (1988) found that neurons remained resistant to CTL-mediated lysis. The authors noted that "neuroblastoma lines [with increased MHC] did not form conjugates with primed T cells." Retrospectively, in light of the need for both MHC and costimulatory signals to result in target lysis, these neuroblastomas may be deficient in either the critical "second signal" or in cell surface adhesion molecule expression needed for binding of the CTL to its target.

The requirement for TCR-MHC recognition can be bypassed using purified granules obtained from CTL or LAK cells. Differing conclusions are drawn by two groups who assessed the susceptibility of primary neurons in culture to purified granules. One group (KEANE et al. 1992) reported that, while leucoagglutinin-treated primary CNS neurons triggered granzyme release from effector cells, neurons remained resistant to the cytolytic effects of these granules, unlike other cultured resident brain cells. They concluded that neurons possess protective mechanisms, beyond blocks in target cell-CTL interaction, that render them refractory to CTL-delivered cytotoxins. A more recent report (RENSING-EHL et al. 1996) readdressed this issue and found that both extracted granules and purified perforin could induce virtually complete lysis of neurons. The major differences between these studies were the age and type of neurons studied; in the paper by Keane and colleagues, whole embryonic brain was used as a source of neurons; in the report by Rensing-Ehl and coworkers, cerebellar granule cells from day-7 neonates served as a source of neurons. These papers underscore the importance of the model system in interpreting these studies: differentiation status, host age, and neuronal subtype are all factors in immune susceptibility.

Finally, if neuronal MHC expression and CTL lysis are linked events, one might expect that reconstitution of class I MHC on the surface of neurons in vivo might render mouse CNS neurons more susceptible to CTL lysis following an inflammatory CNS lesion. However, when CNS neurons were driven to constitutively express a transgene-encoded MHC molecule (RALL et al. 1995) and were subsequently infected with the neurotropic virus LCMV, adoptive transfer of antiviral CTL resulted in recruitment of CTL to the brain parenchyma and more rapid clearance of virus from infected neurons, but no appreciable loss of neurons within infected brains. Thus MHC expression was sufficient to recruit CTL to the CNS parenchyma and to allow virus clearance, but clearance was not accompanied by cell loss (RALL et al. 1995).

How these CTL exert their virucidal, but not cytocidal, effects in vivo is not known. It is conceivable that factors released from neurons or glial cells, such as gangliosides, may interfere with CTL-neuronal interactions. Alternatively, intraparenchymal cytokine production may participate in the clearance of infectious virus from transgenic brains. Inhibition of viral synthesis in the absence of cell death has been documented in the CTL response to HBV infection of hepatocytes (GUIDOTTI et al. 1994) and in human immunodeficiency virus (HIV) (HSEUH et al. 1994). In these systems, inflammatory cytokines such as IFN-γ and TNF-α can downregulate viral expression without concomitant cell death. Studies are underway to test the ability of CTL-elaborated cytokines to resolve neurotropic viral infections in the absence of cell killing.

5 Conclusions

In summary, the paucity of neuronal expression of the class I MHC antigens seems to reflect a reluctance, rather than an inability, to synthesize these proteins. The burden borne by cells which can present foreign peptides in the context of class I MHC is that they will become targets of the cytolytic response. In most cases, such altruistic behavior benefits the host, since CTL lysis of virally infected cells will likely precede virus maturation, thereby limiting the spread of the infection. For some cells, however, such behavior would not benefit the host. In the case of placental trophoblasts that lie at the junction between the maternal and fetal systems, absence of MHC expression helps to protect the developing “graft” from maternal recognition. In the CNS, negligible MHC expression protects essential and nonrenewable neurons from immune-mediated cytolysis.

The potential of neurons to express class I MHC antigens, however, suggests that various stimuli, including cytokines, viral infections, and alterations in neuronal function may allow these cells to become targets for CTL recognition. The mechanisms by which these stimuli cause increased MHC expression and the ultimate consequences of CTL-neuron interactions in the host remain issues of paramount importance to our understanding of immunologically mediated CNS diseases.

Acknowledgements. I would like to thank Diane Lawrence for critically reviewing the manuscript and Sarah Berman for her secretarial support.

References

Bartlett PF, Kerr RSC, Bailey KA (1989) Expression of MHC antigens in the central nervous system. Transplant Proc 21:3163–3165

Bergelson LD (1995) Serum gangliosides as endogenous immunomodulators. Immunol Today 16:483486

Bernards R, Dessain SK, Weinberg RA (1986) N-*myc* amplification causes down-modulation of MHC class I antigen expression in neuroblastoma. Cell 47:667–674

Bianchi L, Gudat F (1980) In: Bianchi L, Gerok W, Sickinger K (eds) Virus and the liver. MTP, Lancaster, pp 197–204

Bilzer T, Stitz L (1994) Immune-mediated brain atrophy: $CD8^+$ T cells contribute to tissue destruction during Borna disease. J Immunol 153:818–823

Brightman MW, Zis K, Anders J (1983) Morphology of cerebral endothelium and astrocytes as determinants of the neuronal microenvironment. Acta Neuropathol Supl 8:21–33

Burke PA, Ozato K (1989) Regulation of major histocompatibility complex class I genes. Year Immunol 4:23–40

Carosella ED, Dausset J, Kirszenbaum M (1996) HLA-G revisited. Immunol Today 17:407–409

Chumbley G, King A, Robertson K, Holmes N, Loke YW (1994) Resistance of HLA-G and HLA-A2 transfectants to lysis by decidual NK cells. Cell Immunol 155:312–322

Cserr HF, Knopf PM (1992) Cervical lymphatics, the blood-brain barrier and the immunoreactivity of the brain: a new view. Immunol Today 13:507–512

Cserr H, Harling-Berg CJ, Knopf PM (1992) Drainage of brain extracellular fluid into blood and deep cervical lymph and its immunological significance. Brain Pathol 2:269–276

David-Watine B, Israël A, Kourilsky P (1990) The regulation and expression of MHC class I genes. Immunol Today 11:286–292

de la Torre JC, Rall G, Oldstone C, Sanna PP, Borrow P, Oldstone MBA (1993) Replication of lymphocytic choriomeningitis virus is restricted in terminally differentiated neurons. J Virol 67:7350–7359

Deschamps J, Mitchell RL, Meijlink F, Kruijer W, Schubert D, Verma IM (1985) Proto-oncogene fos is expressed during development, differentiation, and growth. Cold Spring Harbor Symp Quant Biol 50:733–745

Drew PD, Lonergan M, Goldstein ME, Lampson LA, Ozato K, McFarlin DE (1993) Regulation of MHC class I and β2 microglobulin gene expression in human neuronal cells. Factor binding to conserved cis-acting regulatory sequences correlates with expression of the genes. J Immunol 150:3300–3310

Duguid J, Trzepacz C (1993) Major histocompatibility complex genes have an increased brain expression after scrapie infection. Proc Natl Acad Sci USA 90:114–117

Dyatlovitskaya EV, Bergelson LD (1987) Glycosphingolipids and antitumor immunity. Biochim Biophys Acta 907:125–143

Dyatlovitskaya EV, Koroleva AB, Suskova VS, Rozynov B, Bergelson LD (1991) Influence of ganglioside GM3 and its breakdown products on lymphoblastic transformation and T-suppressor activity. Eur J Biochem 199:643–646

Faulk WP, Temple A (1976) Distribution of beta-2 microglobulin and HLA in chorionic villi of human placentae. Nature 262:799–802

Fazakerley JK, Southern P, Bloom F, Buchmeier MJ (1991) High resolution in situ hybridization to determine the cellular distribution of lymphocytic choriomeningitis virus RNA in the tissues of persistently infected mice: relevance to arenavirus disease and mechanisms of viral persistence. J Gen Virol 72:1611–1625

Feltner DE, Cooper M, Weber J, Israel MA, Thiele CJ (1989) Expression of class I histocompatibility antigens in neuroectodermal tumors is independent of the expression of a transfected neuroblastoma myc gene. J Immunol 143:4292–4299

Ferry BL, Sargent IL, Starkey PM, Redman CWG (1991) Cytotoxic activity against trophoblast and choriocarcinoma cells of LGL from human early pregnancy decidua. Cell Immunol 132:140–149

Fontana A, Fierz W, Wekerle H (1984) Astrocytes present myelin basic protein to encephalitogenic T-cell lines. Nature 307:273–276

Geraghty DE, Koller BH, Orr HT (1987) A human major histocompatibility complex class I gene that encodes a protein with a shortened cytoplasmic segment. Proc Natl Acad Sci USA 84:9145–9149

Gogate N, Swoveland P, Yamabe T, Verma L, Woyciechowska J, Tarnowska-Dziduszko E, Dymecki J, Dhib-Jalbut S (1996) Major histocompatibility complex class I expression on neurons in subacute sclerosing panencephalitis and experimental subacute measles encephalitis. J Neuropathol Exp Neurol 55:435–443

Goldstein GW, Betz AL (1986) The blood-brain barrier. Sci Am 155:74–83

Gombold J, Weiss S (1992) Mouse hepatitis virus A59 increases steady-state levels of MHC mRNAs in primary glial cell cultures and in the murine central nervous system. Microb Pathog 13:493–505

Gopas J, Itzhaky D, Segev Y, Salzberg S, Trink B, Isakov N, Rager-Zisman B (1992) Persistent measles virus infection enhances major histocompatibility complex class I expression and immunogenicity of murine neuroblastoma cells. Cancer Immunol Immunother 34:313–320

Greengard P (1978) Phosphorylated proteins as physiological effectors. Science 199:146–152

Guidotti LG, Ando K, Hobbs MV, Ishikawa T, Runkel L, Schreiber RD, Chisari FV (1994) Cytotoxic T lymphocytes inhibit hepatitis B virus gene expression by a noncytolytic mechanism in transgenic mice. Proc Natl Acad Sci USA 91:3764–3768

Hickey WF, Kimura H (1987) Graft-vs.-host disease elicits expression of class I and class II histocompatibility antigens and the presence of scattered T lymphocytes in rat central nervous system. Proc Natl Acad Sci USA 84:2082–2086

Hickey WF, Osborn JP, Kirby WM (1985) Expression of Ia molecules by astrocytes during acute allergic encephalomyelitis in the Lewis rat. Cell Immunol 91:528–535

Hickey WF, Hsu BL, Kimura H (1991) T-lymphocyte entry into the central nervous system. J Neurosci Res 28:254–260

Hseuh F, Walker C, Blackbourn D, Levy J (1994) Suppression of HIV replication by $CD8^+$ cell clones derived from HIV-infected and uninfected individuals. Cell Immunol 159:271–279

Hunt JS, Orr HT (1992) HLA and maternal-fetal recognition. FASEB J 6:2344–2348

Irani DN, Lin K-I, Griffin DE (1996) Brain-derived gangliosides regulate the cytokine production and proliferation of activated T cells. J Immunol 157:4333–4340

Irani DN, Lin K-I, Griffin DE (1997) Regulation of brain-derived T cells during acute central nervous system inflammation. J Immunol 158:2318–2326

Joly E, Oldstone MBA (1992) Neuronal cells are deficient in loading peptides onto MHC class I molecules. Neuron 8:1185–1190

Joly E, Mucke L, Oldstone MBA (1991) Viral persistence in neurons explained by lack of MHC class I expression. Science 253:1283–1285

Keane RW, Tallent MW, Podack ER (1992) Resistance and susceptibility of neural cells to lysis by cytotoxic lymphocytes and by cytolytic granules. Transplantation 54:520–526

Lampson LA, Fisher CA (1984) Weak HLA and β_2-microglobulin expression of neuronal cell lines can be modulated by interferon. Proc Natl Acad Sci USA 81:6476–6480

Lampson LA, Hickey WF (1986) Monoclonal antibody analysis of MHC expression in human brain biopsies: tissue ranging from "histologically normal" to that showing different levels of glial tumor involvement. J Immunol 136:4054–4062

Lampson LA, Fisher CA, Whelan JP (1983) Striking paucity of HLA-A, B, C and B_2 microglobulin on human neuroblastoma cell lines. J Immunol 130:2471–2478

Lassmann, H, Rossler K, Zimprich F, Vass K (1991) Expression of adhesion molecules and histocompatibility antigens at the blood-brain barrier. Brain Pathol 1:115–123

Lenardo M, Rustgi AK, Schievella AR, Bernards R (1989) Suppression of MHC class I gene expression by N-*myc* through enhancer inactivation. EMBO J 8:3351–3355

Liao N-S, Bix M, Zijlstra M, Jaenisch R, Raulet D (1991) MHC class I deficiency; susceptibility to natural killer (NK) cells and impaired NK activity. Science 253:199–202

Ludwig H, Bode L, Gosztonyi G (1988) Borna disease: a persistent viral infection of the central nervous system. Prog Med Virol 35:107–120

Maehlen J, Daa Schröder H, Klareskog L, Olsson T, Kristensson K (1988) Axotomy induces MHC class I antigen expression on rat nerve cells. Neurosci Lett 92:8–13

Maehlen J, Nennesmo I, Olsson A-B, Olsson T, Daa Schröder H, Kristensson (1989) Peripheral nerve injury causes transient expression of MHC class I antigens in rat motor neurons and skeletal muscles. Brain Res 481:368–372

Main EK, Lampson LA, Hart MK, Kornbluth J, Wilson DB (1985) Human neuroblastoma lines are susceptible to lysis by natural killer cells but not cytotoxic T lymphocytes. J Immunol 135:242–246

Main EK, Monos DS, Lampson LA (1988) IFN-treated neuroblastoma cell lines remain resistant to T cell-mediated allo-killing, and susceptible to non-MHC-restricted cytotoxicity. J Immunol 141:2943–2950

Mason DW, Charleton HM, Jones AJ, Lavy CBD, Puklavek M, Simmonds SJ (1986) The fate of allogeneic and xenogeneic neuronal tissue transplanted into the third ventricle of rodents. Neuroscience 19:685–694

Massa PT, Ozato K, McFarlin DE (1993) Cell type-specific regulation of major histocompatibility complex (MHC) class I gene expression in astrocytes, oligodendrocytes, and neurons. GLIA 8:201207

Miller CA, Carrigan DR (1982) Reversible repression and activation of measles virus infection in neural cells. Proc Natl Acad Sci USA 79:1629–1633

Mucke L, Oldstone MBA (1992) The expression of major histocompatibility complex (MHC) class I antigens in the brain differs markedly in acute and persistent infections with lymphocytic choriomeningitis virus (LCMV). J Neuroimmunol 36:193–198

Murphy C, Nikodem D, Howcroft K, Weissman JD, Singer DS (1996) Active repression of major histocompatibility complex class I genes in a human neuroblastoma cell line. J Biol Chem 271:30992–30999

Narayan O, Herzog S, Feses K, Scheefers H, Rott R (1983) Behavioral disease in rats caused by immunopathological response to persistent Borna virus infection in the brain. Science 220:1401–1404

Neumann H, Cavali A, Jenne DE, Wekerle H (1995) Induction of MHC class I genes in neurons. Science 269:549–552

Neumann H, Schmidt H, Cavali A, Jenne D, Wekerle H (1997) Major histocompatibility complex (MHC) class I gene expression in single neurons of the central nervous system: differential regulation by interferon (IFN)-γ and tumor necrosis factor (TNF)-α. J Exp Med 185:305–316

Nowak MA, Bonhoeffer S, Hill AM, Boehme R, Thomas HC, McDade H (1996) Viral dynamics in hepatitis B virus infection. Proc Natl Acad Sci USA 93:4398–4402

OMalley MB, MacLeish PR (1993) Induction of class I major histocompatibility complex antigens on adult primate retinal neurons. J Neuroimmunol 43:45–58

Pearce BD, Hobbs MV, McGraw TS, Buchmeier MJ (1994) Cytokine induction during T-cell-mediated clearance of mouse hepatitis virus from neurons in vivo. J Virol 68:5483–5495

Pereira RA, Tscharke DC, Simmons A (1994) Upregulation of class I major histocompatibility complex gene expression in primary sensory neurons, satellite cells, and schwann cells of mice in response to acute but not latent herpes simplex virus infection in vivo. J Exp Med 180:841–850

Planz O, Bilzer T, Sobbe M, Stitz L (1993) Lysis of major histocompatibility complex class I-bearing cells in Borna disease virus-induced degenerative encephalopathy. J Exp Med 178:163–174

Rall GF, Oldstone MBA (1995) Virus-neuron-cytotoxic T lymphocyte interactions. Curr Top Microbiol Immunol 202:261–273

Rall GF, Mucke L, Oldstone MBA (1995) Consequences of cytotoxic T lymphocyte interaction with major histocompatibility complex class I-expressing neurons in vivo. J Exp Med 182:1201–1212

Rapoport SI (1976) Blood-brain barrier in physiology and medicine. Raven, New York

Rensing-Ehl A, Malipiero U, Irmler M, Tschopp J, Constam D, Fontana A (1996) Neurons induced to express major histocompatibility complex class I antigen are killed via the perforin and not the Fas (APO-1/CD95) pathway. Eur J Immunol 26:2271–2274

Rodriguez M, Buchmeier MJ, Oldstone MBA, Lampert PW (1983) Ultrastructural localization of viral antigens in the CNS of mice persistently infected with lymphocytic choriomeningtitis virus (LCMV). Am J Pathol 110:95–100

Sanders SK, Giblin PA, Kavathas P (1991) Cell-cell adhesion mediated by CD8 and human histocompatibility leukocyte antigen G, a nonclassical major histocompatibility complex class I molecule on cytotrophoblasts. J Exp Med 174:737–740

Schijns VE, Van der Neut R, Haagmans BL, Bar DR, Schellekens H, Horzinek MC (1991) Tumour necrosis factor-alpha, interferon-gamma and interferon-beta exert antiviral activity in nervous tissue cells. J Gen Virol 72:809–815

Sedgwick JD, Dörries R (1991) The immune system response to viral infection of the CNS. Neurosciences 3:93–100

Simmons A (1989) H-2-linked genes influence the severity of herpes simplex virus infection of the peripheral nervous system. J Exp Med 169:1503–1507

Simmons A, Tscharke DC (1992) Anti-CD8 impairs clearance of herpes simplex virus from the peripheral nervous system: implications for the fate of virally infected neurons. J Exp Med 175:1337–1344

Sobel RA, Ames MB (1988) Major histocompatibility complex molecule expression in the human central nervous system: immunohistochemical analysis of 40 patients. J Neuropathol Exp Neurol 47:19–28

Spriggs MK (1996) One step ahead of the game: viral immunomodulatory molecules. Annu Rev Immunol 14:101–130

Stanwick TL, Anderson RW, Nahmias AJ (1977) Interaction between cyclic nucleotides and herpes simplex viruses: productive infection. Infect Immun 18:342–347

Ting JP-Y, Masafumi T, Macchi M, Frelinger J (1987) The expression and detection of MHC class I antigens on murine neuroblastoma and ependymoblastoma lines. J Neuroimmunol 14:87–98

Tyor WR, Moensch TR, Griffin DE (1989) Characterization of the local and systemic B cell response of normal and athymic nude mice with Sindbis virus encephalitis. J Neuroimmunol 24:207–215

vant Veer LJ, Beijersbergen RL, Bernards R (1993) N-*myc* suppresses major histocompatibility complex class I gene expression through down-regulation of the p50 subunit of NF-κB. EMBO J 12:195–200

Versteeg R, Van Der Minne C, Plomp A, Sijts A, Van Leeuwen A, Schrier P (1990) N-*myc* expression switched off and class I human leukocyte antigen expression switched on after somatic cell fusion of neuroblastoma cells. Mol Cell Biol 10:5416–5423

Vince GS, Johnson PM (1995) Materno-fetal immunobiology in normal pregnancy and its possible failure in recurrent spontaneous abortion? Hum Reprod 10:107–113

Voland JR, Becker C, Hooshmand F (1994) Overexpression of class I MHC in murine trophoblast and increased rates of spontaneous abortion. In: Hunt JS (ed) Immunobiology of reproduction. Springer, Berlin Heidelberg New York, pp 214–235

Ward LA, Massa PT (1995) Neuron-specific regulation of major histocompatibility complex class I, interferon-β, and anti-viral state genes. J Neuroimmunol 58:145–155

Weissman J, Singer DS (1991) A complex regulatory element associated with a major histocompatibility complex class I gene consists of both a silencer and an enhancer. Mol Cell Biol 11:4217–4227

Wekerle H, Linington C, Lassmann H, Meyermann R (1986) Cellular immune reactivity within the CNS. TINS June, pp271–277

Wekerle H, Engelhardt B, Risau W, Meyermann R (1991) Interaction of T lymphocytes with cerebral endothelial cells in vitro. Brain Pathol 1:107–114

Weller RO, Engelhardt B, Phillips MJ (1996) Lymphocyte targeting of the central nervous system: a review of afferent and efferent CNS-immune pathways. Brain Pathol 6:275–288

Whelan JP, Lampson LA (1985) Expression of MHC class I molecules in normal brain, area postrema and olfactory epithelium. Soc Neurosci 11:1105 (abstr 319.1)

Wiegandt H (1971) Glycosphingolipids. Adv Lipid Res 9:2–49

Wong GHW, Bartlett PF, Clark-Lewis I, Battye F, Schrader JW (1984) Inducible expression of H-2 and Ia antigens on brain cells. Nature 310:688–691

Wood GW (1994) Is restricted antigen presentation the explanation for fetal allograft survival? Immunol Today 15:15–18

Yamada S, DePasquale M, Patlak SC, Cserr HF (1991) Albumin outflow into deep cervical lymph from different regions of rabbit brain. Am J Physiol 261:H1197–H1204

Yoffey JM, Courtice FC (1970) In lymphatics, lymph and the lymphomyeloid complex. Academic, New York, pp 309–314

Suppression of MHC Class I Antigen Presentation by Human Adenoviruses

TIM E. SPARER[1] and LINDA R. GOODING[2]

1 Introduction

Adenoviruses are ubiquitous double-stranded DNA viruses that can infect the epithelial surfaces of the gastrointestinal and respiratory tracts (SHENK 1996). Once adenovirus has infected the epithelial surface, it is internalized and uncoated. Its DNA is transported to the nucleus, where it is transcribed. The genes of adenovirus can be divided into early and late transcription units depending on whether they are transcribed before (early) or after (late) the onset of viral DNA replication. The first gene to be expressed is E1A. It produces two alternatively spliced products, 289R and 243R. These products activate transcription of other adenovirus genes as well as both up- and downregulating host cell transcription. One cassette of genes that is activated by E1A is early region 3 (E3). These genes had been deemed "nonessential" because they are not required for viral replication in vitro. Other early-region proteins activated by E1A are E1B (which is involved in viral mRNA transport into the nucleus), E2 (which is involved in viral DNA replication), and E4 (which is associated with virion assembly). Following the onset of viral DNA replication, the late genes are transcribed and translated. These proteins are the structural proteins of the nucleocapsid which are synthesized in the cytoplasm and transported into the nucleus for virion assembly. Virions are released via an unknown mechanism and result in lysis of the cell.

Adenoviruses can be divided into 47 serotypes (Ad1–Ad47) forming six groups (A–F) based on DNA homologies, serum cross-reactivities, and oncongenicity in

[1]St. Mary's Medical School, Respiratory Medicine, Norfolk Place, London W2 1PG, UK

[2]Emory University School of Medicine, Department of Microbiology and Immunology, Atlanta, GA 30322, USA

rodents (Horwitz 1996). Serotypes differ in their ability to infect different epithelial surfaces and in the pathological outcome of infection. For example, Ad8 and Ad19 of group D are associated with keratoconjunctivitis, while groups B and C are associated with mild to severe upper respiratory infections. The most common of these group C viruses are Ad2 and Ad5, which cause acute respiratory infections and can persist for years after initial infection (Brandt et al. 1966; Fox et al. 1969, 1977; Hillis et al. 1973). How adenoviruses persist in the face of a vigorous, virus-neutralizing immune response is not known.

Viruses, especially DNA viruses, have evolved immunomodulatory proteins that are capable of counteracting both the specific and the innate immune responses (reviewed in Gooding 1992; McFadden and Kane 1994; McFadden et al. 1996; Smith 1994), adenoviruses have evolved multiple mechanisms for downregulating the immune response (reviewed in Wold and Gooding 1989; Gooding and Wold 1990). Suppression of immune responses by adenoviruses may be a critical factor that allows them to remain latent in lymphodial tissue (Ginsberg et al. 1990) and/or persist for years after primary infection in spite of a strong anti-adenovirus immune response (Brandt et al. 1966; Fox et al. 1969, 1977; Hillis et al. 1973). In order to avoid detection by cytotoxic T lymphocytes (CTL), one of the main effector cells against viral infections, adenoviruses as a group have evolved two different mechanism for preventing antigen presentation on major histocompatibility complex (MHC) class I (Paabo et al. 1986). Representative serotypes of most genera of adenovirus produce a protein – typified by the E3 protein gp19K of Ad2 and Ad5 – that binds MHC class I within the endoplasmic reticulum (ER), thus preventing its translocation to the cell surface (Paabo et al. 1986; Burgert and Kvist 1985; Tanaka and Tevethia 1988). The highly oncogenic Ad12 utilizes the transactivating protein E1A to reduce mRNA levels of MHC class I and antigen-processing genes in transformed cells (Shemesh et al. 1991, 1993; Rotem Yehudar et al. 1994b).

2 Structure of gp19K and Its Interaction with MHC Class I

The gp19K protein is one of the most abundant early proteins and consists of 142 amino acids after signal sequence cleavage (Fig. 1). Its ER luminal domain contains 107 amino acids (residues 1–107) and is glycosylated at positions 12 and 61 (Kornfeld et al. 1981; Wold et al. 1985). The 20- to 24-amino acid transmembrane domain (residues 108–127) is followed by a 15-amino acid cytoplasmic domain. This final 15 amino acids contains the ER retention signal, which is a combination of linear sequences (Nilsson et al. 1989; Jackson et al. 1990) and conformational motifs (Gabathuler and Kvist 1990). The gp19K protein can bind to class I molecules of many species (Sester and Burgert 1994) and retain them in the ER (Kvist et al. 1978; Signas et al. 1982). The interaction between MHC class I and gp19K is not covalent and does not involve disulfide bonds or carbohydrates, but the gp19K intramolecular disulfide bonds are necessary for binding (Burgert and Kvist 1985). The gp19K protein binds to many different alleles in mouse and humans with

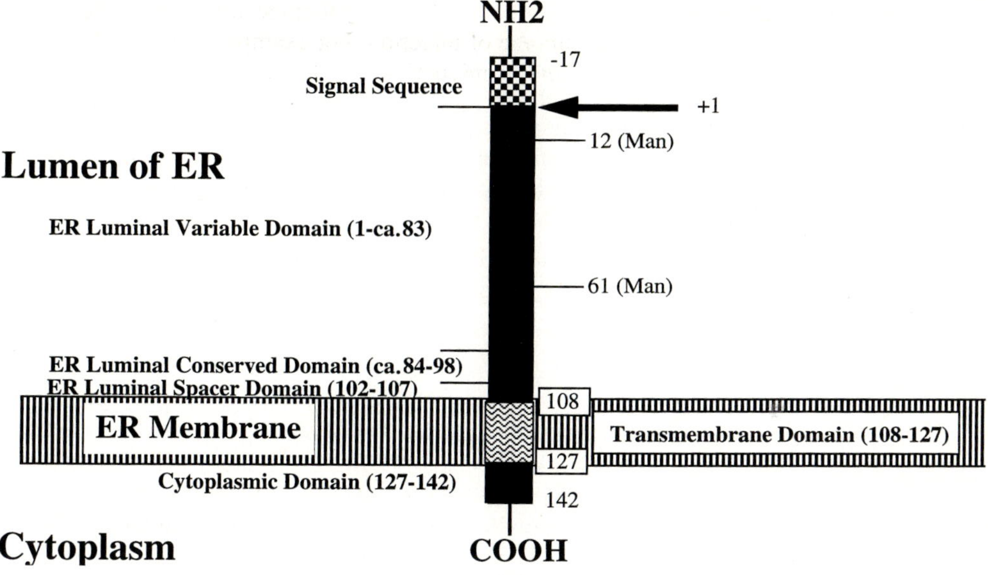

Fig. 1. gp19K from group C (Ad2 and Ad5) adenoviruses. Luminal domains are as defined by HERMISTON et al. (1993b). *ER*, endoplasmic reticulum. See text for details

differing affinities (COX et al. 1990; BEIER et al. 1994). Since gp19K is capable of binding to such diverse class I molecules, it was initially believed that it must bind to the α_3-domain or to a highly conserved region or a conformational epitope elsewhere. Exon shuffling between binding and nonbinding alleles showed that the important region for binding mapped to the α_1- and α_2-helix of the variable domains of the class I molecule (BEIER et al. 1994; FEUERBACH et al. 1994). Using another method, FLOMENBERG et al. (1994) mapped the binding regions on MHC class I using antibodies against HLA class I. They showed that antibodies against the α_3-region, the conserved domain, did not block binding to gp19K, but with antibodies against the N terminus of the α_1-domain or the C terminus of the α_2-domain, gp19K binding was inhibited. The gp19K protein does not bind in the peptide cleft, since CTL can still recognize cells expressing gp19K which is not retained in the ER (COX et al. 1991). It has been proposed that gp19K binds underneath the antigen-binding cleft similar to superantigens binding to MHC class II (FEUERBACH et al. 1994).

To define further the important regions on gp19K for MHC class I binding, extensive mutagenesis of gp19K has been performed. SESTER and BURGERT (1994) reported that the cysteines involved in disufhide linkages are critical for MHC class I binding (cys-11, -28; cys-22, -83), while the other three cysteines have no effect on binding. To investigate the luminal regions, HERMISTON et al. (1993b) generated a series of 14 small (three- to 12-residue) in-frame deletion mutants. Functional studies revealed that all deletions in the luminal domain eliminated MHC class I binding, with the exception of deletions in the membrane proximal region (Δ102–107). These authors proposed to divide the gp19K luminal domain into three subdomains: the variable domain (amino acids 1–84, highly variable among adenovirus serotypes), the conserved domain (amino acids 84–98, highly conserved among adenovirus serotypes), and the spacer domain (amino acids 99–107). The variable domain would interact with the variable regions of the α_1- and α_2-domains on MHC class I and the conserved region could interact with the β_2-microglobulin-binding region (α_3). The spacer domain is probably used for structural purposes, moving gp19K further from the membrane, and is not absolutely required for MHC class I binding.

3 Function of gp19K

The gp19K protein alone functions to retain MHC class I in the ER and to reduce CTL recognition of anti-adenovirus, anti-SV40, and allogeneic CTL (Andersson et al. 1985, 1987; Burgert and Kvist 1985; Rawle et al. 1989; Jefferies and Burgert 1990; Fig. 2). Mutants lacking the six-amino acid ER retention signal still bind MHC class I, and the MHC class I-gp19K complex is translocated to the cell surface, where it can be recognized by CTL (Cox et al. 1991). In some cell lines, there is a decrease in cell surface expression of MHC class I that correlates with a reduction in CTL lysis

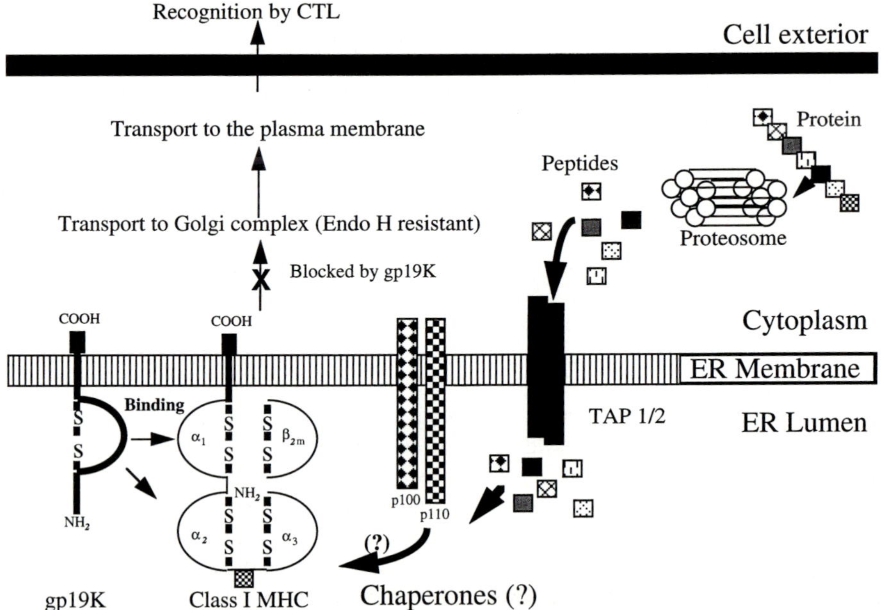

Fig. 2. Steps in presentation of antigen by MHC class I inhibited by adenovirus gp19K. *CTL*, cytotoxic T lymphocytes; *ER*, endoplasmic reticulum; *TAP*, transporter associated with antigen presentation; *β2m*, β2-microglobulin

(ANDERSSON et al. 1985; BURGERT et al. 1987). In other cell lines, there is no reduction in cell surface MHC class I, but there is still a reduction in CTL lysis (COX et al. 1990; ROUTES and COOKE 1990). The reason for this dichotomy is probably that, during virus infection, only newly synthesized class I molecules bear viral peptides and preexisting MHC class I on the surface is irrelevant to antiviral CTL. Hence infected cells can appear to retain MHC class I on the surface and be resistant to MHC class I-restricted antiviral CTL. We find that gp19K is most effective in preventing CTL lysis in cell lines that express low levels of MHC class I. Higher levels of MHC class I expression correlate with intermediate- or low-level inhibition of CTL lysis by gp19K (T.E. Sparer, F.C. Rawle, and L.R. Gooding, unpublished observations). Thus gp19K appears to have a limited capacity for sequestering MHC class I that can be overcome in cells synthesizing higher levels of MHC class I. Similarly, interferon (IFN)-γ treatment of cells, which increases MHC class I expression, also overcomes gp19K-mediated inhibition of CTL lysis (see below). ROUTES and Cooke (1990) have observed that only in cells which are E1A transformed is there a reduction in cell surface MHC class I. Furthermore, these authors found a direct correlation between the amount of E1A expressed and the extent of gp19K-mediated MHC class I downregulation (ROUTES et al. 1993). They propose that, during normal infection, a few cells become E1A transformed, and the resultant resistance to CTL allows these few cells to become persistently infected. In contrast, KORNER and BURGERT (1994) found that expression of E3 proteins in lymphoid cells, the cell type that harbors adenovirus during persistence, caused extensive inhibition of MHC class I expression in the absence of E1A. These authors also found that the NF-κB sites within the E3 promoter can be activated with either tumor necrosis factor (TNF) or other processes such as cell activation, increasing transcription of the gp19K gene and resulting in a reduction of CTL lysis (KORNER et al. 1992). Recently it has been shown that gp19K ER retention is necessary for the activation of NF-κB transcription factor (PAHL et al. 1996). When gp19K is retained in the ER, the ER becomes "overloaded" and activates a pathway leading to NF-κB activation. NF-κB stimulates the transcription of the class I genes. This could be a form of antiviral response by an infected cell to overcome gp19K. On the other hand, gp19K's NF-κB-binding sites within the E3 promoter would compensate for the increase in MHC class I expression. Perhaps in a persistent infection in some cells MHC class I would be overexpressed and would overcome gp19K retention, while in others NF-κB activation of E3 or lack of MHC class I in the infected cell would allow the cell to resist CTL lysis.

In addition to retention of MHC class I, it is possible that gp19K also acts by inhibiting chaperone proteins in the ER. In some gp19K-transfected cell lines, two other proteins are coprecipitated with gp19K and MHC class I (FEUERBACH and BURGERT 1993). These proteins, with a molecular mass of 100 and 110 kDa, respectively, are transmembrane glycoproteins with disulfide bridges that are induced by starvation, but not by heat shock. The proteins can be dissociated when peptide is added, which indicates that they are chaperone proteins. These proteins could be involved in MHC folding or assembly and/or peptide binding. Even without gp19K present, some gp100/110 is associated with MHC class I (K^d), indicating that the proteins may be associated with normal peptide loading. These chaperones are not calnexin (88 kDa), because, unlike gp100/110, calnexin is a nonglycoslyated phosphoprotein; gp100/110 are not identical to gp96, because they are not induced by heat shock. One possibility is that they could be related to the 105-kDa protein that associates with D^b molecules of RMA-S lysates, as described by TOWNSEND et al. (1990). It remains to be established whether gp19K could be preventing egress from the Golgi apparatus by preventing peptide binding or transport or by forcing MHC into a conformation that cannot release gp100/110. It could be that a total reduction in MHC class I on the cell surface may not be beneficial for adenoviruses, since the lack of cell surface MHC class I could make the cell vulnerable to lysis by natural killer (NK) cells, although one study did not find a correlation between adenovirus infection/cell surface MHC class I and NK lysis (BOSSE et al. 1991).

The role of gp19K in vivo is less clearly defined. It could elicit its effect on either the afferent (generation) or the efferent (effector) phase of the immune response. The presence of gp19K in the immunizing virus did not alter CTL generation in the mouse by either intraperitoneal or intranasal inoculation (RAWLE et al. 1989; SPARER et al. 1996). In contrast, Lee et al. (1995) claim to have prevented CTL generation by inoculating mice with an adenovirus vector overexpressing gp19K. Recently, ILAN et al. (1997) have shown that adenovirus-specific CTL generation and humoral response are prevented when an adenovirus vector overexpresses all of the E3 proteins. These data are confounded by the inclusion of the entire E3 region. It is possible that the effects that were observed could be amplified by the other immunregulatory proteins contained in E3. Since human adenovirus is a nonproductive infection in mice (i.e., relatively low levels of gp19K expression), one possible explanation for these discrepancies is that only overexpression of gp19K prevents CTL expansion. In contrast, COX et al. (1994) could not find any differences in vaccinia-specific CTL when gp19K was overexpressed by a vaccinia vector. Even this does not answer the question completely, however. The vaccinia response could be driving a much stronger immune response that could overcome the gp19K effect (e.g., IFN-γ), or vaccinia-infected cells might not be the same as those infected by adenovirus; therefore, gp19K cannot perform its usual function.

Perhaps it is not surprising that gp19K does not function in vivo as it does in vitro. After all, a reduction of 50%–90% in CTL-specific lysis, the most that can be attributed to gp19K expression in vitro, means that there are cells that can still be recognized by adenovirus-specific CTL. The cells that are recognized could stimulate CTL priming in vivo. In addition, it is possible that in vivo priming is carried out by cells which can process exogenous adenovirus antigens. Rock and coworkers have shown that both a subset of macrophages and cloned dendritic cells can present exogenous antigens in the context of class I molecules to CTL (ROCK et al. 1993; SHEN et al. 1997). If adenovirus antigens are processed in noninfected macrophages, CTL epitopes could be presented without interference by gp19K.

If gp19K cannot prevent CTL generation, perhaps its in vivo role dampens the efferent (effector) phase of the immune response. Even if adenovirus cannot prevent CTL priming, it could dampen cellular destruction. Initially, in a cotton rat model of adenovirus pneumonia, it appeared that gp19K was involved in the reduction of adenovirus-induced immune pathology (i.e., efferent function) (GINSBERG et al. 1990). Cotton rats inoculated intranasally with adenovirus lacking gp19K showed greater perivascular/peribroncholiar infiltrates than rats infected with the parent virus.

Others have shown that islet cells from E3 transgenic mouse donors survive longer than nontransgenic cells (EFRAT et al. 1995). Since CTL play a major role in islet cell rejection, these data imply that gp19K could prevent recognition of transplanted cells. Whether the increased longevity of the graft is due to other E3 proteins, such as the TNF resistance proteins, cannot be excluded. In contrast, in mice infected with a recombinant vaccinia virus overexpressing gp19K, there are no changes in NK activity, $CD4^+/CD8^+$ ratios, or vaccinia pathogenesis (COX et al. 1994; GRUNHAUS et al. 1994). Once again, this could be due to an overly aggressive anti-vaccinia response. However, in our hands, deletion of gp19K from the infecting adenovirus had no effect on pulmonary infiltration or CTL responses in a mouse pneumonia model (SPARER et al. 1996). Hence, at present, the utility of gp19K for subverting CTL effects in acute infection remains uncertain.

Why did gp19K not decrease the efferent arm (i.e., destruction of infected cells and/or lung pathology)? First, it is possible that adenovirus infects cells that overexpress MHC class I and can override the gp19K effect, or gp19K could be underexpressed in lung cells, thus not sufficiently reducing MHC class I antigen presentation. Second, cytokines such as IFN-γ are capable of increasing MHC class I expression and/or transporter proteins, which would overcome the gp19K effect (ROUTES 1992; FLOMENBERG et al. 1996; SPARER et al. 1996). In either of these cases of acute adenovirus infection, the role of gp19K in persistence is not addressed. In a natural infection, the survival of a few lymphoid cells in which gp19K has been overexpressed due to activation of NF-κB could prevent CTL recognition (ROUTES et al. 1993; KORNER and BURGERT 1994). Currently, there is no model for adenovirus persistence in vivo, but the fact that Ad7 gp19K is highly conserved in clinical isolates (HERMISTON et al. 1993a) underscores the importance of gp19K in natural infection. In a natural infection, the function of gp19K could permit persistence of a few adenovirus-infected cells, while allowing acute infection pathogenesis or an adenovirus-specific immune response.

4 Adenovirus Type 12 E1A Downregulation of MHC Class I

All subgenera of adenovirus modulate MHC class I expression via E3 gp19K except for group A (PAABO et al. 1986, 1989). Group A (i.e., Ad12) uses another early region protein, E1A, to modulate MHC class I expression. The E1A of Ad12 is a transactivating protein that induces transcription of class I genes in acutely infected mouse cells (ROSENTHAL et al. 1985), but downregulates the transcription of MHC class I (SCHRIER et al. 1983; EAGER et al. 1985) and transporter/proteosome genes (ROTEM YEHUDAR et al. 1994a, 1996) in Ad12-transformed cells. Although both Ad5 and Ad12 are capable of transforming cells, only Ad12-transformed cells can induce tumors in rodents. The downregulation of class I molecules may be responsible for the increase in tumorigenicity in this group (BERNARDS et al. 1983; SCHRIER et al. 1983; VASAVADA et al. 1986; ROTEM YEHUDAR et al. 1994a,b, 1996), although other groups do not see a correlation between a decrease in MHC class I and tumorigenesis

(PEREIRA et al. 1995). Soddu and Lewis (1992) have even shown that high levels of MHC class I actually increase tumorigenicity in BALB/c mice. Downregulation of MHC class I transcription occurs in transformed rat (SCHRIER et al. 1983), mouse (EAGER et al. 1985), and human cells (PAABO et al. 1989) by several means without affecting β_2-microglobulin mRNA levels.

Several studies have shown that Ad12 E1A might function via downregulation of the murine enhancer element. The enhancer element in mouse H-2K^b class I contains three elements: R1, R1', and R2. The R1 region is activated by NF-κB binding, and over 70% of the enhancer activity comes from this region (BURKE et al. 1989). The R1 enhancer activity is shut down in Ad12-transformed cells, but not in Ad5-transformed ones. The nonconserved portion between the two maps to the first exon of between CR2 and CR3 (PEREIRA et al. 1995). The inhibition of NF-κB binding is specific for the transcriptionally active form of the protein such that NF-κB p50–p65 heterodimer is prevented from binding to R1, while the transcriptionally inactive p50–p50 homodimer is unaffected (KUSHNER et al. 1996; LIU et al. 1996). This is in contrast to the observation made by SCHOUTEN et al. (1995), who observed a decrease in levels of p50 due to a decrease in the processing of the p50 precursor (p105). The inhibition of p50–p65 binding to R1 occurs in the nucleus and not in nuclear translocation, implying that downregulation is due to a nuclear localizing factor (LIU et al. 1996). E1A could be the NF-κB-inhibitory factor, but purified recombinant E1A is not capable of directly preventing NF-κB binding in vitro (LIU et al. 1996). It is possible that E1A utilizes a cellular factor to bind to NF-κB p50–p65.

In transformed mouse cells, Ad12 E1A downregulates MHC class I by decreasing NF-κB binding, but E1A also induces binding of the negative regulator COUP-TF, which binds to the R2 of the class I enhancer (KRALLI et al. 1992; KUSHNER et al. 1996). In addition, preliminary evidence suggests a downregulation at the third enhancer-binding site R1'. It appears that Ad12 E1A uses global downregulation at all three enhancer sites in order to downregulate transcription of MHC class I. Others have shown that, in order to obtain full suppression of class I, E1A must bind to the upstream negative regulatory repeats (CAA(A)CAAA) in the H-2^{bm} promoter (OZAWA et al. 1993). From these data, it appears that the E1A downregulation of MHC class I is complex and varied, depending on the regulatory elements involved in the different species.

The 289R E1A product causes downregulation of the amount of cell surface MHC class I not only because the amount of MHC class I produced is decreased, but also because MHC class I egress is halted in the ER (SHEMESH et al. 1991; SHEMESH and EHRLICH 1993). Ad12 E1A also downregulates transcription of members of the proteosome complex (LMP2 and LMP7) and the peptide transporter (TAP-2) (ROTEM YEHUDAR et al. 1994a,b, 1996). The proteosome complex is responsible for cleavage of endogenously synthesized proteins that are then transported into the ER, where they are loaded onto MHC class I. ROTEM YEHUDAR et al. (1996) showed that E1A causes inhibition of transport of newly synthesized swine and H-2D^b class I molecules, even though the levels of mRNA are normal. They showed that transfection or infection with vaccinia expressing TAP genes only partially restored MHC class I transport. Expression of the LMP2 and LMP7 of the proteosome complex along with the TAP-2 gene completely restored the egress of swine class I molecules, but not

H-2D^{b}. Levels of mouse H-2D^{b} molecules were not completely restored until IFN-γ was added. These data indicate that adenovirus blockage of the transporter and proteosome genes can prevent MHC class I progression through the ER, but total reconstitution in some cases requires the addition of IFN-γ.

Does the decrease in MHC class I alter the immune response? The answer to this question is controversial. Ad12 downregulation of MHC class I was shown to decrease influenza-specific CTL recognition in Ad12-transformed cells lines, but not in Ad5-transformed cell lines in vitro (YEWDELL et al. 1988). In vivo, Ad12-transformed cells transfected with MHC class I are no longer tumorigenic (TANAKA et al. 1985). Further, KAST et al. (1989) have shown that induction of adenovirus-specific CTL can lead to tumor eradication, even without increasing the amount of MHC class I on tumor cells. These data support the hypothesis that the decrease in MHC class I and the immune response allows for tumor survival. In contrast to these findings, SODDU and LEWIS (1992) showed that, in spite of low levels of class I, Ad12-transformed cells can be recognized by allospecific CTL and can function to elicit Ad12-specific protection against tumors; thus, even though there are low levels of MHC class I on the surface, these are sufficient to mediate functions. Similarly, PEREIRA et al. (1995) did not find a correlation between downregulation of MHC class I, allospecific CTL recognition and tumor induction in Ad12/5 hybrid cell lines. Perhaps the discrepancies can be explained by differences in the adenovirus-transformed cell lines used in these studies.

Even though the different subtypes infect different mucosal surfaces and cause a variety of different diseases, all adenoviruses combat the immune response by interfering with class I antigen presentation. Downregulation of the class I antigen-processing and -presentation pathway is a common theme among viruses. The herpesviruses have developed proteins that can interfere with TAP (FRUH et al. 1995; HILL et al. 1995) or with MHC class I processing via the EBNA1 repeats (LEVITSKAYA et al. 1995). Recently, the gp40 protein from murine cytomegalovirus (CMV) has been shown to inhibit MHC class I egress from the *cis*-Golgi (ZIEGLER et al. 1997). Although both gp19K and gp40 retain MHC class I, there are differences in cellular location, retention signals, and coprecipitation with MHC class I. As more viral immunoregulatory proteins are discovered, this may present an opportunity for discovering other processes present in normal antigen presentation. In the non-group C subtypes of adenovirus, there are many other E3 proteins which have no functions assigned to them (BASLER and HORWITZ 1996; YEH et al. 1996). Whether these proteins can counteract MHC class I processing and presentation via different means remains to be investigated.

References

Andersson M, Paabo S, Nilsson T, Peterson PA (1985) Impaired intracellular transport of class I MHC antigens as a possible means for adenoviruses to evade immune surveillance. Cell 43:215–222

Andersson M, McMichael A, Peterson PA (1987) Reduced allorecognition of adenovirus-2 infected cells. J Immunol 138:3960–3966

Basler CF, Horwitz MS (1996) Subgroup B adenovirus type 35 early region 3 mRNAs differ from those of the subgroup C adenoviruses. Virology 215:165–177
Beier DC, Cox JH, Vining DR, Cresswell P, Engelhard VH (1994) Association of human class I MHC alleles with the adenovirus E3/19K protein. J Immunol 152:3862–3872
Bernards R, Schrier PI, Houweling A, Bos JL, van der Eb AJ, Zijlstra M, Melief CJ (1983) Tumorigenicity of cells transformed by adenovirus type 12 by evasion of T-cell immunity. Nature 305:776–779
Bosse D, Ades E (1991) Studies of adenovirus subtypes and down-regulation of HLA class I expression: correlations to natural-killer-mediated cytolysis. Pathobiology 59:313–315
Brandt CD, Wassermann FE, Fox JP (1966) The Virus Watch Program. IV. Recovery and comparison of two serological varieties of adenovirus type 5. Proc Soc Exp Biol Med 123:513–518
Burgert HG, Kvist S (1985) An adenovirus type 2 glycoprotein blocks cell surface expression of human histocompatibility class I antigens. Cell 41:987–997
Burgert HG, Kvist S (1987) The E3/19K protein of adenovirus type 2 binds to the domains of histocompatibility antigens required for CTL recognition. EMBO J 6:2019–2026
Burgert HG, Maryanski JL, Kvist S (1987) "E3/19K" protein of adenovirus type 2 inhibits lysis of cytolytic T lymphocytes by blocking cell-surface expression of histocompatibility class I antigens. Proc Natl Acad Sci USA 84:1356–1360
Burke PA, Hirschfeld S, Shirayoshi Y, Kasik JW, Hamada K, Appella E, Ozato K (1989) Developmental and tissue-specific expression of nuclear proteins that bind the regulatory element of the major histocompatibility complex class I gene. J Exp Med 169:1309–1321
Cox JH, Yewdell JW, Eisenlohr LC, Johnson PR, Bennink JR (1990) Antigen presentation requires transport of MHC class I molecules from the endoplasmic reticulum. Science 247:715–718
Cox JH, Bennink JR, Yewdell JW (1991) Retention of adenovirus E19 glycoprotein in the endoplasmic reticulum is essential to its ability to block antigen presentation. J Exp Med 174:1629–1637
Cox JH, Buller RM, Bennink JR, Yewdell JW, Karupiah G (1994) Expression of adenovirus E3/19K protein does not alter mouse MHC class I-restricted responses to vaccinia virus. Virology 204:558–562
Eager KB, Williams J, Breiding D, Pan S, Knowles B, Appella E, Ricciardi RP (1985) Expression of histocompatibility antigens H-2K, -D, and -L is reduced in adenovirus-12-transformed mouse cells and is restored by interferon gamma. Proc Natl Acad Sci U S A 82:5525–5529
Efrat S, Fejer G, Brownlee M, Horwitz MS (1995) Prolonged survival of pancreatic islet allografts mediated by adenovirus immunoregulatory transgenes. Proc Natl Acad Sci USA 92:6947–6951
Feuerbach D, Burgert HG (1993) Novel proteins associated with MHC class I antigens in cells expressing the adenovirus protein E3/19K. EMBO J 12:3153–3161
Feuerbach D, Etteldorf S, Ebenau Jehle C, Abastado JP, Madden D, Burgert HG (1994) Identification of amino acids within the MHC molecule important for the interaction with the adenovirus protein E3/19K. J Immunol 153:1626–1636
Flomenberg P, Gutierrez E, Hogan KT (1994) Identification of class I MHC regions which bind to the adenovirus E3-19k protein. Mol Immunol 31:1277–1284
Flomenberg P, Piaskowski V, Truitt RL, Casper JT (1996) Human adenovirus-specific CD8+ T-cell responses are not inhibited by E3-19K in the presence of gamma interferon. J Virol 70:6314–6322
Fox JP, Brandt CD, Wassermann FE, Hall CE, Spigland I, Kogon A, Elveback LR (1969) The virus watch program: a continuing surveillance of viral infections in metropolitan New York families. VI. Observations of adenovirus infections: virus excretion patterns, antibody response, efficiency of surveillance, patterns of infections, and relation to illness. Am J Epidemiol 89:25–50
Fox JP, Hall CE, Cooney MK (1977) The Seattle Virus Watch. VII. Observations of adenovirus infections. Am J Epidemiol 105:362–386
Fruh K, Ahn K, Djaballah H, Sempe P, van Endert PM, Tampe R, Peterson PA, Yang Y (1995) A viral inhibitor of peptide transporters for antigen presentation. Nature 375:415–418
Gabathuler R, Kvist S (1990) The endoplasmic reticulum retention signal of the E3/19K protein of adenovirus type 2 consists of three separate amino acid segments at the carboxy terminus. J Cell Biol 111:1803–1810
Ginsberg HS, Horswood RL, Chanock RM, Prince GA (1990) Role of early genes in pathogenesis of adenovirus pneumonia [published erratum appears in Proc Natl Acad Sci USA 1991 Jan 15;88(2):681]. Proc Natl Acad Sci USA 87:6191–6195
Gooding LR (1992) Virus proteins that counteract host immune defenses. Cell 71:5–7
Gooding LR, Wold WS (1990) Molecular mechanisms by which adenoviruses counteract antiviral immune defenses. Crit Rev Immunol 10:53–71

Grunhaus A, Cho S, Horwitz MS (1994) Association of vaccinia virus-expressed adenovirus E3-19K glycoprotein with class I MHC and its effects on virulence in a murine pneumonia model. Virology 200:535–546

Hermiston TW, Hellwig R, Hierholzer JC, Wold WS (1993a) Sequence and functional analysis of the human adenovirus type 7 E3-gp19K protein from 17 clinical isolates. Virology 197:593–600

Hermiston TW, Tripp RA, Sparer T, Gooding LR, Wold WS (1993b) Deletion mutation analysis of the adenovirus type 2 E3-gp19K protein: identification of sequences within the endoplasmic reticulum lumenal domain that are required for class I antigen binding and protection from adenovirus-specific cytotoxic T lymphocytes. J Virol 67:5289–5298

Hill A, Jugovic P, York I, Russ G, Bennink J, Yewdell J, Ploegh H, Johnson D (1995) Herpes simplex virus turns off the TAP to evade host immunity. Nature 375:411–415

Hillis WD, Cooper MR, Bang FB (1973) Adenovirus infections in West Bengal. I. Persistence of viruses in infants and young children. Indian J Med Res 61:980–988

Horwitz M (1996) Adenoviruses. In: Fields BN, Knipe DM, Howley PM (eds) Virology. 3rd edn. Raven, Philadelphia, pp 2149–2171

Ilan Y, Droguett G, Chowdhury NR, Li Y, Sengupta K, Thummala NR, Davidson A, Chowdhury JR, Horwitz MS (1997) Insertion of the adenoviral E3 region into a recombinant viral vector prevents antiviral humoral and cellular immune responses and permits long-term gene expression. Proc Natl Acad Sci USA 94:2587–2592

Jackson MR, Nilsson T, Peterson PA (1990) Identification of a consensus motif for retention of transmembrane proteins in the endoplasmic reticulum. EMBO J 9:3153–3162

Jefferies WA, Burgert HG (1990) E3/19K from adenovirus 2 is an immunosubversive protein that binds to a structural motif regulating the intracellular transport of major histocompatibility complex class I proteins. J Exp Med 172:1653–1664

Kast WM, Offringa R, Peters PJ, Voordouw AC, Meloen RH, van der Eb AJ, Melief CJ (1989) Eradication of adenovirus E1-induced tumors by E1A-specific cytotoxic T lymphocytes. Cell 59:603–614

Korner H, Burgert HG (1994) Down-regulation of HLA antigens by the adenovirus type 2 E3/19K protein in a T-lymphoma cell line. J Virol 68:1442–1448

Korner H, Fritzsche U, Burgert HG (1992) Tumor necrosis factor alpha stimulates expression of adenovirus early region 3 proteins: implications for viral persistence. Proc Natl Acad Sci USA 89:11857–11861

Kornfeld R, Wold WS (1981) Structures of the oligosaccharides of the glycoprotein coded by early region E3 of adenovirus 2. J Virol 40:440–449

Kralli A, Ge R, Graeven U, Ricciardi RP, Weinmann R (1992) Negative regulation of the major histocompatibility complex class I enhancer in adenovirus type 12-transformed cells via a retinoic acid response element. J Virol 66:6979–6988

Kushner DB, Pereira DS, Liu X, Graham FL, Ricciardi RP (1996) The first exon of Ad12 E1A excluding the transactivation domain mediates differential binding of COUP-TF and NF-kappa B to the MHC class I enhancer in transformed cells. Oncogene 12:143–151

Kvist S, Ostberg L, Persson H, Philipson L, Peterson PA (1978) Molecular association between transplantation antigens and cell surface antigen in adenovirus-transformed cell line. Proc Natl Acad Sci USA 75:5674–5678

Lee MG, Abina MA, Haddada H, Perricaudet M (1995) The constitutive expression of the immunomodulatory gp19k protein in E1-, E3- adenoviral vectors strongly reduces the host cytotoxic T cell response against the vector. Gene Ther 2:256–262

Levitskaya J, Coram M, Levitsky V, Imreh S, Steigerwald-Mullen PM, Klein G, Kurilla MG, Masucci MG (1995) Inhibition of antigen processing by the internal repeat of the Epstein-Barr virus nuclear antigen-1. Nature 6533:685–688

Liu X, Ge R, Ricciardi RP (1996) Evidence for the involvement of a nuclear NF-kappa B inhibitor in global down-regulation of the major histocompatibility complex class I enhancer in adenovirus type 12-transformed cells. Mol Cell Biol 16:398–404

McFadden G, Kane K (1994) How DNA viruses perturb functional MHC expression to alter immune recognition. Adv Cancer Res 63:117–209

McFadden G, Graham K, Barry M (1996) New strategies of immune modulation by DNA viruses. Transplant Proc 28:2085–2088

Nilsson T, Jackson M, Peterson PA (1989) Short cytoplasmic sequences serve as retention signals for transmembrane proteins in the endoplasmic reticulum. Cell 58:707–718

Ozawa K, Hagiwara H, Tang X, Saka F, Kitabayashi I, Shiroki K, Fujinaga K, Israel A, Gachelin G, Yokoyama K (1993) Negative regulation of the gene for H-2Kb class I antigen by adenovirus 12-E1A is mediated by a CAA repeated element. J Biol Chem 268:27258–27268

Paabo S, Nilsson T, Peterson PA (1986) Adenoviruses of subgenera B, C, D, and E modulate cell-surface expression of major histocompatibility complex class I antigens. Proc Natl Acad Sci USA 83:9665–9669

Paabo S, Severinsson L, Andersson M, Martens I, Nilsson T, Peterson PA (1989) Adenovirus proteins and MHC expression. Adv Cancer Res 52:151–163

Pahl HL, Sester M, Burgert HG, Baeuerle PA (1996) Activation of transcription factor NF-kappaB by the adenovirus E3/19K protein requires its ER retention. J Cell Biol 132:511–522

Pereira DS, Rosenthal KL, Graham FL (1995) Identification of adenovirus E1A regions which affect MHC class I expression and susceptibility to cytotoxic T lymphocytes. Virology 211:268–277

Rawle FC, Tollefson AE, Wold WS, Gooding LR (1989) Mouse anti-adenovirus cytotoxic T lymphocytes. Inhibition of lysis by E3 gp19K but not E3 14.7K. J Immunol 143:2031–2037

Rock KL, Rothstein L, Gamble S, Fleischacker C (1993) Characterization of antigen-presenting cells that present exogenous antigens in association with class I MHC molecules. J Immunol 150:438–446

Rosenthal A, Wright S, Quade K, Gallimore P, Cedar H, Grosveld F (1985) Increased MHC H-2K gene transcription in cultured mouse embryo cells after adenovirus infection. Nature 315:579–581

Rotem Yehudar R, Shechter H, Ehrlich R (1994a) Transcriptional regulation of class-I major histocompatibility complex genes transformed in murine cells is mediated by positive and negative regulatory elements. Gene 144:265–270

Rotem Yehudar R, Winograd S, Sela S, Coligan JE, Ehrlich R (1994b) Downregulation of peptide transporter genes in cell lines transformed with the highly oncogenic adenovirus 12. J Exp Med 180:477–488

Rotem Yehudar R, Groettrup M, Soza A, Kloetzel PM, Ehrlich R (1996) LMP-associated proteolytic activities and TAP-dependent peptide transport for class 1 MHC molecules are suppressed in cell lines transformed by the highly oncogenic adenovirus 12. J Exp Med 183:499–514

Routes JM (1992) IFN increases class I MHC antigen expression on adenovirus-infected human cells without inducing resistance to natural killer cell killing. J Immunol 149:2372–2377

Routes JM, Cook JL (1990) Resistance of human cells to the adenovirus E3 effect on class I MHC antigen expression. Implications for antiviral immunity. J Immunol 144:2763–2770

Routes JM, Metz BA, Cook JL (1993) Endogenous expression of E1A in human cells enhances the effect of adenovirus E3 on class I major histocompatibility complex antigen expression. J Virol 67:3176–3181

Schouten GJ, van der Eb AJ, Zantema A (1995) Downregulation of MHC class I expression due to interference with p105-NF kappa B1 processing by Ad12E1A. EMBO J 14:1498–1507

Schrier PI, Bernards R, Vaessen RT, Houweling A, van der Eb AJ (1983) Expression of class I major histocompatibility antigens switched off by highly oncogenic adenovirus 12 in transformed rat cells. Nature 305:771–775

Sester M, Burgert HG (1994) Conserved cysteine residues within the E3/19K protein of adenovirus type 2 are essential for binding to major histocompatibility complex antigens. J Virol 68:5423–5432

Shemesh J, Ehrlich R (1993) Aberrant biosynthesis and transport of class I major histocompatibility complex molecules in cells transformed with highly oncogenic human adenoviruses. J Biol Chem 268:15704–15711

Shemesh J, Rotem Yehudar R, Ehrlich R (1991) Transcriptional and posttranscriptional regulation of class I major histocompatibility complex genes following transformation with human adenoviruses. J Virol 65:5544–5548

Shen Z, Reznikoff G, Dranoff G, Rock KL (1997) Cloned dendritic cells can present exogenous antigens on both MHC class I and class II molecules. J Immunol 6:2723–2730

Shenk T (1996) Adenoviridae: the viruses and their replication. In: Fields BN, Knipe DM, Howley PM (eds) Virology, 3rd edn. Raven, Philadelphia, pp 2111–2148

Signas C, Katze MG, Persson H, Philipson L (1982) An adenovirus glycoprotein binds heavy chains of class I transplantation antigens from man and mouse. Nature 299:175–178

Smith GL (1994) Virus strategies for evasion of the host response to infection. Trends Microbiol 2:81–88

Soddu S, Lewis AM Jr (1992) Driving adenovirus type 12-transformed BALB/c mouse cells to express high levels of class I major histocompatibility complex proteins enhances, rather than abrogates, their tumorigenicity. J Virol 66:2875–2884

Sparer TE, Tripp RA, Dillehay DL, Hermiston TW, Wold WS, Gooding LR (1996) The role of human adenovirus early region 3 proteins (gp19K, 10.4K, 14.5K, and 14.7K) in a murine pneumonia model. J Virol 70:2431–2439

Tanaka Y, Tevethia SS (1988) Differential effect of adenovirus 2 E3/19K glycoprotein on the expression of H-2Kb and H-2Db class I antigens and H-2Kb- and H-2Db-restricted SV40-specific CTL-mediated lysis. Virology 165:357–366

Tanaka K, Isselbacher KJ, Khoury G, Jay G (1985) Reversal of oncogenesis by the expression of a major histocompatibility complex class I gene. Science 228:26–30

Townsend A, Elliott T, Cerundolo V, Foster L, Barber B, Tse A (1990) Assembly of MHC class I molecules analyzed in vitro [published erratum appears in Cell 1990 Sep 21;62(6):following 1233]. Cell 62:285–295

Vasavada R, Eager KB, Barbanti Brodano G, Caputo A, Ricciardi RP (1986) Adenovirus type 12 early region 1A proteins repress class I HLA expression in transformed human cells. Proc Natl Acad Sci USA 83:5257–5261

Wold WS, Gooding LR (1989) Adenovirus region E3 proteins that prevent cytolysis by cytotoxic T cells and tumor necrosis factor. Mol Biol Med 6:433–452

Wold WS, Cladaras C, Deutscher SL, Kapoor QS (1985) The 19-kDa glycoprotein coded by region E3 of adenovirus. Purification, characterization, and structural analysis. J Biol Chem 260:2424–2431

Yeh HY, Pieniazek N, Pieniazek D, Luftig RB (1996) Genetic organization, size, and complete sequence of early region 3 genes of human adenovirus type 41. J Virol 70:2658–2663

Yewdell JW, Bennink JR, Eager KB, Ricciardi RP (1988) CTL recognition of adenovirus-transformed cells infected with influenza virus: lysis by anti-influenza CTL parallels adenovirus-12-induced suppression of class I MHC molecules. Virology 162:236–238

Ziegler H, Thale R, Lucin P, Muranyi W, Flohr T, Hengel H, Farrell H, Rawlinson W, Koszinowski UH (1997) A mouse cytomegalovirus glycoprotein retains MHC class I complexes in the ERGIC/cis-Golgi compartments. Immunity 6:57–66

Herpesvirus Evasion of the Immune System

DAVID C. JOHNSON and ANN B. HILL

L-220, Department of Molecular Microbiology & Immunology, Oregon Health Sciences University, Portland, OR 97201, USA

1 Introduction

It is now well established that animal viruses evade host immunity by many different mechanisms. In an immunocompetent host, the evolutionary pressure brought to bear on viruses is obviously immense, and thus their existence will ultimately dependent upon one form or another of immune subterfuge, subversion, suppression, or abrogation. However, there must also be a well-established balance between the ability of a virus to go around host immunity and the host's ability to adapt to the virus and ultimately contain it. In the event that this balance is not established, either the host or the virus will be eliminated. It is also important to note that viruses have had a very long time to study the immune system, many millions of years longer than immunologists. Thus, much can be learned by examining the effects of viruses on the immune system and their evasion strategies, not only about the intricate interplay between viruses and the immune system, but also about the workings of the immune system itself.

Smaller viruses, such as myxoviruses, picornaviruses, and retroviruses, evade the immune system by changing their coats, restricting the expression of dominant epitopes, or infecting important opponents. By contrast, large DNA viruses, such as the adenoviruses, herpesviruses, and poxviruses, have the luxury of large coding capacities and thus express an elaborate array of proteins that have specific effects on immune recognition and effector functions.

Herpesviruses are particularly interesting examples of the interplay between virus and host, because their life cycle is characterized by long-term persistence or lifelong latency. Persistence usually includes expression of most viral polypeptides, so that viral progeny are produced and shed, usually at low levels. An example is human cytomegalovirus (HCMV), a β-herpesvirus that replicates relatively slowly and that can be continuously or intermittently shed by immunocompetent adults and infants (reviewed in Britt and Alford 1996; Mocarski 1996). HCMV probably persists in monocytes and macrophages and in the salivary gland, and persistence appears to be related to the state of differentiation of host cells, so that infection is abortive in some cells but lytic in others. During abortive replication, viral gene expression may be relatively limited and, as a result, immune responses may also be dampened. HCMV and its murine cousin, MCMV, may also establish infections that may be more appropriately defined as latent infections. For example, MCMV can be reactivated, by coculture with permissive cells, from the spleen, salivary glands, prostate, and testes, even though there is little or no gene expression in the tissues. Latency normally involves expression of a limited subset of the viral genes, if any at all, and for several viruses latency is restricted to a narrow set of tissues. During latency, the genome is normally present as a closed circle rather than in the linear form found in virions. A classic example of latency is that of herpes simplex virus (HSV), which establishes a latent infection in sensory neurons that innervate the mucosal epithelium where the primary infection occurs (reviewed Roizman and Sears 1996; Whitley 1996). During latency, HSV proteins are not expressed. Periodic reactivation from latency, induced by stress, ultraviolet (UV) light, or other stimuli, leads to HSV replication in the neuron and reinfection at the mucosa. Epstein-Barr virus (EBV) establishes a latent infection in B cells, where the viral genome is continuously maintained as a closed circle and with very restricted gene expression.

Latency itself is an important mechanism for herpesvirus immune evasion. HSV is largely invisible to the immune system during its latent phase, since the nervous system is an immunopriviledged site and viral proteins are not expressed. This is an ideal scenario in which the virus hides and waits until robust immunity subsides. EBV is latent in B cells, cells with much more exposure to the immune system, but nonetheless remains camouflaged by virtue of limited gene expression and by inhibiting presentation of viral antigens. However, all the herpesviruses must eventually reactivate from latency and replicate, and viral progeny must be spread to other hosts; during this process, the viruses come under fire from a fully primed immune system. Therefore, this cycle of latency, replication, and virus spread, followed again by latency, presents obvious problems as well as benefits in terms of dealing with host immunity. To replicate in this environment, herpesviruses have devised many fascinating mechanisms to evade host immunity, among these methods of blocking antigen presentation pathways. In this chapter, we review the molecular mechanisms that we currently know herpesviruses use to manipulate antigen presentation. In order to consider the biological significance of these mechanisms, we have included a brief review of current knowledge of the immune response to each virus, focusing in particular on the $CD8^+$ cytotoxic T lymphocyte (CTL) response.

2 Herpes Simplex Viruses

2.1 Immunity

HSV is the best studied of the α-herpesvirus family. Primary HSV infection in the mucosal epithelium is characterized by strong immune responses, including nonspecific, humoral, and cell-mediated immunity. From the limited information available, it is not absolutely clear whether humoral or cell-mediated responses predominate in clearance of virus, but most of the evidence suggests that cell-mediated immunity is more important (reviewed in Schmid and Rouse 1992; York and Johnson 1994).

Mice are the most widely used animal models of HSV infection, but there are clearly qualitative and quantitative differences between the immunity seen in mice and in humans, and the relative contribution of different immune cell types may be skewed in mice. Moreover, there are major differences in the biology of virus infection in mice versus humans. Latency in mice does not lead to spontaneous recurrences, as is the case in humans. HSV rarely spreads beyond the sensory ganglia to aggressively infect the central nervous system (CNS) in humans, yet this is common in mice (reviewed in Roizman and Sears 1996; Whitley 1996). However, studies of humans can also be difficult to interpret, because most involve relatively few patients, the data represents only a snapshot in the course of lifelong infection, and control groups are often not used.

Strong humoral responses are produced in response to HSV infection, and antibodies generated during primary infection can effectively neutralize viruses; however, there is little or no evidence that antisera can prevent or ameliorate HSV infection in humans (reviewed in KOHL 1992). In mice, antibodies can protect against infection and may restrict neuronal spread in some instances. However, antibodies become less effective in clearing virus after infection of even a single cell, because HSV can spread efficiently and rapidly from cell to cell without entering the extracellular milieu and because, after establishment of latency, there are no targets for the antibodies. Indeed, HSV can reactivate from latency and cause recurrent infections in spite of high antibody titers (ZWEERINK and STANTON 1981), and antibody titers do not predict the time to recrudescence or severity of secondary lesions (DOUGLAS and COUCH 1970).

Innate (nonspecific) and acquired (specific) cellular immunity appear to be crucial for virus clearance during ongoing infection in mice and humans, although these responses may be less effective in preventing the initial (primary) infection and in limiting virus spread during the early stages of infection in the mucosa (reviewed in COREY and SPEAR 1986; SCHMID and ROUSE 1992). Macrophages and natural killer (NK) cells are the first cells to be involved in HSV clearance, and these cells limit spread while specific immune responses develop (reviewed in DOMKE-OPTIZ and SWATZKYM 1990). However, macrophages and NK cells are not sufficient to prevent infection and spread of HSV in mice (reviewed in SCHMID and ROUSE 1992). Nevertheless, studies of patients with defects in NK cells have suggested that NK cells are an important component of the anti-HSV immune response (BIRON et al. 1989; JAWAHAR et al. 1996).

Studies in mice and humans have demonstrated that T cells are important or essential to the process of containing and clearing primary and recurrent HSV infections. Patients treated with immunosuppressive agents prior to organ transplantation or with depressed T cell function frequently develop severe HSV disease (QUINNAN et al. 1984; COREY and SPEAR 1986). In mice, adoptive transfer of anti-HSV T lymphocytes protected recipients and curtailed the progression of HSV infection (reviewed in SCHMID and ROUSE 1992). Both $CD4^+$ and $CD8^+$ T cells could protect mice from HSV challenge, and immunodepletion experiments similarly indicated important roles for both cell types in vivo (MARTIN et al. 1987; NASH et al. 1987; reviewed in SCHMID and ROUSE 1992).

$CD8^+$ CTL can protect mice from primary, acute HSV infection and limit the establishment of latent infection (BONNEAU and JENNINGS 1989). However, in some laboratories, low frequencies of $CD8^+$ CTL have been observed, and the $CD8^+$ CTL response to HSV in mice has been characterized as "weak" (SCHMID and ROUSE 1992). There has also been evidence for anti-HSV, $CD4^+$ CTL in mice (NASH et al. 1987; KOLAITIS et al. 1990), and these CTL may compensate for the low numbers of $CD8^+$ CTL. Early studies of human CTL clones also provided evidence for $CD4^+$ CTL (YASUKAWA and ZARLING 1984a; SCHMID 1988). It is not clear how effective $CD4^+$ CTL may be in vivo if host cells do not express class II, although cytokines may induce expression of class II proteins in the mucosa. Studies from several laboratories indicated that humans produce both $CD4^+$ and $CD8^+$ CTL responses directed at HSV-infected target cells, but the frequencies of the $CD8^+$ CTL were low (YASUK-

AWA and ZARLING 1984a; SCHMID 1988; TORPEY et al. 1989; TIGGES et al. 1992). In some of these studies, bulk cultures were restimulated with UV-inactivated HSV (TORPEY et al. 1989), although in other cases restimulation was with peripheral blood mononuclear cells (PBMC) infected with HSV (TIGGES et al. 1992). Thus the low frequencies of $CD8^+$ CTL that were observed may well be related to methods of restimulation. More recently, there was a report of relatively high frequencies of $CD8^+$ CTL restimulated using phytohemagglutinin (PHA)-stimulated, HSV-infected PBMC (POSAVAD et al. 1996). Therefore, it is not clear at this point whether the low frequencies of $CD8^+$ CTL observed in humans in previous studies were related to methods of detecting these cells or to the inhibition of antigen presentation that occurs after HSV infection (see below).

2.2 Complement and Immunoglobulin Receptors

HSV expresses receptors for complement and for IgG, and these may effect some degree of resistance to humoral immune responses or certain types of cellular responses (reviewed in YORK and JOHNSON 1994). HSV glycoprotein C, denoted gC, binds complement factor C3 fragments (SEIDEL-DUGAN et al. 1897; CINES et al. 1982; SMILEY and FRIEDMAN 1985; EISENBERG et al. 1987; KUBOTA et al. 1987; MCNEARNEY et al. 1987; TAL-SINGER et al. 1991). HSV complement receptors can protect virions and virus-infected cells from complement-mediated neutralization either in the presence or absence of antibodies, but there is no evidence to date that this occurs in vivo (MCNEARNEY et al. 1987; HIDAKA et al. 1991). The HSV IgG receptors are composed of a complex of two glycoproteins, gE and gI, and bind the Fc domain of both monomeric and antigen-complexed IgG (JOHNSON and FEENSTRA 1987; JOHNSON et al. 1988). Early work suggested that the Fc receptors on the surface of virions might interfere with binding of cytotoxic antibodies or lymphocytes (COSTA and RABSON 1975). It has also been suggested that the Fc receptors might allow "double binding" or "bipolar bridging" of IgG, so that IgG bound to antigen (through its antigen-binding domain) on the surface of infected cells would also be bound at the Fc domain (LEHNER et al. 1975; FRANK and FRIEDMAN 1989). This binding of the Fc domain might preclude complement binding or interactions with Fc receptors on immune cells. There is evidence that Fc receptors on infected cells can provide resistance to IgG and complement in vitro (FRANK and FRIEDMAN 1989; DUBIN et al. 1991), but there is no evidence to date that this alters the course of disease in vivo (reviewed in YORK and JOHNSON 1994).

2.3 Inhibition of Natural Killer, Macrophage, and T Cell Function by Cell-to-Cell Spread

Soon after infection with HSV, human fibroblasts become sensitive to lysis by NK cells (FITZGERALD-BOCARSLY et al. 1989). However, late in the infection, the HSV-infected cells become relatively resistant to the effects of NK cells or lymphokine-activated killer (LAK) cells, and, after contact with the virus-infected targets, the effectors can be rendered unable to lyse other targets (CONFER et al. 1990). This effect was initially attributed to "disarming" of the NK cells, which, after contacting HSV-infected fibroblasts, were inactivated in a reversible fashion. However, subsequent studies demonstrated that there was a requirement for spread of HSV into the NK cells after contact with infected fibroblasts, so that expression of HSV early (E) genes occurs in the NK cells (YORK and JOHNSON 1993). A similar inhibitory effect was observed with $CD8^+$ CTL placed in contact with HSV-infected targets (POSAVAD et al. 1994). Inhibition of NK or T lymphocytes required direct contact with infected cells, infection of the effectors by extracellular virus was inefficient, and NK cells were irreversibly altered within a few hours of contact with infected targets (CONFER et al. 1990; YORK and JOHNSON 1993). HSV is well adapted to spreading directly from cell to cell in solid tissues, and it appears that the virus utilizes this property to spread from infected target cells into lymphocytes in order to turn off the effectors. It is not clear how HSV alters the lymphocytes, although it is likely that these effects are mediated through structural components of the virus or E gene products. Whether this effect is relevant in vivo is not clear.

2.4 Inhibition of MHC Class I Antigen Presentation

Evidence has been presented that HSV infection of human fibroblasts and keratinocytes leads to resistance to HSV-specific, $CD8^+$ CTL lysis, although virus-infected B cells can be lysed (POSAVAD and ROSENTHAL 1992; KOELLE et al. 1993; POSAVAD et al. 1994; YORK et al. 1994). Fibroblasts and keratinocytes are normally hosts for the virus in vivo, and CTL lysis of B cells is likely to be largely irrelevant in vivo. This inhibition was observed early after infection, well before infectious progeny were produced, and was due to inhibition of MHC class I antigen presentation because incubation with appropriate peptide antigens allowed the HSV-infected fibroblasts to be lysed by $CD8^+$ CTL (TIGGES et al. 1992; KOELLE et al. 1993). Inhibition of antigen presentation may explain defects in $CD8^+$ CTL observed in vivo (discussed above) and the preference for structural components of the virus (TIGGES et al. 1992), which would enter the class I pathway immediately after infection. Therefore, HSV infection leads to a rapid cessation of class I antigen presentation. By contrast, mouse fibroblasts can be lysed by anti-HSV, $CD8^+$ CTL, and thus antigen presentation is not blocked in these cells.

HSV infection of cells leads to rapid degradation of host mRNA, which is largely caused by a structural component of the virion known as the virion host shutoff (vhs) protein. HSV-1 vhs protein is much less active than vhs of HSV-2 (FENWICK and EVERETT 1990). HSV-1 and HSV-2 mutants lacking vhs reduce cell surface expression of MHC class I less efficiently and with slower kinetics than is observed with wild type HSV-1 or HSV-2, but $CD8^+$ CTL lysis of fibroblasts infected with these mutants is still difficult to detect early after infection (POSAVAD et al. 1994; TIGGES et al. 1996). The vhs protein clearly decreases synthesis of MHC class I within 2–6 h of infection with HSV-2 and more slowly after infection with HSV-1 (HILL et al. 1994; YORK et al. 1994; TIGGES et al. 1996); however, in most experiments, MHC class I is expressed early in the infection. Therefore, vhs cannot account for the abrogation of antigen presentation in the first hours of infection or where vhs-negative mutants are used. Furthermore, experiments involving an HSV recombinant, F-US5MHC, that expresses mouse MHC class I (K^b) proteins demonstrated that expression of class I molecules in human fibroblasts during the course of virus infection did not allow presentation of viral antigens (YORK et al. 1994). Therefore, viral proteins (other than vhs) that are produced early in infection, apparently inhibit MHC class I presentation, although vhs plays some role in blocking de novo synthesis of MHC class I, especially at late stages of infection (TIGGES et al. 1996).

2.5 ICP47, An Inhibitor of TAP

MHC class I proteins in HSV-infected human fibroblasts were found to be retained in the endoplasmic reticulum (ER) or *cis*-Golgi compartment, apparently in a peptide-empty form (HILL et al. 1994; YORK et al. 1994). Mutants unable to express an HSV protein, ICP47, were unable to cause MHC class I retention, and ICP47, expressed using recombinant adenovirus vectors, blocked class I antigen presentation (YORK et al. 1994). ICP47 is one of the immediate-early (IE) HSV proteins, which can be expressed in the absence of other HSV polypeptides; the other IE proteins upregulate E gene expression. ICP47 is a small (88-amino acid) protein that has no obvious signal or transmembrane sequences. Therefore, obvious cellular targets of ICP47 were cytosolic components of the class I antigen presentation pathway. Indeed, ICP47 bound to the transporter associated with antigen presentation, TAP, and inhibited TAP-mediated peptide translocation across the ER membrane (FRUH et al. 1995; HILL et al. 1995). More detailed biochemical analysis of the interaction between ICP47 and TAP indicated that ICP47 bound to the cytoplasmic surface of TAP and inhibited peptide binding without affecting the binding of ATP (AHN et al. 1996a; TOMAZIN et al. 1996). Peptides, at relatively high concentrations, inhibited ICP47 binding to TAP, and binding of ICP47 was to a region or domain of TAP that included the peptide-binding site (Tomazin et al. 1996). ICP47 binding to TAP was of high affinity compared to that of peptides, and it bound in a more stable fashion than peptides. Moreover, it remained accessible to proteases added from the cytoplasmic surface and was not transported into the ER membrane or lumen (TOMAZIN et al. 1996). Together, these results demonstrate that ICP47 blocks peptide binding to TAP and subsequent transport by binding tightly to the peptide-binding domain of TAP.

It was interesting that ICP47 could be boiled for 10 min without affecting its inhibitory activity (AHN et al. 1996a; TOMAZIN et al. 1996). More recently, the protein was synthesized as an 87-amino acid peptide (missing the N-terminal methionine), and this molecule had similar inhibitory capacity to ICP47 produced in bacteria (GALOCHA et al. 1997). Therefore, ICP47 has the capacity to renature after heating or in vitro synthesis, or alternatively the protein only takes on an active conformation during binding to TAP. Consistent with the latter hypothesis, circular dichrosim studies revealed no evidence of ordered structure for uncomplexed ICP47 in solution (BEINART et al. 1997). Fragments of HSV-1 ICP47 (ICP47-1) encompassing residues 2–53 and 1–35 were active in inhibiting TAP in permeabilized cells (GALOCHA et al. 1997), suggesting that the 1–35 region contains the core region for binding to TAP.

However, it should be noted that the most highly conserved region of ICP47-1 and HSV-2 ICP47 (ICP47-2) is in the region between residues 33 and 47, which is largely outside the region that is sufficient to block TAP (residues 2–35). There is also significant homology between HSV-1 and HSV-2 ICP47 at the N terminus in the region found to be sufficient for function. Since the stability of ICP47–TAP interaction was not characterized in these studies, residues outside the 2–35 region may affect binding in some consequential way. Moreover, assays involving peptide transport in permeabilized cells may not comprehensively duplicate how ICP47 functions in infected cells, and ICP47 may have additional targets.

ICP47 appears to be able to effectively inhibit presentation of the vast majority of the 75 HSV polypeptides (YORK et al. 1994) by being among the very first proteins produced in infected cells and by binding with high affinity to TAP. Indeed, we recently found that human fibroblasts infected with an HSV-1 ICP47 mutant were effectively lysed by $CD8^+$ CTL (65% cell lysis), yet cells infected with wild-type HSV were not lysed above background levels (% lysis) observed with uninfected cells (P. Jugovic and D.C. Johnson, unpublished). Moreover, expression of ICP47 for long periods can reduce MHC class I expression by over 99% in transfected cells (S. Riddell, personal communication). At the earliest moments of HSV infection, one might expect that structural components of the virion might well bypass ICP47's inhibitory effects and enter the class I pathway, and those rare $CD8^+$ CTL cloned from patients have specificity for these proteins (TIGGES et al. 1992). One intriguing consequence of ICP47's effects might be increased sensitivity to NK cell lysis. Indeed, there are numerous studies demonstrating that HSV-infected human cells become targets for NK cells, and IE gene expression is sufficient for this effect (FITZGERALD-BOCARSLY et al. 1991; KAUFMAN et al. 1992).

In contrast to what is seen with human fibroblasts, HSV-infected mouse cells were much less sensitive to the effects of ICP47-1. Concentrations of ICP47-1 reaching 10 μM introduced into permeabilized mouse fibroblasts did not obviously inhibit TAP-mediated transport (TOMAZIN et al. 1996). In similar assays, ICP47-2 was somewhat more effective, although concentrations of 10 μM or more were required in order to obtain partial inhibition (A. Hill, R. Tomazin, P. Jugovic, and D.C. Johnson, unpublished). In another assay, mouse TAP expressed in insect cell microsomes was inhibited by 40% using 10 μM ICP47-1 (AHN et al. 1996). By comparison, ICP47-1 at concentrations of only 0.15 μM inhibited 50% of the TAP in human fibroblasts (AHN et al. 1996; TOMAZIN et al. 1996). Therefore, it appears that both ICP47-1 and ICP47-2 are poor inhibitors of TAP and antigen presentation in mouse cells. Nevertheless, one must explain weak $CD8^+$ T cell responses in mice and observations that $CD4^+$ T cells are more important than $CD8^+$ T cells in some experimental infections of mice (MANICKAN and ROUSE 1995). One could hypothesize that ICP47 affects antigen presentation in mice, even though there are not conspicuous effects on antigen presentation in mouse fibroblasts. Consistent with this, HSV-2-infected mouse cells were found to be less susceptible to HSV-specific CTL and, by using HSV-1×HSV-2 intertypic recombinants, this resistance to CTL mapped to the Us region of the HSV-2 genome (CARTER et al. 1984), which includes the ICP47 gene (MCGEOCH et al. 1985). Previously, we found that the ICP47-2 protein inhibited mouse TAP somewhat better than ICP47-1. Therefore, it is possible that this difference accounts for the resistance of HSV-2-infected cells to $CD8^+$ CTL and may suggest that there are inhibitory effects of ICP47 in mouse cells.

Recently, we characterized the replication and pathogenesis of an HSV-1 ICP47-negative mutant in mice. The ICP47-negative mutant was able to initiate an infection in the mouse cornea, but displayed markedly reduced ability to cause pathology and encephalitis in the nervous system when compared to wild-type HSV-1 (GOLDSMITH et al. 1998). The mutant caused a primary infection in the epithelial layer of the cornea and spread to the sensory ganglia, but instead of replicating well in the nervous system and subsequently causing encephalitis, as with wild-type HSV-1, the virus was largely cleared. An ICP47-rescued version of this virus behaved similarly to wild-type HSV-1. However, in mice depleted of $CD8^+$ T cells (using anti-CD8 antibodies), the ICP47-negative mutant caused pathology similar to that of wild-type HSV-1. There is evidence that $CD8^+$ T cells play a central role in restricting virus replication in sensory ganglia (SIMMONS and TSCHARKE 1992), yet these T cells play a less important role in limiting virus during the primary infection in the cornea (reviewed in SCHMID and ROUSE 1992). Therefore, ICP47 appears to alter the course of viral pathogenesis in an organ in which HSV replication is contained by $CD8^+$ T cells. This implies that ICP47 confers resistance to $CD8^+$ T cells in these mice in some unknown cell type, e.g., neurons or other cells in nervous system. To date, we have not established whether this is related to inhibition of the MHC class I antigen presentation pathway or to some other effect on the immune system. One would have predicted that an ICP47-negative mutant would not behave differently from wild-type HSV in mice. These results provide an interesting example of the dangers inherent in extrapolating from in vitro biochemical data to animals.

It should also be noted that the effects of ICP47 in vivo are not likely to be absolute. Human fibroblasts treated with interferon (IFN)-γ can be lysed by HSV-specific CTL (TIGGES et al. 1992), and the HSV-induced downregulation of MHC class I in human keratinocytes was reversed by treatment with IFN-γ (MIKLOSKA et al. 1996). IFN-γ upregulates the expression of TAP and, with higher TAP present, it appears that ICP47 cannot reach quantities sufficient to block TAP. These results may explain previous observations that $CD8^+$ T cells appear in human HSV lesions only after $CD4^+$ T cells have been present for several days and cytokines such as IFN-γ are present (CUNNINGHAM and MERIGAN 1983). Thus we might expect that these $CD8^+$ T cells could recognize HSV-infected cells later in the course of lesion formation and after IFN-γ (produced by $CD4^+$ T cells or NK cells) has upregulated TAP. It is well known that IFN-γ levels strongly influence the severity of clinical disease (CUNNINGHAM and MERIGAN 1983; BURCHETT et al. 1992; YAMAMOTO et al. 1993). Therefore, there appears to be a balance established between the early efforts of HSV to block the immune system, so that progeny can be produced, and later efforts by the host to overcome the immune escape and shutdown virus replication.

3 Cytomegaloviruses

The CMVs are the best studied of the β-herpesviruses, viruses which infect a variety of cells in vivo, including macrophage/monocytes, endothelial cells, epithelial cells, and a long list of other cell types. These are large viruses with coding capacity for over 200 proteins, more than twice that of HSV. It is already clear that HCMV and MCMV contain rich repertoires of proteins that interfere with the MHC class I antigen presentation pathway. The multiplicity of mechanisms they employ raises questions of its own: why do the viruses need so many genes, and how do they coordinate these genes to achieve the desired manipulation of the immune system?

CMVs are ancient and ubiquitous members of the herpesvirus family. Most or all mammalian species possess CMV, and most individuals acquire CMV infection, usually in early life. CMV are highly species specific; for instance, individual Old World primate species possess distinct CMV and, although rhesus CMV can be propagated in human fibroblasts in vitro, there is little evidence for sustained cross-species infection in vivo. CMV that are phylogenetically more distant cannot even infect cells from the other species in vitro.

Current evidence, based on genomic sequences of several CMV, is consistent with the notion that the cytomegaloviruses have cospeciated with their hosts (McGeoch 1992). Thus there has been a protracted period of coevolution between these viruses and their individual host species' immune systems. Genes in the central part of the MCMV genome and in the unique long (U_L) component of the HCMV genome, regions thought to encode functions necessary for virus replication, are more homologous, whereas genes nearer ends of the genome and in the U_S region of HCMV, where most of the immunomodulatory genes are located, are less homologous.

Nevertheless, HCMV and MCMV appear very similar in many aspects of their biology. They both infect similar organs (liver and spleen in primary infection, lung in recurrent infection, and salivary gland as a site of persistence), and both viruses can be recovered from most organs during infective episodes and infect a similar spectrum of cell types (epithelial, endothelial, monocyte/macrophage). Both viruses have a similar relationship with their host (causing severe disease upon immunosuppression of a host that was previously free of symptomatic infection). There are also some differences, due to host or viral factors or both. For example, unlike MCMV, HCMV can cross the placenta and cause congenital infection. Moreover, HCMV and MCMV differ in their replication cycle in permissive cells. HCMV replication is unusually protracted, especially in cells such as macrophages, where replication can take many days. MCMV can usually replicate in a much shorter period of time, producing infectious virions in less than 24 h. The slower replication cycle of HCMV may necessitate differences in the mechanisms of interference with antigen processing. Cells which produce viral progeny in a relatively short period may be less vulnerable to attack by certain components of the immune system, e.g., CTL, than those which replicate virus more slowly.

3.1 Murine Cytomegalovirus

3.1.1 Immune Control and the Characteristics of the Immune Response

MCMV has been studied as a model of CMV infection for two decades and has many advantages over HCMV, given that an animal model is available and the virus tends to replicate better in cultured cells. As might be expected for a virus this large which replicates in a large number of cell types, immune control of MCMV is multifaceted and complex. Important control of MCMV infection is mediated through both the innate and acquired immune systems (reviewed in KOSZINOWSKI 1991). Here, we will summarize aspects of the $CD8^+$ T cell and NK responses, both of which are affected by viral interference with the MHC class I antigen presentation pathway.

$CD8^+$ cytotoxic T cells are readily generated by mice in response to infection by MCMV. Splenocytes or draining lymph nodes from infected mice, restimulated with MCMV in vitro, differentiate into effector CTL. MCMV gene expression can be manipulated by drug treatment of cells to allow enhanced expression of IE genes, i.e., those expressed in the absence of de novo synthesis of viral proteins. When cells expressing IE proteins were offered to MCMV-primed CTL lines, a dominant CTL response in H-2^d mice was found to be directed against an epitope derived from the IE protein pp89 (KOSZINOWSKI et al. 1987). However, when E gene transcription was allowed to progress, the pp89 epitope was no longer presented, despite the continued synthesis of pp89 (REDDEHASE et al. 1986; DEL VAL et al. 1989). These experiments provided the first example of inhibition of antigen presentation by MCMV (and other herpesviruses). The molecular basis of this is described below. After E gene expression was allowed, however, a second population of MCMV-specific CTL was able to lyse the infected cells in both H-2^d and H-2^b mice (DEL VAL et al. 1989; CAMPBELL et al. 1992). The antigen or antigens recognized by these “E-specific” CTL have not yet been identified, nor has the means by which they avoid the interference with antigen presentation, which effectively precludes pp89 presentation. The pp89-specific response was effective in virus control, as adoptively transferred pp89-specific CTL could protect against a lethal MCMV infection, and immunization with pp89, expressed using a recombinant vaccinia, protected mice against subsequent challenge with a lethal dose of MCMV (REDDEHASE et al. 1987; JONJIC et al. 1988; REDDEHASE et al. 1988; DEL VAL et al. 1991). Therefore, both MCMV IE- and E-specific CTL are generated and the IE-specific CTL can protect hosts, but MCMV E gene products inhibit presentation of IE proteins to $CD8^+$ T cells.

NK cells recognize and lyse some cells that have lost MHC class I expression (YOKOMAYA 1995), and these cells therefore play an important role in detecting cells that are infected with virus, where the virus blocks the class I pathway. As with other herpesviruses, NK cells play a crucial role in controlling MCMV in vivo (SCALZO et al. 1992; ORANGE et al. 1995; TAY et al. 1995; ORANGE and BIRON 1996). There are at least two genetic susceptibility loci to MCMV in mice, one of which is called Cmv-1 and maps to the region on chromosome 6 that contains NK receptors (SCALZO et al. 1995). Mice differing in this locus have splenic CMV titers which differ 10^3- to 10^4-fold during acute infection, yet the titers of resistant mice approach those of susceptible mice if NK cells are depleted (SCALZO et al. 1992). Two murine NK receptors, Ly-49A and NKR-P1, that recognize MHC class I and impart a negative signal to NK cells are encoded in the chromosome 6 NK locus to which the Cmv-1 locus is tightly linked. The molecular basis of the NK-linked resistance to MCMV in the mouse is a subject of active investigation. The nature of NK recognition in general is a complex subject currently receiving considerable attention (for comprehensive recent coverage, see PARHAM 1997), and the impact of MCMV on NK recognition will no doubt prove very interesting.

3.1.2 Retention of Class I Molecules in a Pre-Golgi Compartment by gpm152 (gp40)

The molecular details of how MCMV inhibits class I presentation of IE proteins have begun to be understood in recent years. Investigations of the synthesis and expression of MHC class I in H-2^d fibroblasts indicated that, after E genes were expressed, MHC class I slowly disappeared from the cell surface and most newly synthesized MHC class I was retained inside the cell in an endoglycosidase H-sensitive form (DEL VAL et al. 1992). At the time when the pp89 epitope was no longer recognized by CTL, pp89 peptide capable of sensitizing CTL could still be extracted from L^d molecules within infected cells, indicating that the retained class I molecules were loaded with peptide (DEL VAL et al. 1992). Moreover, other proteins, β-galactosidase expressed from an IE promoter (DEL VAL et al. 1989), or an SV40 T antigen encoded by the cell (CAMPBELL et al. 1992) were not presented when MCMV E genes were expressed. Thus the antigen presentation defect caused by E gene products is not selective for pp89, affecting both another antigen expressed as an IE gene and a constitutively synthesized cellular protein. Nevertheless, as noted above, presentation of E gene products occurs during the early phase of MCMV infection, perhaps because there is more abundant expression of these proteins.

A deletion mutant, Δ94.5, which contains a 9.4-kb deletion, mapped the region encoding the gene responsible for ER retention of class I to the *Hin*dIII-E region of the MCMV genome, i.e., in cells infected with Δ94.5, acquisition of endoglycosidase H resistance by class I molecules is normal (THALE et al. 1995). By injecting DNA fragments into cells and scoring for intracellular MHC class I retention, the MCMV gene m152 was found to mediate MHC class I retention (ZIEGLER et al. 1997). A recombinant vaccinia virus expressing m152 caused both retention of class I MHC and a major reduction in CTL recognition of pp89.

MCMV m152 encodes a type I membrane glycoprotein with a predicted molecular mass of 42 kDa (ZIEGLER et al. 1997). The gene produces two protein products of 37 kDa and 40 kDa, respectively, both of which are glycosylated and reduce to a single band on sodium dodecyl sulfate polyacrylamide gel electrophoresis (SDS-PAGE) after treatment with endoglycosidase H. Here, we will refer to the products of the m152 ORF as gpm152. A pulse-chase experiment of gpm152 expressed using a vaccinia vector revealed that, after 3 h of chase, most of the protein was no longer recovered using an antipeptide antibody raised to part of the intraluminal domain of the protein. The small fraction of the glycoprotein which was recovered after 3 h was endoglycosidase H resistant and ran on SDS gels at a position intermediate between gp37 and gp40. Thus gpm152 apparently had a short half-life and was probably degraded, but a fraction of the protein was processed in the Golgi compartment. In contrast, the class I MHC that was retained by gpm152 had a longer half-life. Confocal microscopy of MHC class I in MCMV-infected cells revealed that it colocalized with the ER-Golgi intermediate compartment marker p58. Staining of gpm152 revealed a similar pattern, although there was not exact colocalization, and gpm152 was also identified in lysosomal compartments. Coprecipitation of gpm152 and class I MHC was not observed in either standard or mild detergent lysates. A mutant of gpm152 lacking the putative cytosolic tail was still able to cause retention of class I MHC. Thus, although it is clear that gpm152 causes MHC class I retention, the mechanism by which it does so is unknown.

Surprisingly, even though synthesis and transport of class I was normal in cells infected with Δ94.5 (the mutant lacking the region containing m152), with prolonged expression of E genes there was still a progressive loss of sensitivity to lysis by pp89-specific CTL, concomitant with decreased surface MHC class I (THALE et al. 1995). This does not appear to be due to decreased synthesis or transport of MHC class I. Therefore, it was postulated that there is a mechanism or mechanisms for downregulating class I at the cell surface encoded outside the 9.4-kb fragment deleted from Δ94.5. Finally, it should also be noted that m152 is a member of a clustered family of genes within the MCMV genome. The function of other members of this family is not known, but by analogy with HCMV it is possible that some may affect antigen presentation.

3.1.3 Association of gpm04 (gp34) with MHC Class I Molecules in the Endoplasmic Reticulum

The MCMV gene m04 encodes a second protein that affects the MHC class I pathway, a 34-kDa glycoprotein previously known as gp34. Here, we will refer to this glycoprotein as gpm04. It was identified as a protein coprecipitating with class I MHC in MCMV-infected cells (KLEIJNEN et al. 1997) and exists in two glycosylation states within an infected cell. The first form has three endoglycosidase H-sensitive N-linked glycans and is presumably resident in a pre-Golgi compartment. Most of the gpm04 found in infected cells is found in this form. The second form is associated with MHC class I and contains two endoglycosidase H-resistant glycans and one endoglycosidase H-sensitive glycan. Since some of this second form can be iodinated on the cell surface, it is presumed that it resides stably bound to MHC class I at the cell surface. However, to date there is no information as to how gpm04 affects the function of the MHC class I molecules to which it is bound.

Synthesis of gpm04 begins within 2 h of infection of fibroblasts, and the protein is quite stable. It is expressed in stochiometric excess over MHC class I, leading to the suggestion that it may displace gpm152, allowing egress of some MHC class I to the surface. Since surface MHC class I might import resistance to NK cells, it has been speculated that gpm04 could act as a decoy to prevent NK lysis of infected cells (KLEIJNEN et al. 1997). Alternatively, gpm04 may act to destabilize MHC class I on the surface of infected cells (THALE et al. 1995). However, to date, there is no data that support either hypothesis. In the MCMV genome, the gpm04 gene is a member of a family (the m02 family) of type-1 membrane glycoproteins comprising 15 members, five of which share with m04 two conserved cysteine residues and a proline-rich region in the carboxy-terminal third of the protein (RAWLINSON et al. 1996). Nothing is known about the function of these other proteins, nor of the proteins they encode, but, again by analogy with HCMV, it is possible that several members of this family alter antigen presentation.

3.1.4 Transcriptional Downregulation of MHC Class I in Transformed Cells

Using an SV40-transformed fibroblast line, it was noted that MCMV infection led to a decreased synthesis of class I molecules (CAMPBELL and SLATER 1994). This was found to be due to an inhibition of transcription. Further studies revealed that this effect was limited to transformed cell lines; in primary fibroblast lines, the synthesis of MHC class I was actually stimulated by MCMV infection (A. Campbell, personal communication). Thus the relevance of this phenomenon in vivo remains unclear. It is distinctly possible that, in vivo, a population of cells infected with MCMV behaves as do the transformed cells, so that there is downregulation of class I genes.

3.1.5 Class I Homologue

The sequence of the MCMV genome revealed an open reading frame (m144) predicted to encode a protein with significant homology to MHC class I proteins (RAWLINSON et al. 1996). This was of particular interest, because HCMV is also known to encode a class I homologue, the UL18 gene (discussed below). However, m144 does not share a common position in the MCMV genome with that of UL18 in HCMV. As predicted, each viral gene product is more homologous to the MHC protein of its own species than to the other viral protein. The greatest homology between the two proteins is in the α_3-domains, and both proteins bind β_2-microglobulin (BROWNE et al. 1990; P. Bjorkman, personal communication). Both viral gene products have lost the predicted CD8-binding site. At present, the best guess for the function of both viral gene products is that they serve to provide resistance to NK cells, and there is some evidence for this with the HCMV product (see below).

3.2 Human Cytomegalovirus

3.2.1 Immunity

Given what we now know about HCMV's immune evasion, it is perhaps surprising that one of the first demonstrations of the importance of cellular immunity in controlling herpesviruses in humans came from studies of HCMV, in which it has been shown that CTL are critical for control of HCMV after bone marrow transplantation (REUSSER et al. 1991). It became apparent, however, that the nature of antigen recognition by HCMV-specific CTL was unusual. Many HCMV-specific CTL clones could recognize cells infected with UV-irradiated virus, i.e., viruses capable of entering target cells and uncoating, but not expressing viral proteins (RIDDELL et al. 1991). The major tegument protein of HCMV, pp65, was identified as the major target antigen, and it was obvious that delivery of the protein to the cytosol by the infecting virion was sufficient for target sensitization (LI et al. 1994). Recent work indicates that pp65 provides immunodominant CTL epitopes in many and perhaps most individuals (BERENCSI et al. 1997). A second important target of the anti-HCMV $CD8^+$ CTL response is the major IE protein (GILBERT et al. 1993; BERENCSI et al. 1997). The IE protein is a more conventional antigen for presentation by the MHC class I pathway, since it is a nonstructural protein and its presentation depends on de novo synthesis of the protein very early in virus infection.

As with MCMV, NK cells are thought to play an important role in controlling HCMV disease. An adolescent with selective and complete NK cell deficiency suffered severe infections with a number of herpesviruses, including HCMV (BIRON et al. 1989).

3.2.2 Rapid Degradation of MHC Class I in Infected Cells: Independent Mediation of this Effect by US2 and US11

Cells infected with HCMV showed a progressive decline in cell surface levels of MHC class I and concomitant decreased susceptibility of peptide-pulsed targets to lysis by CTL (WARREN et al. 1994). Metabolic labeling and pulse-chase analysis of HCMV-infected cells revealed that class I heavy chain was synthesized in normal amounts but then very rapidly degraded (BEERSMA et al. 1993; WARREN et al. 1994; YAMASHITA et al. 1994). The half-life of β_2-microglobulin was normal, and other membrane glycoproteins, such as transferrin receptor, were not affected. Degradation appeared to be relatively specific for human class I alleles; in mouse cells expressing the human class I allele HLA-B27, only the human class I was degraded (BEERSMA et al. 1993). The region of the HCMV genome responsible for this effect was identified by utilizing a panel of HCMV mutants with deletions in the U_S component. These studies identified at least two loci independently capable of mediating the rapid destruction of newly synthesized class I, a region including US2 and other genes, and a second region including the US11 gene (JONES et al. 1995).

The mechanism by which US11 and US2 cause rapid destruction of MHC class I was recently studied using US2- and US11-transfected U373 cell lines (WIERTZ et al.

1996a,b). Pulse-chase studies of the US11 transfectant revealed that the half-life of MHC class I in these cells was less than 1 min. However, breakdown of MHC class I was inhibited in the presence of proteasome inhibitors (LLnL, Cbz-LLL, or lactacystin), and a lower molecular weight form of MHC class I accumulated. This intermediate was a deglycosylated form of the heavy chain, and there was evidence favoring deglycosylation by the action of *N*-glycanase, known to be present in the cytoplasm. Cell fractionation experiments confirmed that the heavy-chain intermediate was found in the cytosol (WIERTZ et al. 1996a). Thus US11 caused newly synthesized MHC class I which had been cotranslationally glycosylated to be extruded into the cytosol, where it was deglycosylated by *N*-glycanase and then degraded by the proteasome. The US11 glycoprotein is an ER resident, type I membrane glycoprotein, and there was no evidence for direct binding of US11 to class I, although it would seem likely that there is a transient interaction before degradation.

US2 appears to act in a similar manner as US11. US2-transfected U373 cells showed rapid degradation of the class I heavy chain following extrusion from the ER membrane (WIERTZ et al. 1996b). The process with US2 was less efficient than in US11 transfectants, and perhaps for this reason it was possible to provide a more detailed analysis of intermediates in the degradation pathway. In the US2-transfected cells, the class I heavy chain associated with β_2-microglobulin and acquired a folded conformation, yet was deglycosylated and degraded. Most intriguingly, deglycosylated heavy chain was found associated with the Sec61 complex. Sec61 is the mammalian homologue of a yeast protein known to be important in cotranslational translocation of proteins into the ER and is thought to act as a proteinaceous pore through the membrane. Furthermore, the US2 glycoprotein itself could be found to coprecipitate with class I and was also subject to translocation into the cytosol, deglycosylation, and proteasomal degradation. Therefore, US2 and US11 apparently trigger a common pathway used by cells for protein import into the ER and probably also for export of proteins, so that MHC class I is reverse translocated back out of the ER through the Sec-61 membrane pore.

3.2.3 Retention of MHC Class I in the Endoplasmic Reticulum by US3

The US3 open reading frame, which is adjacent to the US2 gene in the HCMV genome, encodes another protein which binds to class I molecules, but causes retention in the ER (AHN et al. 1996b; JONES et al. 1996). In both infected and transfected cells, US3 specifies two protein products, with molecular weights of 22 kDa and 18 kDa, respectively. The 22-kDa protein contains N-linked oligosaccharides and coprecipitates with MHC class I in digitonin (but not NP40) cell lysates. The US3 glycoprotein has a half-life of about 1 h and is mostly found in an endoglycosidase H-sensitive form, although a small fraction of the proteins manages to leave the ER, especially in experiments involving long chase periods (JONES et al. 1996). In long chase experiments, a fraction of class I heavy chains associate with β_2-microglobulin in US3-transfected U373 cells and appear to be loaded with peptides from thermostability experiments (AHN et al. 1996b). However, these class

I complexes remain endoglycosidase H sensitive for protracted periods. Therefore, it appears that gpUS3 associates with MHC class I before peptide loading, but the MHC class I is not prevented from acquiring peptides. Confocal microscopy involving the US3 transfectants revealed an intense perinuclear staining for class I heavy chains, supporting the notion that class I is efficiently retained in the ER by US3 (JONES et al. 1996). This apparently contrasts with a different subcellular localization of class I seen in cells expressing MCMV (ZIEGLER et al. 1997), although in other ways the action of the genes seems analogous. US3 is expressed as an IE gene in HCMV-infected human fetal fibroblasts and U373 cells and is therefore likely to exert its action before US2 and US11 in these cells. It has been proposed that US3 may assist the function of US2 and US11 by retaining MHC class I for subsequent destruction (AHN et al. 1996b).

3.2.4 Interference of US6 in the Function of TAP

A fourth gene in the HCMV US region, US6, has been found to decrease MHC class I expression in transiently transfected cells (AHN et al. 1997; HENGEL et al. 1997; KOSZINOWSKI et al. 1997; P. Cresswell, personal communication). Peptide transport assays involving permeabilized cells revealed that TAP-mediated peptide transport was blocked in HCMV-infected cells. The US6 gene was found to be responsible for this effect. Unlike ICP47, the US6 glycoprotein causes its effect by interacting with TAP at the luminal interface and can be coprecipitated with TAP, tapasin, MHC class I, and calnexin as part of a large multimeric complex.

3.2.5 Inhibition of Natural Killer Cells by Class I Homologue UL18

The sequence of the HCMV genome revealed an open reading frame, UL18, which is homologous to human class I molecules (BECK and BARRELL 1988). The UL18 protein, expressed by using a vaccinia virus vector, binds β_2-microglobulin (BROWNE et al. 1990), and this heterodimeric complex binds peptides (FAHNESTOCK et al. 1995). Nevertheless, the protein has been difficult to detect in HCMV-infected cells. It has been speculated that this viral protein acts as a decoy for NK cells. To test this hypothesis, UL18 was transfected into the HLA-A,B,C-negative human cell line 721.221, which is normally a target for HLA-C-inhibited human NK clones. Cell surface expression of gpUL18 was extremely low; nevertheless, the transfectant was more resistant to lysis by NK cells of a number of specificities (REYBURN et al. 1997). From these data, it was concluded that the UL18 protein confers resistance to NK cells. Inclusion of saturating amounts of antibodies directed against the inhibitory receptor NK1R did not inhibit lysis, and the investigators concluded that the UL18 product must interact with another NK receptor. The function of the UL18 product clearly requires further investigation.

3.2.6 Prevention by pp65 of the Recognition of Immediate-Early Protein

One further selective effect on antigen presentation has been described in HCMV-infected cells. As mentioned above, CTL clones specific for the major IE protein of HCMV were isolated from immune donors, and their specificity was demonstrated by using cells infected with a vaccinia vector expressing IE. HCMV-infected cells, however, were very poorly lysed (GILBERT et al. 1993). This was attributed to a selective effect of the virion tegument protein pp65 on the ability of IE to be correctly processed for presentation (GILBERT et al. 1996). In cells coinfected with vaccinia vectors expressing the IE protein and pp65, no presentation of IE was seen, whereas cells infected with the vector expressing IE alone were lysed. pp65 apparently causes IE to be phosphorylated, probably indirectly because pp65 produced in bacteria has no kinase activity (S. Riddell, personal communication). A mutant form of pp65 without the capacity to cause phosphorylation of IE did not block antigen presentation of IE (GILBERT et al. 1996). It was suggested that phosphorylation of IE precludes its access to the class I antigen presentation pathway, perhaps at the level of the proteasome. Since this effect should be very early in the infection, the relative roles of pp65 and the IE protein US3 in preventing presentation of IE in cells infected with HCMV remain unclear. Nevertheless, the vaccinia experiments have demonstrated that coexpression of pp65 and IE decreases the presentation of IE epitopes. One of the more curious aspects of this story is the existence of IE-specific CTL in donors infected with HCMV. If IE is not effectively presented in HCMV-infected cells, how were these CTL primed in vivo?

3.3 Why Do Cytomegaloviruses Express Multiple Genes Which Inhibit Antigen Processing?

MCMV and HCMV clearly express a substantial number of proteins that inhibit or have the potential to inhibit the MHC class I antigen presentation pathway, yet there are still $CD8^+$ CTL responses directed to these viruses. These CTL are known to play a critical role in controlling CMV infections. It is possible that the multiple genes act additively or synergistically to improve the overall efficacy of the blockade. This may be important, since in some cells these viruses are slow to induce viral gene expression so that, by expressing more genes, inhibtion may be more complete. There are a number of alternative hypotheses. First, the CMV genes may be regulated differentially in different cell types. The CMV infect many different cell types in vivo, and it is easy to imagine that the virus would desire different degrees of immune control in cells in which progeny virus is produced versus cells in which persistent infection occurs. Thus we might expect differential expression of the genes responsible for immune evasion, either quantitatively or kinetically, in different host cells. There is clearly evidence for host cell control of viral transcription, including US11 (KERRY et al. 1997). Second, there is evidence for differential susceptibility of class I alleles to the effects of CMV immunomodulatory proteins. Many of the CMV gene products described above probably interact directly with MHC class I and, since class I molecules are polymorphic, not all may be affected to the same extent. A possible purpose of seemingly redundant functions, such as those performed by US2 and US11, would be to more effectively cover the range of class I alleles. The susceptibility of various mouse class I proteins to US2 and US11 was investigated, and differences were found in their susceptibility to the two viral proteins (MACHOLD et al. 1997). It seems likely that more subtle differences may also exist in the susceptibility of human class I alleles to these viral gene products. The foregoing answers to this question must also consider the evolutionary argument that even a minute selective advantage gained by individual members of this family of proteins will be sufficiently beneficial to allow the virus a selective advantage in the face of $CD8^+$ T cells over millions of years.

4 Epstein-Barr Virus

EBV is a member of the γ-herpesvirus family which maintains a lifelong latent infection in B lymphocytes and productively infects stratified epithelium in the nasopharynx (reviewed in KIEFF 1996; RICKINSON and KIEFF 1996). EBV encodes about 100 proteins that are expressed during lytic infection (BAER et al. 1984), but during latent infection of B lymphocytes only a relatively small number of latent proteins are expressed. In B cells immortalized in vitro, known as lymphoblastoid cell lines (LCL), EBV nuclear antigens (EBNA)-1 to -6 and latent membrane proteins LMP1, LMP2A, and LMP2B are expressed (reviewed KIEFF 1996). EBV is also associated with several human cancers, i.e., Burkitt's lymphoma (BL), nasopharyngeal carcinoma (NPC), and Hodgkin's disease (HD). In BL, EBNA-1 is the only latent protein expressed, whereas in BL and HD there is expression of LMP proteins as well (reviewed in MOSS et al. 1996; RICKINSON and KIEFF 1996). EBNA-1 to -6 and LMP proteins are also expressed in infectious mononucleosis (IM) and post-transplantation lymphoproliferative disease. Latently infected cells can be switched into lytically infected cells in vitro by the use of phorbol esters, cross-linking of surface IgM, or treatment with calcium ionophore (LUKA et al. 1979). During lytic infection, viral proteins can be subdivided into early antigens (EA), membrane antigens (MA), and viral capsid antigens (VCA) (THORLEY-LAWSON et al. 1977).

4.1 Immune Responses

Humoral responses to EBV during the acute phase of IM consist largely in IgG and IgM directed to MA and VCA (reviewed in KHANNA et al. 1995). Interestingly, the anti-MA responses appear to be largely restricted to gp85, rather than gp350/220, yet anti-gp-350/220 antibodies are much more effective in neutralizing EBV (HENLE and HENLE 1980). Antibody responses to EBNA-1 and EBNA-2 appear later, during the convalescent phase. The late appearance and specificity (anti-gp85 rather than anti-gp350/220) of anti-EBV humoral responses suggests that they do not play a primary role in limiting virus spread (PEARSON et al. 1979; reviewed in KHANNA et al. 1995).

Anti-EBV cellular responses function to limit acute IM infections and to maintain or restrict latent infections. EBV is frequently reactivated and shed in more substantial amounts from the oropharynx after immunosuppression (reviewed in KHANNA et al. 1995; RICKINSON and KIEFF 1996). In vitro and in vivo studies have provided ample evidence that T lymphocytes control expansion of latently infected B cells (THORLEY-LAWSON et al. 1977; THORLEY-LAWSON 1980). B cells in culture will proliferate in response to infection with EBV, and this proliferation can be controlled by T lymphocytes, primarily $CD8^+$ CTL reactivated in vitro (MOSS et al. 1978). In IM, there is dramatic expansion of T cells and establishment of a population of $CD46RO^+$ cells, considered to be memory T cells, that can later be stimulated to become $CD8^+$ and $CD4^+$ CTL. Apparently, these CTL play a primary role in limiting virus spread. $CD4^+$ and $CD8^+$ CTL recognize both latent and lytic antigens in IM (STRANG and RICKINSON 1987; TOMKINSON et al. 1989). CTL epitopes have been described in eight of the nine latent antigens, with most of the reactivity to EBNA-3, -4, and -6. Curiously, EBNA-1, which is expressed in all latently infected B cells and in BL, appears not to act as a target for $CD8^+$ T cells based on the inability to generate or identify anti-EBNA-1 $CD8^+$ CTL, either as clones or in polyclonal culture. However, an anti-EBNA-1 $CD4^+$ T cell clone has been described (MOSS et al. 1994). There is evidence for control of EBV by IFN-γ, released by $CD4^+$ T cells (ANDERSON et al. 1983; BEJARANO et al. 1990), and the relative contribution of cytolysis versus cytokines in controlling EBV is not well understood.

4.2 Resistance of Epstein-Barr Virus Cytotoxic T Lymphocytes and Inhibition of Antigen Presentation

Since EBV antigens are expressed in latently infected cells and in EBV-associated tumors and yet the cells persist in the blood for long periods, one might expect that there are mechanisms in play to prevent recognition by T lymphocytes. Indeed, at least four different reasons for the defects in T cell recognition of BL and other EBV-positive cells have been described: (1) limited expression of viral antigens, (2) downregulation of antigen presentation genes and cell adhesion molecules, (3) loss of dominant CTL epitopes, and (4) failure of latent antigens to undergo processing and antigen presentation.

4.2.1 Limited Expression of Viral Antigens

Many BL cells express EBNA-1 and no other EBV antigens (SAMPLE et al. 1991; MARCHINI et al. 1992a,b), and this very restricted form of gene expression has obvious advantages in terms of avoiding recognition by T cells. Since EBNA-1 is both necessary and sufficient for EBV DNA replication in mitotic cells (YATES et al. 1984; MIDDLETON and SUGDEN 1994), EBNA-1 is observed in all EBV-infected B cells that persist and in all malignancies associated with EBV. It was widely believed that

EBNA-1 does not include MHC class I epitopes or that these are masked in some fashion (reviewed in Moss et al. 1996). However, more recent observations on *cis*-acting sequences in EBNA-1 that preclude antigen presentation (see Sect. 4.2.4) probably explain, at least in part, defects in recognition of EBNA-1-expressing cells and the persistence in latently infected and BL cells. But this does not explain how other EBV latent proteins avoid recognition, so that tumors, e.g., NP and HD, that express LMP1 and LMP2 are not rejected. It is possible that these proteins acquire properties that make their presentation less efficient. Consistent with this, LMP1 derived from an NPC cell was not recognized by CTL when expressed in a mouse model, whereas another LMP1 from B95.8 cells was immunogenic (Hu et al. 1991). Changes in LMP1 from HD cells compared with "wild-type" EBV have been reported (Knecht et al. 1993), suggesting that LMP1 is altered during tumorigenesis.

4.2.2 Downregulation of Antigen Presentation and Cell Adhesion Molecules

Although BL cells do express MHC class I proteins at the cell surface and can present to allospecific CTL clones (Torsteindottir et al. 1986), there have been reports that the surface MHC class I in these cells is conformationally aberrant, perhaps as a result of inefficient peptide loading (Khanna et al. 1994; Rowe et al. 1995). There are low levels of TAP, and perhaps other components of the class I pathway, which may result from deficiencies in transcription of MHC genes in these cells (Khanna et al. 1994). Combined with these defects, there is a marked reduction in expression of lymphocyte function-associated antigen (LFA)-1, LFA-3, and intercellular adhesion molecule (ICAM)-1 in BL cells, which might be expected to reduce CTL adherence (Gregory et al. 1988; Khanna et al. 1993).

4.2.3 Loss of Dominant Cytotoxic T Lymphocyte Epitopes

Masucci and colleagues described a dominant CTL epitope in EBNA-4 presented by HLA-A11, which is a relatively infrequent haplotype in Caucasian and central African populations (Gavioli et al. 1993). In populations from the lowlands of Papua New Guinea or China, where HLA-A11 is very frequent, EBV isolates rarely contained the epitope; instead, the epitope was altered, abrogating CTL recognition and binding to HLA-A11 (de Campos-Lima et al. 1993, 1994). Therefore, like other viruses (Pircher et al. 1990; Phillips et al. 1991), EBV has apparently evolved to avoid expression of immunodominant CTL epitopes. However, this mechanism may not be universal, because an EBNA-3 HLA-B35 epitope is largely conserved in West African populations in which the B35 is relatively frequent (Lee et al. 1995).

4.2.4 Failure to Undergo Processing and Antigen Presentation

Recently, a novel and interesting effect on the antigen presentation pathway was observed with the EBNA-1 protein (Levitskaya et al. 1995). EBNA-1 is comprised of two unique domains, an N-terminal domain (amino acids 1–89) and a C-terminal domain (amino acids 327–641), separated by a region containing Gly-Ala repeats that varies in length in different EBV (BAER et al. 1984). Recombinant EBNA-1 proteins devoid of Gly-Ala repeats could be recognized by $CD8^+$ CTL clones (specific for CTL epitopes introduced into the protein), but EBNA-1 proteins with the Gly-Ala repeat were not recognized (LEVITSKAYA et al. 1995). Moreover, when the Gly-Ala repeat was transferred to another protein, EBNA-4, which is a good CTL antigen, the recombinant EBNA-4 proteins were not recognized by EBNA-4-specific $CD8^+$ CTL clones. However, since other class I antigens could be presented in EBNA-1-expressing cells, the class I pathway was not itself blocked. Therefore, this Gly-Ala sequence apparently acts in *cis* to prevent presentation of EBNA-1 (and other proteins into which it is inserted) by class I molecules. This result may explain the inability to generate $CD8^+$ CTL clones specific for EBNA-1 and the persistence of cells expressing EBNA-1 in vivo. The mechanism by which EBNA-1 avoids presentation has not yet been described in the literature, but recent results suggest that EBNA-1 proteins containing the Gly-Ala repeat are more resistant to in vitro proteasome-mediated degradation than are forms of the protein without the Gly-Ala repeat (M. Masucci, personal communication).

5 Conclusions

In the long term, each of these herpesviruses attempts to replicate and spread to other hosts, often in the face of a fully primed, robust immune response. However, each must also avoid undue damage to the host, as the survival of the virus depends on this, while at the same time ensuring that the host does not eradicate the virus. Different herpesviruses accomplish these aims by different means. HSV establishes a latent infection in the nervous system, a safe haven of long-lived, nondividing cells which are not subject to the normal immune surveillance mechanisms. Periodic reactivation of HSV leads to production of progeny that can spread infection to other hosts. ICP47 probably functions early during replication in epithelial cells to prevent presentation to $CD8^+$ T cells so that virus can be produced, but later in the disease the effects of ICP47 are lessened by IFN-γ. The CMV infect a variety of cell types, including macrophages/monocytes, which are professional antigen-presenting cells. Persistence in monocytes/macrophages may be related to the stage of differentatiation, so that during infection of less-differentiated cells only a limited set of viral antigens are expressed. However, to achieve production of infectious progeny, there must be replication either in differentiated macrophages or in other cells, followed again by immune control. It is likely that the ability of CMV to avoid immune response in macrophages/monocytes requires immunomodulatory functions that are not required in other cell types. Thus it appears that CMV carries a heavy load of genes that function in immune evasion. EBV persists in B cells, which are also subject to intensive immune regulation and replicates primarily in nasopharyngeal epithilium. However, as with HSV, there is a latent state in which very limited gene expression occurs. One of these genes, EBNA-1, has evolved an ingenious ability to avoid the class I antigen presentation pathway in latently infected or tumor cells. Therefore, each of these viruses has very different requirements for its immunomodulatory arsenal, replicating in different cells types and using different strategies to persist in the host. With the tools of molecular virology and immunology, it should be possible to characterize viral genes that inhibit the immune system without a great deal of difficulty. However, the task of understanding how these genes impact the overall picture of virus pathogenesis and persistence will be more difficult.

References

Ahn K, Meyer TH, Uebel S, Sempe P, Djaballah H, Yang Y, Peterson PA, Fruh K, Tampe R (1996a) Molecular mechanism and species specificity of TAP inhibition by herpes simplex virus ICP47. EMBO J 15:3247–55

Ahn K, Angulo A, Ghazal P, Peterson PA, Yang Y, Fruh K (1996b) Human cytomegalovirus inhibits antigen presentation by a sequential multistep process. Proc Natl Acad Sci USA 93:10990–10995

Ahn K, Gruhler A, Galocha B, Jones TR, Wiertz EJHJ, Ploegh HL, Peterson PA, Yang Y, Fruh K (1997) The ER-luminal domain of the HCMV glycoprotein US6 inhibits peptide translocation by TAP Immunity 6:613–621

Anderson U, Britton S, de Ley M, Bird G (1983) Evidence for the ontogenetic precedence of suppressive T cell functions in the human neonate. Eur J Immunol 13:613–619

Baer R, Bankier AT, Biggin MD, Desinger PL, Farrell PJ, Gibson TJ, Halfull G, Hudson G, Satchwell C, Sequin C, Fuffnell P, Barrell B (1984) DNA sequence and expression of the B95-8 Epstein-Barr virus genome. Nature (London) 310:207–211

Beck S, Barrell BG (1988) Human cytomegalovirus encodes a protein homologous to MHC class I antigens. Nature 331:269–272

Beersma MFC, Bijlmakers MJE, Ploegh H (1993) Human cytomegalovirus down-regulates HLA class I expression by reducing the stability of class I heavy chains. J Immunol 151:4455–4464

Beinart D, Neumann L, Vebel S, Tampé R (1997) Structure of the viral TAP-inhibitor ICP47 induced by membrane association. Biochemistry 36:4694–4700

Bejarano MT, Masucci MG, Morgan A, Morein B, Klein G, Klein E (1990) Epstein-Barr virus (EBV) antigens processed and presented by B cells, B blasts, and macrophages trigger T-cell-mediated inhibition of EBV-induced B-cell transformation. J Virol 64:1398–1401

Berencsi K, Endresz V, Gyulai Z, Pincus S, Cox WI, Plotkin SA, Gonczol E (1997) Cytotoxic T lymphocyte responses to human cytomegalovirus (HCMV). Sixth International Cytomegalovirus Workshop, 5–9 March, Perdido Beach, Alabama (abstr 151)

Biron CA, Byron KS, Sullivan JL (1989) Severe herpesvirus infections in an adolescent without natural killer cells. N Engl J Med 320:1731–1735

Bonneau RH, Jennings SR (1989) Modulation of acute and latent herpes simplex virus infection in C57BL/6 mice by adoptive transfer of immune lymphocytes with cytotoxic activity. J Virol 63:1480–1484

Britt WJ, Alford CA (1996) Cytomegalovirus. In: Fields BN, Knipe DM, Howley PM et al (eds) Fields virology. Raven–Lippincott, Philadelphia, pp 2493–2523

Browne H, Smith G, Beck S, Minson T (1990) A complex between the MHC class I homologue encoded by human cytomegalovirus and beta 2 microglobulin. Nature 347:770–772

Burchett SK, Corey L, Mohan KM, Westall J, Ashley R, Wilson CB (1992) Diminished interferon-g and lymphocyte proliferation in neonatal and postpartum primary herpes simplex virus infection. J Infect Dis 165:813–818

Campbell AE, Slater JS (1994) Down-regulation of major histocompatibility complex class I synthesis by murine cytomegalovirus early gene expression. J Virol 68:1805–1811

Campbell AE, Slater JS, Cavanaugh VJ, Stenberg RM (1992) An early event in murine cytomegalovirus replication inhibits presentation of cellular antigens to cytotoxic T lymphocytes. J Virol 66:3011–3017
Carter VC, Jennings SR, Rice PL, Tevethia SS (1984) Mapping of a herpes simplex virus type 2-encoded function that affects the susceptibility of herpes simplex virus-infected target cells to lysis by herpes simplex virus-specific cytotoxic T lymphocytes. J Virol 49:766–771
Cines DB, Lyss AP, Bina M (1982) Fc and C3 receptors induced by herpes simplex virus on cultured human endothelial cells. J Clin Invest 69:123–128
Confer DL, Vercellotti GM, Kotasek D, Goodman JL, Ochoa A, Jacob HS (1990) Herpes simplex virus-infected cells disarm killer lymphocytes. Proc Natl Acad Sci USA 87:3609–3613
Corey L, Spear PG (1986) Infections with herpes simplex viruses (2). N Engl J Med 314:749–757
Costa JC, Rabson AS (1975) Role of Fc receptors in herpes simplex virus infection. Lancet 1:77–78
Cunningham AL, Merigan TC (1983) Gamma interferon production appears to predict time of recurrence of herpes labialis. J Immunol 130:2397–2400
de Campos-Lima PO, Gavioli R, Zhang QJ, Wallace LE, Dolcetti R, Rowe M, Rickinson AB, Masucci MG (1993) HLA-A11 epitope loss isolates of Epstein-Barr virus from a highly A11+ population. Science 260:98–100
de Campos-Lima PO, Levitsky V, Brooks J, Lee SP, Hu LF, Rickinson AB, Masucci MG (1994) T cell responses and virus evolution: loss of HLA A11-restricted CTL epitopes in Epstein-Barr virus isolates from highly A11-positive populations by selective mutation of anchor residues. J Exp Med 179:1297–1305
Del Val M, Munch K, Reddehase MJ, Koszinowski UH (1989) Presentation of CMV immediate-early antigen to cytolytic T lymphocytes is selectively prevented by viral genes expressed in the early phase. Cell 58:305–315
Del Val M, Schlicht HJ, Volkmer H, Messerle M, Reddehase MJ, Koszinowski UH (1991) Protection against lethal cytomegalovirus infection by a recombinant vaccine containing a single nonameric T-cell epitope. J Virol 65:3641–3646
Del Val M, Hengel H, Hacker H, Hartlaub U, Ruppert T, Lucin P, Koszinowski UH (1992) Cytomegalovirus prevents antigen presentation by blocking the transport of peptide-loaded major histocompatibility complex class I molecules into the medial-Golgi compartment. J Exp Med 176:729–738
Domke-Optiz I, Swatzkym R (1990) Natural resistance to herpes simplex virus infections: the macrophage-interferon axis. In: Boston AL (ed) Herpes viruses, the immune system, and AIDS. Kluwer, Dordrecht, pp 171–202
Douglas RG, Couch RB (1970) A prospective study of chronic herpes virus infection and recurrent herpes labialis in humans. J Immunol 104:289–295
Dubin G, Socolof E, Frank I, Friedman HM (1991) Herpes simplex virus type 1 Fc receptor protects infected cells from antibody-dependent cellular cytotoxicity. J Virol 65:7046–7050
Eisenberg RJ, Ponce de Leon M, Friedman HM, Fries LF, Frank MM, Hastings JC, Cohen GH (1987) Complement component C3b binds directly to purified glycoprotein C of herpes simplex virus types 1 and 2. Microb Pathog 3:423–435
Fahnestock MJ, Johnson JI, Feldman RM, Neveu JM, Lane WS, Bjorkman PJ (1995) The MHC class I homologue encoded by human cytomegalovirus binds endogenous peptides. Immunity 3:583–590
Fenwick ML, Everett RD (1990) Transfer of UL41, the gene controlling virion-associated host shutoff, between strains of herpes simplex virus. J Gen Virol 71:411–418
Fitzgerald-Bocarsly P, Feldman M, Curl S, Schnell J, Denny T (1989) Positively selected Leu-11a (CD16+) cells require the presence of accessory cells or factors for the lysis of herpes simplex virus-infected fibroblasts but not herpes simplex virus-infected Raji. J Immunol 143:1318–1326
Fitzgerald-Bocarsly P, Howell DM, Pettera L, Tehrani S, Lopez C (1991) Immediate-early gene expression is sufficient for induction of natural killer cell-mediated lysis of herpes simplex virus type 1-infected fibroblasts. J Virol 65:3151–3160
Frank I, Friedman HM (1989) A novel function of the herpes simplex virus type 1 Fc receptor: participation in bipolar bridging of antiviral immunoglobulin G. J Virol 63:4479–4488
Fruh K, Ahn K, Djaballah H, Sempe P, van Endert PM, Tampe R, Peterson PA, Yang Y (1995) A viral inhibitor of peptide transporters for antigen presentation. Nature 375:415–418
Galocha B, Hill AB, Cook R, Barnett B, Nolan A, Raimondi A, McGeoch D, Ploegh HL (1997) Identification of the active site of ICP47, an herpes simplex virus encoded inhibitor of TAP. J Exp Med 185:1565–1572

Gavioli R, Kurilla MG, de Campos-Lima PO, Wallace LE, Dolcetti R, Murray RJ, Rickinson AB, Masucci MG (1993) Multiple HLA A11-restricted cytotoxic T-lymphocyte epitopes of different immunogenicities in the Epstein-Barr virus-encoded nuclear antigen 4. J Virol 67:1572–1578

Gilbert MJ, Riddell SR, Li CR, Greenberg PD (1993) Selective interference with class I major histocompatibility complex presentation of the major immediate-early protein following infection with human cytomegalovirus. J Virol 67:3461–3469

Gilbert MJ, Riddell SR, Plachter B, Greenberg PD (1996) Cytomegalovirus selectively blocks antigen processing and presentation of its immediate-early gene product. Nature 383:720–722

Goldsmith K, Chen W, Johnson DC, Henducks RL (1998) ICP47 enhances herpes simplex virus neurovirulence by the blocking $CD8^+$ T cell response. J Exp Med (in press)

Gregory CP, Murray R, Edwards CF and ABR (1988) Down regulation of cell adhesion molecules LFA-3 and ICAM-1 in Epstein-Barr virus-positive Burkitt's lymphoma underlies tumor cell escape from virus-specific T cell surveillance. J Exp Med 167:1811–1824

Hengel H, Koopman JO, Flohr T, Murangi W, Goulmy E, Hammerling GJ, Koszinowski UH, Momburg F (1997) A viral ER resident glycoprotein inactivates the MHC encoded peptide transporter. Immunity 6:623–632

Henle W, Henle G (1980) Epidemiologic aspects of (EBV) -associated diseases. Ann NY Acad Sci 354:326–331

Hidaka Y, Sakai Y, Toh Y et al (1991) Glycoprotein C of herpes simplex virus type 1 is essential for the virus to evade antibody-independent complement-mediated virus inactivation and lysis of virus-infected cells. J Gen Virol 72:915–921

Hill AB, Barnett BC, McMichael AJ, McGeoch DJ (1994) HLA class I molecules are not transported to the cell surface in cells infected with herpes simplex virus types 1 and 2. J Immunol 152:2736–41

Hill A, Jugovic P, York I, Russ G, Bennink J, Yewdell J, Ploegh H, Johnson D (1995) Herpes simplex virus turns off the TAP to evade host immunity. Nature 375:411–415

Hu LF, Minarovits J, Cao SL, Contreras-Salazar B, Rymo L, Falk K, Klein G, Ernberg I (1991) Variable expression of latent membrane protein in nasopharyngeal carcinoma can be related to methylation status of the Epstein-Barr virus BNLF-1 5'-flanking region. J Virol 65:1558–1567

Jawahar S, Moody C, Chan M, Finberg R, Geha R, Chatila T (1996) Natural killer (NK) cell deficiency associated with an epitope-deficient Fc receptor type IIIA (CD16-II). Clin Exp Immunol 103:408–413

Johnson DC, Feenstra V (1987) Identification of a novel herpes simplex virus type 1-induced glycoprotein which complexes with gE and binds immunoglobulin. J Virol 61:2208–2216

Johnson DC, Frame MC, Ligas MW, Cross AM, Stow ND (1988) Herpes simplex virus immunoglobulin G Fc receptor activity depends on a complex of two viral glycoproteins, gE and gI. J Virol 62:1347–1354

Jones TR, Hanson LK, Sun L, Slater JS, Stenberg RM, Campbell AE (1995) Multiple independent loci within the human cytomegalovirus unique short region down-regulate expression of major histocompatibility complex class I heavy chains. J Virol 69:4830–4841

Jones TR, Wiertz EJ, Sun L, Fish KN, Nelson JA, Ploegh HL (1996) Human cytomegalovirus US3 impairs transport and maturation of major histocompatibility complex class I heavy chains. Proc Natl Acad Sci USA 93:11327–11333

Jonjic S, del Val M, Keil GM, Reddehase MJ, Koszinowski UH (1988) A nonstructural viral protein expressed by a recombinant vaccinia virus protects against lethal cytomegalovirus infection. J Virol 62:1653–1658

Kaufman DS, Schoon RA, Leibson PJ (1992) Role for major histocompatibility complex class I in regulating natural killer cell-mediated killing of virus-infected cells. Proc Natl Acad Sci USA 89:8337–8341

Kerry JA, CD, and V, Stanley TL, Jones TR, Stenberg RM (1997) Role of cellular transcription factors in human cytomegalovirus replication. Sixth International Cytomegalovirus Workshop, 5–9 March 1997, Perdido Beach, Alabama (abstr 2)

Khanna R, Burrows SR, Suhrbier A, Jacob CA, Griffin H, Misko IS, Sculley TB, Rowe M, Rickinson AB, Moss DJ (1993) EBV peptide epitope sensitization restores human cytotoxic T cell recognition of Burkitt's lymphoma cells. Evidence for a critical role for ICAM-2. J Immunol 150:5154–5162

Khanna R, Burrows SR, Argaet V, Moss DJ (1994) Endoplasmic reticulum signal sequence facilitated transport of peptide epitopes restores immunogenicity of an antigen processing defective tumour cell line. Int Immunol 6:639–645

Khanna R, Burrows SR Moss DJ (1995) Immune regulation in Epstein-Barr virus-associated diseases. Microbiol Rev 59:387–405

Kieff E. (1996) Epstein-Barr virus and its replication. In: Fields BN, Knipe DM, Howlet PM et al. (eds) Fields virology. Raven–Lippincott, Philadelphia, pp 2343–2396

Kleijnen MF, Huppa JB, Lucin P, Mukherjee S, Farrell H, Campbell AE, Koszinowski UH, Hill AB, Ploegh HL (1997) A mouse cytomegalovirus glycoprotein, gp34, forms a complex with folded class I MHC molecules in the ER which is not retained but is transported to the cell surface. EMBO J 16:685–694

Knecht H, Bachmann E, Brousset P, Sandvej K, Nadal D, Bachmann F, Odermatt BF, Delsol G, Pallesen G (1993) Deletions within the LMP1 oncogene of Epstein-Barr virus are clustered in Hodgkin's disease and identical to those observed in nasopharyngeal carcinoma. Blood 82:2937–2942

Koelle DM, Tigges MA, Burke RL, Symington FW, Riddell SR, Abbo H, Corey L (1993) Herpes simplex virus infection of human fibroblasts and keratinocytes inhibits recognition by cloned CD8+ cytotoxic T lymphocytes. J Clin Invest 91:961–968

Kohl S (1992) The role of antibody in herpes simplex virus infection in humans. Curr Top Microbiol Immunol 179:75–88

Kolaitis G, Doymaz M, Rouse BT (1990) Demonstration of MHC class II-restricted cytotoxic T lymphocytes in mice against herpes simplex virus. Immunology 71:101–106

Koszinowski UH (1991) Molecular aspects of immune recognition of cytomegalovirus. Transplant Proc 23:70–73

Koszinowski UH, Reddehase MJ, Keil GM, Schickedanz J (1987) Host immune response to cytomegalovirus: products of transfected viral immediate-early genes are recognized by cloned cytolytic T lymphocytes. J Virol 61:2054–2058

Koszinowski UH, Koopman JO, Flohr T, Muranyi W, Goulmy E, Hämmerling GJ, Momburg F, Hengel H (1997) Inactivation of the MHC encoded peptide transporters by a cytomegaloviral ER resident glycoprotein. Sixth International Cytomegalovirus Workshop, 5–9 March 1997, Perdido Beach, Alabama, p 9

Kubota Y, Gaither TA, Cason J, O'Shea JJ Lawley TJ (1987) Characterization of the C3 receptor induced by herpes simplex virus type 1 infection of human epidermal, and A431 cells. J Immunol 138:1137–1142

Lee SP, Morgan S, Skinner J, Thomas WA, Jones SR, Sutton J, Khanna R, Whittle HC, Rickinson AB (1995) Epstein-Barr virus isolates with the major cytotoxic T cell epitope are prevalent in a highly B35.01-positive African population. Eur J Immunol 25:102–110

Lehner T, Wilton JM, Shillitoe EJ (1975) Immunological basis for latency, recurrences and putative oncogenicity of herpes simplex virus. Lancet 2:60–62

Levitskaya J, Coram M, Levitsky VS, I, Steigerwald-Mullen PM, G, K, MG, K, MG, M (1995) Inhibition of antigen processing by the internal repeat region of the Epstein-Barr virus nuclear antigen-1. Nature 375:685–688

Li CR, Greenberg PD, Gilbert MJ, Goodrich JM, Riddell SR (1994) Recovery of HLA-restricted cytomegalovirus (CMV)-specific T-cell responses after allogeneic bone marrow transplant: correlation with CMV disease and effect of ganciclovir prophylaxis. Blood 83:1971–1979

Luka J, Kallin B, Klein G (1979) Induction of the Epstein-Barr virus (EBV) cycle in latently infected cells by n-butyrate. Virology 94:228–231

Machold RP, Wiertz EJHJ, Jones TR, Ploegh HL (1997) The HCMV gene products US11 and US2 differ in their ability to attack allelic forms of murine major histocompatibility complex (MHC) class I heavy chains. J Exp Med 185:363–366

Manickan E, Rouse BT (1995) Roles of different T-cell subsets in control of herpes simplex virus infection determined by using T-cell-deficient mouse-models. J Virol 69:8178–8179

Marchini A, Cohen JI, Wang F, Kieff E (1992a) A selectable marker allows investigation of a nontransforming Epstein-Barr virus mutant. J Virol 66:3214–3219

Marchini A, Longnecker R, Kieff E (1992b) Epstein-Barr virus (EBV)-negative B-lymphoma cell lines for clonal isolation and replication of EBV recombinants. J Virol 66:4972–4981

Martin S, Moss B, Berman PW, Laskey LA, Rouse BT (1987) Mechanisms of antiviral immunity induced by a vaccinia virus recombinant expressing herpes simplex virus type 1 glycoprotein D: cytotoxic T cells. J Virol 61:726–734

McGeoch DJ (1992) Molecular evolution of the large DNA viruses of eukaryotes. Semin Virol 3:399–409

McGeoch DJ, Dolan A, Donald S, Rixon FJ (1985) Sequence determination and genetic content of the short unique region in the genome of herpes simplex virus type I. J Mol Biol 181:1–13

McNearney TA, Odell C, Holers VM, Spear PG, Atkinson JP (1987) Herpes simplex virus glycoproteins gC-1 and gC-2 bind to the third component of complement and provide protection against complement-mediated neutralization of viral infectivity. J Exp Med 166:1525–1535

Middleton T, Sugden B (1994) Retention of plasmid DNA in mammalian cells is enhanced by binding of the Epstein-Barr virus replication protein EBNA1. J Virol 68:4067–4071

Mikloska Z, Kesson AM, Penfold ME, Cunningham AL (1996) Herpes simplex virus protein targets for CD4 and CD8 lymphocyte cytotoxicity in cultured epidermal keratinocytes treated with interferon-gamma. J Infect Dis 173:7–17

Mocarski ES Jr (1996) Cytomegaloviruses and their replication. In: Fields BN, Knipe DM, Howley PM et al (eds) Fields virology. Raven–Lippencott, Philadelphia, pp 2447–2492

Moss DJ, Rickinson AB, Pope JH (1978) Long-term T cell-mediated immunity to Epstein-Barr virus in man. I. Complete regression of virus-induced transformation in cultures of seropositive donor leukocytes. Int J Cancer 22:662–668

Moss DJ, Burrows SR, Suhrbier A, Khanna R (1994) Potential antigenic targets on Epstein-Barr virus-associated tumours and the host response. Ciba Found Symp 187:4–13

Moss DJ, Schmidt D, Elliott S, Suhrbier A, Burrows S, Khanna R (1996) Strategies involved in developing an effective vaccine for EBV-associated diseases. Cancer Res 69:211–243

Nash AA, Jayasuriya A, Phelan J, Cobbold SP, Waldmann H, Prospero T (1987) Different roles for L3T4+ and Lyt 2+ T cell subsets in the control of an acute herpes simplex virus infection of the skin and nervous system. J Gen Virol 68:825–833

Orange JS, Biron CA (1996) Characterization of early IL-12, IFN-alphabeta, and TNF effects on antiviral state and NK cell responses during murine cytomegalovirus infection. J Immunol 156:4746–4756

Orange JS, Wang B, Terhorst C, Biron CA (1995) Requirement for natural killer cell-produced interferon gamma in defense against murine cytomegalovirus infection and enhancement of this defense pathway by interleukin 12 administration. J Exp Med 182:1045–1056

Parham P (ed) NK cells, MHC class I antigens and missing self. Immunol Rev 155

Pearson GR, Qualtiere LF, Klein G, Norin T, Bal IS (1979) Epstein-Barr virus-specific antibody-dependent cellular cytotoxicity in patients with Burkitt's lymphoma. Int J Cancer 24:402–406

Phillips RE, Rowland-Jones S, Nixon DF, Gotch FM, Edwards JP, Ogunlesi AO, Elvin JG, Rothbard A, Bangham CR, Rizza CR et al (1991) Human immunodeficiency virus genetic variation that can escape cytotoxic T cell recognition Nature 354:453–459

Pircher D, Moskopidis D, Rohrer U, Burki K, Hengartner H, Zinkernagel RM (1990) Viral excape by selection of cytotoxic T cell-resistant virus variants in vivo. Nature (Lond) 346:629

Posavad CM, Rosenthal KL (1992) Herpes simplex virus-infected human fibroblasts are resistant to and inhibit cytotoxic T-lymphocyte activity. J Virol 66:6264–6272

Posavad CM, Newton JJ, Rosenthal KL (1994) Infection and inhibition of human cytotoxic T lymphocytes by herpes simplex virus. J Virol 68:4072–4074

Posavad CM, Koelle DM, Corey L (1996) High frequency of CD8+ cytotoxic T-lymphocyte precursors specific for herpes simplex viruses in persons with genital herpes. J Virol 70:8165–8168

Quinnan GV, Masur H, Rook AH, Armstrong G, Frederick WR, Epstein J, Manischeqitz MS, Mecher AM, Jackson L, Ames J, Smith HA, Parker M, Pearson GR, Panillo J, Mitchell C, SE, S (1984) Herpesvirus infections in the acquired immune deficiency syndrome. JAMA 252:72–77

Rawlinson WD, Farrell HE, Barrell BG (1996) Analysis of the complete DNA sequence of murine cytomegalovirus. J Virol 70:8833–8849

Reddehase MJ, Fibi MR, Keil GM, Koszinowski UH (1986) Late-phase expression of a murine cytomegalovirus immediate-early antigen recognized by cytolytic T lymphocytes. J Virol 60:1125–1129

Reddehase MJ, Mutter W, Munch K, Buhring HJ, Koszinowski UH (1987) CD8-positive T lymphocytes specific for murine cytomegalovirus immediate-early antigens mediate protective immunity. J Virol 61:3102–3108

Reddehase MJ, Jonjic S, Weiland F, Mutter W, Koszinowski UH (1988) Adoptive immunotherapy of murine cytomegalovirus adrenalitis in the immunocompromised host: CD4-helper-independent antiviral function of CD8-positive memory T lymphocytes derived from latently infected donors. J Virol 62:1061–1065

Reusser P, Riddell SR, Meyers JD, Greenberg PD (1991) Cytotoxic T-lymphocyte response to cytomegalovirus after human allogeneic bone marrow transplantation: pattern of recovery and correlation with cytomegalovirus infection and disease. Blood 78:1373–1380

Reyburn H, O, M, Valez-Gomez M, Sheu EG, Pazmany L, Davis DM, Strominger JL (1997) Human NK cells: their ligands, receptors and functions. Immunol Rev 155:119–125

Rickinson AB, Kieff E (1996) Epstein-Barr virus. In: Fields BN, Knipe DM, Howley PM et al (eds) Fields virology. Raven–Lippincott, Philadelphia, pp 2397–2446

Riddell SR, Rabin M, Geballe AP, Britt WJ, Greenberg PD (1991) Class I MHC-restricted cytotoxic T lymphocyte recognition of cells infected with human cytomegalovirus does not require endogenous viral gene expression. J Immunol 146:2795–2804

Roizman B, Sears AE (1996) Herpes simplex viruses and the replication. In: Fields BN, Knipe DM, Howley PM et al (eds) Fields virology. Raven–Lippincott, Philadelphia, pp 2231–2295

Rowe M, Khanna R, Jacob CA, Argaet V, Kelly A, Powis S, Belich M, Croom-Carter D, Lee S, Burrows SR et al (1995) Restoration of endogenous antigen processing in Burkitt's lymphoma cells by Epstein-Barr virus latent membrane protein-1: coordinate up-regulation of peptide transporters and HLA-class I antigen expression. Eur J Immunol 25:1374–1384

Sample J, Brooks L, Sample C, Young L, Rowe M, Gregory C, Rickinson A, Kieff E (1991) Restricted Epstein-Barr virus protein expression in Burkitt lymphoma is due to a different Epstein-Barr nuclear antigen 1 transcriptional initiation site. Proc Natl Acad Sci USA 88:6343–6347

Scalzo AA, Fitzgerald NA, Wallace CR, Gibbons AE, Smart YC, Burton RC, Shellam GR (1992) The effect of the Cmv-1 resistance gene, which is linked to the natural killer cell gene complex, is mediated by natural killer cells. J Immunol 149:581–589

Scalzo AA, Lyons PA, Fitzgerald NA, Forbes CA, Yokoyama WM, Shellam GR (1995) Genetic mapping of Cmv1 in the region of mouse chromosome 6 encoding the NK gene complex-associated loci Ly49 and musNKR-P1. Genomics 27:435–441

Schmid DS (1988) The human MHC-restricted cellular response to herpes simplex virus type 1 is mediated by CD4+, CD8- T cells and is restricted to the DR region of the MHC complex. J Immunol 140:3610–3616

Schmid DS, Rouse BT (1992) The role of T cell immunity in control of herpes simplex virus. Curr Top Microbiol Immunol 179:57–74

Seidel-Dugan C, Ponce de Leon M, Friedman HM, Eisenberg RJ, Cohen GH (1897) Identification of C3b-binding regions on herpes simplex virus type 2 glycoprotein C. J Virol 64:1897–1906

Simmons A, Tscharke DC (1992) Anti-CD8 impairs clearance of herpes simplex virus from the nervous system: implications for the fate of virally infected neurons. J Exp Med 175:1337–1344

Smiley ML, Friedman HM (1985) Binding of complement C3b to glycoprotein C is modulated by sialic acid on herpes simplex virus type I-infected cells. J Virol 55:857–851

Strang G, Rickinson AB (1987) Multiple HLA class I-dependent cytotoxicities constitute the non-HLA-restricted response in infectious mononucleosis. Eur J Immunol 17:1007–1013

Tal-Singer R, Seidel-Dugan C, Fries L, Huemer HP, Eisenberg RJ, Cohen GH, Friedman HM (1991) Herpes simplex virus glycoprotein C is a receptor for complement component iC3b. J Infect Dis 164:750–753

Tay CH, Welsh RM, Brutkiewicz RR (1995) NK cell response to viral infections in beta 2-microglobulin-deficient mice. J Immunol 154:780–789

Thale R, Szepan U, Hengel H, Geginat G, Lucin P, Koszinowski UH (1995) Identification of the mouse cytomegalovirus genomic region affecting major histocompatibility complex class I molecule transport. J Virol 69:6098–6105

Thorley-Lawson DA (1980) The suppression of Epstein-Barr virus in vitro occurs after infection but before transformation of the cell. J Immunol 124:745–751

Thorley-Lawson DA, Chess L, Strominger JL (1977) Suppression on in vitro Epstein-Barr virus infection. A new role for adult human lymphocytes. J Exp Med 146:495–508

Tigges MA, Koelle D, Hartog K, Sekulovich RE, Corey L, Burke RL (1992) Human CD8+ herpes simplex virus-specific cytotoxic T-lymphocyte clones recognize diverse virion protein antigens. J Virol 66:1622–1634

Tigges MA, Leng S, Johnson DC, Burke RL (1996) Human herpes simplex virus (HSV)-specific CD8+ CTL clones recognize HSV-2-infected fibroblasts after treatment with IFN-gamma or when virion host shutoff functions are disabled. J Immunol 156:3901–3910

Tomazin R, Hill AB, Jugovic P, York I, van Endert P, Ploegh HL, Andrews DW, Johnson DC (1996) Stable binding of the herpes simplex virus ICP47 protein to the peptide binding site of TAP. EMBO J 15:3256–3266

Tomkinson BE, Maziarz R Sullivan JL (1989) Characterization of the T cell-mediated cellular cytotoxicity during acute infectious mononucleosis. J Immunol 143:660–670

Torpey DJ, Lindsley MD, Rinaldo C (1989) HLA-restricted lysis of herpes simplex virus-infected monocytes and macrophages mediated by CD4+ and CD8+ T lymphocytes. J Immunol 142:1325–1332
Torsteindottir S, Masucci MG, Ehlin-Hendricksson G, Brautbar G, Basset GB, Klein G, Klein E (1986) Differentiation-dependent sensitivity of human B-cell-derived lines to major histocompatibility complex-restricted T-cell cytotoxicity. Proc Natl Acad Sci USA 83:5620–5624
Warren AP, Ducroq DH, Lehner PJ, Borysiewicz LK (1994) Human cytomegalovirus-infected cells have unstable assembly of major histocompatibility complex class I complexes and are resistant to lysis by cytotoxic T lymphocytes. J Virol 68:2822–2829
Whitley RJ (1996) Herpes simplex viruses. In: Fields BN, Knipe DM, Howley PM et al (eds) Fields virology. Raven–Lippincott, Philadelphia, pp 2297–2342
Wiertz EJ, Jones TR, Sun L, Bogyo M, Geuze HJ, Ploegh HL (1996a) The human cytomegalovirus US11 gene product dislocates MHC class I heavy chains from the endoplasmic reticulum to the cytosol. Cell 84:769–779
Wiertz EJ, Tortorella D, Bogyo M, Yu J, Mothes W, Jones TR, Rapoport TA, Ploegh HL (1996b) Sec61-mediated transfer of a membrane protein from the endoplasmic reticulum to the proteasome for destruction. Nature 384:432–438
Yamamoto T, Osaki T, Yoneda K, Ueta E (1993) Immunological investigation of adult patients with primary herpes simplex virus-1 infection. J Oral Pathol Med 22:263–267
Yamashita Y, Shimokata S, Saga S, Mizuno S, Tsurumi T, Nishiyama Y (1994) Rapid degradation of the heavy chain of class I major histocompatibility complex antigens in the endoplasmic reticulum of human cytomegalovirus-infected cells. J Virol 68:7933–7943
Yasukawa M, Zarling JM (1984a) Human cytotoxic T cell clones directed against herpes simplex virus-infected cells. I. Lysis restricted by HLA class II MB and DR antigens. J Immunol 133:422–427
Yates J, Warren N, Reisman D, Sugden B (1984) A cis-acting element from the Epstein-Barr viral genome that permits stable replication of recombinant plasmids in latently infected cells. Proc Natl Acad Sci USA 81:3806–3810
Yokomaya WM (1995) Natural killer cell receptors specific for major histocompatibility complex class I molecules. Proc Natl Acad Sci USA 92:3081–3085
York IA, Johnson DC (1993) Direct contact with herpes simplex virus-infected cells results in inhibition of lymphokine-activated killer cells because of cell-to-cell spread of virus. J Infect Dis 168:1127–1132
York IA, Johnson DC (1994) Inhibition of humoral and cellular immune recognition by herpes simplex viruses. In: McFadden G(ed) Viroceptors, virokines and related immune modulators encoded by DNA viruses. Landes, Austin, pp 89–110
York IA, Roop C, Andrews DW, Riddell SR, Graham FL, Johnson DC (1994) A cytosolic herpes simplex virus protein inhibits antigen presentation to CD8+ T lymphocytes. Cell 77:525–535
Ziegler H, Thäle R, Lucin P, Muranyi W, Flohr T, Hengel H, Farrell H, Rawlinison W, Koszinowski UH (1997) A mouse cytomegalovirus protein retains MHC class I complexes in the ERCIC/cis-Golgi compartments. Immunity 6:57–66
Zweerink HJ, Stanton LW (1981) Immune response to herpes simplex virus infections: virus-specific antibodies in sera from patients with recurrent facial infections. Infect Immun 31:624–630

Intracellular Transport of Molecules Engaged in the Presentation of Exogenous Antigens

TOMMY W. NORDENG[1], JEAN-PIERRE GORVEL[2], and ODDMUND BAKKE[1]

[1]Division of Molecular Cell Biology, Department of Biology, University of Oslo, 0316 Oslo, Norway

[2]Centre d'Immunologie, INSEM-CNRS de Marseille-Luminy, Case 906, 13288 Marseille Cedex 09, France

1 Introduction

Major histocompatibility complex (MHC) class I and class II molecules bind peptides and present them to T cells. Whereas class I molecules predominantly bind peptides generated from endogenously synthesized cytoplasmic proteins*, class II molecules mainly present peptides derived from internalized degraded material (for reviews about class I- and class II-presenting pathways, see GERMAIN and MARGULIES 1993; NEEFJES and MOMBURG 1993; BUUS 1994).

After synthesis and translocation into the endoplasmic reticulum (ER), class II molecules associate with the invariant chain (Ii) and the resulting complex traverses the Golgi complex and accumulates in endosomal compartments. Here, Ii is proteolytically degraded (BLUM et al. 1988), leaving the class II molecules free to bind peptides derived from endocytosed antigen. The peptide class II complexes are subsequently transported to the cell surface for presentation to $CD4^+$ T cells.

During the past 10 years, our knowledge has considerably improved about the molecular mechanisms of protein antigen processing and presentation in the context of MHC molecules. Their polymorphism and high level of expression have allowed extensive biochemical, structural, and genetic studies on pathogen-derived peptides presented to T lymphocytes. Recently, novel molecules related to MHC antigens have been discovered. Their function is to present "nonclassical" pathogenic antigens to T cells, and they can be divided in two groups: the nonclassical MHC class I proteins (class Ib molecules) and the class I-related molecules (the CD1 family) encoded outside of the MHC. Whereas the class Ib molecules present nonconventional peptide antigen by TAP (transporter associated with antigen presentation)-dependent mechanisms, the CD1 molecules bind non-peptide antigen, probably somewhere in the class II peptide-loading pathway. In this review, in addition to MHC class II-mediated presentation of exogenous antigen, we will also discuss a parallel role for CD1 molecules.

2 Assembly and Initial Transport of MHC Class II and Invariant Chain

2.1 Structure

In humans, class II molecules are encoded by the polymorphic HLA-DR, HLA-DQ, and HLA-DP genes and expressed as noncovalent heterodimers of two transmembrane polypeptides, the α-chain (35 kDa) and the β-chain (27 kDa) (for a review, see KAUFMAN et al. 1984). Whereas the luminal domains of class II molecules have an

By "cytoplasm" we mean the content of a cell within the plasma membrane, but outside the nucleus and all membrane bounded compartments.

intrinsic ability to associate (KJÆR-NIELSEN et al. 1990; WETTSTEIN et al. 1991), interactions by the transmembrane domains promote the formation of correctly assembled complexes (COSSON et al. 1992). After synthesis and translocation into the ER, the α- and β-chains associate with a third transmembrane glycoprotein, Ii (SUNG et al. 1981; KVIST et al. 1982), forming a nonameric $(\alpha\beta Ii)_3$-subunit complex (ROCHE et al. 1991b).

Ii is a type II transmembrane protein, and a number of Ii isoforms exist, defined by the primary amino acid sequence. The major form is a glycoprotein of 31–33 kDa (p33). In humans, an alternative form of Ii, p35, containing an amino-terminal cytoplasmic extension of 16 residues, results from initiation of translation at an alternative AUG codon upstream of the one used to generate p33 (STRUBIN et al. 1986; O'SULLIVAN et al. 1987). Additional species of the p33 and p35 isoforms result from alternative splicing of the Ii mRNA where the additional exon 6b is introduced. This gives rise to the p41 and p43 forms, respectively, containing a 64-amino acid insertion in the luminal domain (STRUBIN et al. 1986; O'SULLIVAN et al. 1987; KOCH 1988).

2.2 Expression of MHC Class II and Invariant Chain

Coordinate expression of MHC class II and Ii molecules has been reported in human tissue (QUARANTA et al. 1984; VOLC-PLATZER et al. 1984) and in tissue culture cells (LONG 1985). However, whereas transcription of Ii is controlled by elements common to those controlling MHC class II transcription (ZHU et al. 1990; BROWN et al. 1991), distinct elements have been identified (WRIGHT et al. 1995) with tissue-specific activity (ZHU et al. 1990). These elements probably enable distinct regulation of transcription from the Ii gene relative to MHC class II genes. In fact, Ii has been observed in almost absence of MHC class II in some cell lines (LONG et al. 1984; ACCOLLA et al. 1985; LENARDO et al. 1989) and vice versa (MOMBURG et al. 1986).

Expression of both Ii and class II molecules can be increased in a variety of cell lines by stimulation with cytokines such as tumor necrosis factor (TNF)-α (PESSARA et al. 1990; KOLK et al. 1993), interferon (IFN)-γ (COLLINS et al. 1984; EADES et al. 1990; BROWN et al. 1991, 1993; KOLK et al. 1992), and interleukin (IL)-4 (NOELLE et al. 1986; POLLA et al. 1986). However, treatment with mitomycin C or phorbol esters is reported to upregulate Ii synthesis in fibroblasts not expressing MHC class II (RAHMSDORF et al. 1983, 1986), and adding granulocyte-macrophage colony-stimulating factor (GM-CSF) to macrophage cell lines leads to elevated Ii expression, whereas class II molecules are not induced (KLAGGE et al. 1997). In contrast, treatment with the heavy metal lead (Pb) increases MHC class II expression (MCCABE et al. 1990), possibly by a post-translational mechanism (MCCABE et al. 1991; GUO et al. 1996). Whereas Ii expression can be increased by a similar mechanism in mouse cells (MCCABE et al. 1991), lead has no effect on Ii expression in human cells (GUO et al. 1996). This demonstrates the existence of species- or cell type-specific factors regulating MHC class II expression at a post-transcriptional level (MAFFEI et al. 1989) and yet another mechanism for modulation of Ii versus MHC class II expression. In antigen-presenting cells (APC), Ii is usually produced in excess of class II molecules (KVIST et al. 1982; NGUYEN et al. 1989), and overproduction can be further increased by activation of protein kinase C (PKC) (SHIH et al. 1995). Interestingly, PKC activation has also been demonstrated to affect proteolysis of Ii (BAROIS et al. 1997), and in transfected cell lines, inhibited Ii protein degradation in an expression level-dependent fashion (T.W. Nordeng and O. Bakke, manuscript in preparation).

2.3 Interaction of Invariant Chain with MHC Class II α- and β-Chains

A region corresponding almost exclusively to exon 3 of Ii (residues 82–107) plays a critical role in the interaction of Ii with αβ (FREISEWINKEL et al. 1993; ROMAGNOLI et al. 1994). It is now well established that the CLIP region occupies the peptide-binding groove of class II molecules, thereby promoting correct folding and inhibiting premature binding of resident peptides (ROMAGNOLI et al. 1994; BANGIA et al. 1995; GAUTAM et al. 1995; GHOSH et al. 1995; LEE et al. 1995; LIANG et al. 1995, 1996; MALCHEREK et al. 1995; MORKOWSKI et al. 1995; SETTE et al. 1995; WU et al. 1996). CLIP can be subdivided into two functional regions: the C-terminal segment (residues 92–105) occupies the peptide-binding groove, whereas the N-terminal segment (residues 81–98) binds outside the groove and seems to be important for a fast off-rate for CLIP (KROPSHOFER et al. 1995a,b). However, the exact nature of CLIP binding is clearly different for each member of the polymorphic MHC class II family, as the affinities for individual αβ dimers vary dramatically (BANGIA et al. 1995; MALCHEREK et al. 1995; SETTE et al. 1995).

A model explaining how a single region of Ii may bind to the large number of peptide-binding domains has been proposed by LEE and MCCONNELL (1995). They predicted that bound CLIP will assume the same conformation as that adopted by antigenic peptides and that CLIP may bind in a single, general way across products of class II alleles. Recent work by WEENINK et al. (1997) lends support to this theory. They propose that there is a general backbone of a periodic nature within the CLIP sequence that minimizes deleterious contacts and allows promiscuous binding to class II molecules. Thus, the CLIP property may lie in avoiding high-affinity binding rather than in fitting into a polymorphic groove.

A role for additional regions of Ii other than CLIP in maintaining the association has been reported (VOGT et al. 1995; NEWCOMB et al. 1996), but such interactions are not effective on their own in the assembly process (FREISEWINKEL et al. 1993; BIJLMAKERS et al. 1994; ROMAGNOLI et al. 1994).

2.4 Invariant Chain Association Helps MHC Class II Molecules to Exit the Endoplasmic Reticulum

Ii acts as a “chaperone” for class II molecules in their maturation process (CLAESSON-WELSH et al. 1985; LAYET et al. 1991; SCHAIFF et al. 1991; ANDERSON et al. 1992); the binding of Ii aids, but is not an absolute requirement for, the transport of class II molecules out of the ER (LAYET et al. 1991; ANDERSON et al. 1992; NIJENHUIS et al. 1994). Similarly, in the presence of MHC class II, Ii is transported more efficiently from the ER to the Golgi complex (SIMONIS et al. 1989; LAMB et al. 1991), indicating that exit from the ER is mutually facilitated by αβ-Ii assembly. Studies based on cross-linking of proteins suggest that class II and Ii molecules are released from the ER as a 3×3 subunit complex (ROCHE et al. 1991b), but the precise order of assembly is not clear. αβ-dimers either assemble stepwise to preexisting Ii trimers (ROCHE et al. 1991b), or separate α- and β-chains associate with Ii trimers one at a time to form a nonameric complex (LAMB et al. 1992; ANDERSON et al. 1994). However, the contribution of Ii to subunit assembly differs for allelic variants of αβ, suggesting that sequential associations of α, β, and Ii may be affected by polymorphic differences (BIKOFF et al. 1995). αβ expressed in cells lacking Ii transiently forms large aggregates in the ER (BONNEROT et al. 1994; MARKS et al. 1995), suggesting that assembly may be important to avoid interactions of empty MHC class II binding sites to exposed segments in the ER (ZHONG et al. 1996).

Calnexin, a ubiquitous ER phosphoprotein, associates with unassembled subunits of multimeric complexes (DEGEN et al. 1991; AHLUWALIA et al. 1992; GALVIN et al. 1992; HOCHSTENBACH et al. 1992; SCHREIBER et al. 1994) as well as viral (HAMMOND et al. 1994) and secretory monomeric (OU et al. 1993) glycoproteins. It has been shown that calnexin also associates with Ii and the α- and β-chains and remains associated with the assembling αβ-Ii complex until the final class II subunit is added to form the nonameric complex (ANDERSON et al. 1994; SCHREIBER et al. 1994). As dissociation of calnexin parallels egress of αβ-Ii from the ER, these results suggest that calnexin retains and stabilizes both free subunits and partially assembled MHC class II-Ii complexes until the nonamer is complete. However, the molecular requirements for association of αβ-Ii complexes with calnexin remain uncertain, as neither replacement of the transmembrane region of the DRβ-subunit with a guanosine phosphatidylinositol (GPI) anchor nor deglycosylation of the complex constituents abolishes the association (ARUNACHALAM et al. 1995). When expressed in the absence of Ii, class II molecules are found also to aggregate with the ER resident chaperone BiP (BONNEROT et al. 1994). Thus it is likely that αβ-Ii interaction also mediates dissociation of these molecules from other chaperones known to bind and to retain misfolded or partially folded proteins in the ER.

Although association of several peptides to MHC class II alone is not efficient in the ER (ERICSON et al. 1994), probably due to the low pH required for binding (JENSEN 1991), another function of Ii seems to be the prevention of premature peptide binding to class II molecules (BUSCH et al. 1996). The p35 and p43 forms of Ii are retained in the ER (LOTTEAU et al. 1990; LAMB et al. 1991) by a calnexin-independent mechanism (ARUNACHALAM et al. 1995), and a double-arginine motif in the prolonged amino-terminal segment of the cytoplasmic tail has been identified as the ER retention signal (SCHUTZE et al. 1994). p35 Ii inhibits presentation of endogenous antigen (DODI et al. 1994; LONG et al. 1994; VEENSTRA et al. 1996), and in vitro studies utilizing isolated αβ-Ii trimers have shown that class II molecules only bind peptide in the absence of Ii (ROCHE et al. 1990; NEWCOMB et al. 1993; ERICSON et al. 1994). Using antigen transgenic and Ii knockout mice, BODMER and colleagues (1994) found Ii to limit the diversity of endogenous peptides bound to class II molecules.

2.5 Multimerization of Invariant Chain

In the absence of association with class II molecules, human Ii forms a trimer, and an 18-kDa fragment of Ii produced by proteinase K, lacking the cytoplasmic and the transmembrane region, is also trimeric (Marks et al. 1990). The region between amino acids 163 and 183 (exon 6) is essential for trimerization of Ii (Bijlmakers et al. 1994; Gedde-Dahl et al. 1997). This carboxy-terminal luminal region is thus essential for trimerization, but, as other studies have shown (Amigorena et al. 1995; Newcomb et al. 1996), it is not required for maintaining the nonameric complex.

The exact function of Ii trimerization is currently uncertain. Oligomerization is not a prerequisite for Ii-MHC class II association as Ii mutants unable to trimerize still bind to MHC class II (Bertolino et al. 1995). The association of p33/p41Ii with p35/p43Ii in trimers and higher molecular weight aggregates contributes to the retention of the p33/p41 forms in the ER (Arunachalam et al. 1994), where they undergo degradation to smaller, amino-terminally cleaved forms (Nguyen et al. 1989; Marks et al. 1990). Ii retention in the ER could be a way of promoting correct conformation of $(\alpha\beta Ii)_3$. Another possible function of Ii trimerization could be the formation of a motif made up of adjacent chains which could be recognized by the protein-sorting machinery. Arneson and Miller (1995) have demonstrated that the Ii trimer must consist of at least two chains containing the sorting signals in order to be transported efficiently to endosomes. This observation is compatible with the reported inability of cells transfected with truncated Ii constructs lacking the region important for trimerization to present certain peptides (Bertolino et al. 1995). Coupling between trimerization and transport is further supported by the observation that Ii-induced vacuolation of endosomes is restricted to the same region (Gedde-Dahl et al. 1997).

2.6 The First Transport Steps

Once assembled, the nonameric complex is transported to the Golgi network, where it undergoes glycosylation and terminal sialylation. A small portion of Ii (2%–5%) may acquire a chondroitin sulfate moiety to a particular serine side chain (residue 201 in p33) (Sant et al. 1985a,b). This proteoglycan version of Ii can interact with CD44 on T cells and stimulates the response of these molecules (Naujokas et al. 1993). In the absence of Ii, αβ is poorly glycosylated (Elliott et al. 1994). Conversely, in the absence of αβ, Ii fails to gain resistance to endoglycosidase H (EndoH) (Pieters et al. 1993), suggesting that Ii alone poorly acquires complex-type sugars or is not transported to the endocytic pathway via the Golgi complex (Chervonsky et al. 1995). However, the Ii glycosylation pattern does not differ between the MHC class II-negative mutant lymphoblastoid cell line T2 and its MHC class II-positive parental cell line T1 (Henne et al. 1995), demonstrating that association with MHC class II is not a general prerequisite for Ii transport via the Golgi network.

3 MHC Class II in the Endocytic Pathway

3.1 Retention of MHC Class II Molecules in the Antigen-Processing Pathway

Newly synthesized class II molecules are retained in the endocytic pathway for 1–3 h before they appear at the cell surface (NEEFJES et al. 1990). In endosomes, Ii is sequentially degraded by proteases (BLUM et al. 1988; NGUYEN et al. 1989; MARKS et al. 1990; PIETERS et al. 1991; NEWCOMB et al. 1993; XU et al. 1994), an event required for proper peptide-MHC class II complex formation (ROCHE et al. 1991a; NEEFJES et al. 1992; DAIBATA et al. 1994; MARIC et al. 1994; AMIGORENA et al. 1995).

By subcellular fractionation of a human fibroblast cell line stably transfected with Ii and HLA-DR, we have shown that material endocytosed in the fluid phase is retained in early endosomes together with Ii and class II molecules (GORVEL et al. 1995). At a high level of expression, Ii has been found to accumulate in an unusual cohort of large endosomal vesicles in transfected cells (PIETERS et al. 1993; ROMAGNOLI et al. 1993; ARUNACHALAM et al. 1994; POND et al. 1995; SUGITA et al. 1995; GEDDE-DAHL et al. 1997; STANG et al. 1997), where vacuolation is proportional to the expression of level of Ii (T.W. Nordeng, S. Kjølsrud, B. Bremnes, and O. Bakke, manuscript in preparation). The rate of endocytic flow is lowered from this compartment (ROMAGNOLI et al. 1993; GORVEL et al. 1995). However, the delay is not caused by the vacuolation itself, as the transport rate of membrane-bound and fluid-phase ligands through the endocytic pathway of both M1 cells (GORVEL et al. 1995) and polarized Madin-Darby canine kidney (MDCK II) cells (T.W. Nordeng and O. Bakke, manuscript in preparation) is lowered, although these cells fail to vacuolate in response to Ii expression.

In conclusion, these findings suggest that Ii regulates the movement of both endocytosed antigen and newly synthesized MHC class II-Ii complexes in the endosomal pathway and thus contributes to the efficiency of peptide capture by MHC class II where antigen is limiting. The phenotypic alterations induced by Ii, however, have only been seen so far in transfected cells. The effect of Ii is thus a biological phenomena, but it still remains to be seen whether the same or similar mechanisms are active in "professional" APC such as B cells, macrophages, and dendritic cells. Interestingly, however, isolated Langerhans cells expressing Ii are found to contain large lucent acidic vacuoles with the characteristics of early endosomes (KÄMPGEN et al. 1991; PURE et al. 1990; STÖSSEL et al. 1990).

3.2 Processing Events in the MHC Class II Presentation Route

In general, for MHC class II presentation, antigen must be internalized by endocytosis for subsequent processing within endosomes/lysosomes. Exogenous antigens are endocytosed by APC by different mechanisms. In B lymphocytes, for instance, receptor-mediated endocytosis by surface Ig is an efficient way of specific uptake. However, fluid-phase endocytosis, micropinocytosis, or phagocytosis also allows solutes and particles to be internalized by APC. The antigen then reaches a first compartment located at the cell periphery, the early endosome (pH, approximately 6.3), from which the antigen can be recycled back to the cell surface or it may be transported to later endocytic organelles and eventually to lysosomes for ultimate degradation (pH, 4.5–5.0).

Acidic pH in the endocytic pathway is maintained to a large extent by the vacuolar H^+-ATPase. This multimeric membrane protein is located in endosomes, lysosomes, the trans-Golgi network (TGN), and clathrin-coated vesicles. Different inhibitors are known to block vacuolar acidification. These are acidotropic weak bases such as chloroquine and ammonium chloride (THORENS et al. 1986), carboxylic ionophores such as monensin and nigericin (TARTAKOFF 1983), and the macrolide antibiotics bafilomycin A_1 and concanamycin B (BOWMAN et al. 1988; YOSHIMORI et al. 1991; WOO et al. 1992). The latter two reagents inhibit the vacuolar H^+-ATPase directly at nanomolar concentrations on intact cells, and it has been demonstrated that bafilomycin A_1 prevents the formation of carrier vesicles in BHK cells (CLAGUE et al. 1994). Intracellular acidification of TGN is directly associated with the transfer of an ADP-ribosylation factor (ARF) from the cytoplasm to membrane vesicles (ZEUZEM et al. 1992). Several other studies have also emphasized the idea that acidification in the endocytic pathway contributes to protein trafficking (KLIONSKY et al. 1992; YILLA et al. 1993; CHAPMAN et al. 1994; BÉNAROCH et al. 1995; ESCOLA et al. 1996). In APCs, it has been demonstrated that low pH is not only important for antigen processing, but also for MHC class II transport steps and antigen presentation (BÉNAROCH et al. 1995; ESCOLA et al. 1996).

Antigen processing includes both proteolysis, apparently mediated by a spectrum of different proteases (VIDARD et al. 1992), and disulfide reduction (JENSEN 1991; HAMPL et al. 1992). Disulfide reduction seems to be mediated in high-density, lysosome-like compartments (COLLINS et al. 1991), and immunoelectron microscopy has shown that lysosome-like compartments in some cells contain high levels of class II molecules (PETERS et al. 1991; HARDING et al. 1992). Processing of endocytosed antigen is blocked at 18°C (HARDING et al. 1990), also suggesting involvement of late endosomes or lysosomes, as transport from early endosomes to later compartments is blocked at this temperature. Furthermore, liposome-encapsuled antigens are efficiently processed only after lysosomal targeting (HARDING et al. 1991a,b). These observations suggest a role for lysosomes in class II antigen processing.

Processing of Ii is necessary to achieve peptide loading on mature class II molecules (ROCHE et al. 1991a; DAIBATA et al. 1994). During intracellular transport of MHC class II-Ii complexes, Ii is sequentially degraded from the luminal C-terminal side, but is still associated with class II molecules (BLUM et al. 1988; NGUYEN et al. 1989; MARKS et al. 1990; PIETERS et al. 1991; NEWCOMB et al. 1993; XU et al. 1994) in a nonameric complex (AMIGORENA et al. 1995). Although it has been established that processing of Ii and exogenous antigen involves proteolysis, it has been difficult to resolve the involvement of the specific proteases (for a review, see BERG et al. 1995). Both cathepsin D and cathepsin B are necessary for Ii degradation (DAIBATA et al. 1994; MIZUOCHI et al. 1994), and it seems that aspartyl proteases (cathepsin D and E) may be involved in the early steps of Ii degradation (MARIC et al. 1994; KAGEYAMA et al. 1996), whereas cysteine proteases (cathepsin B and L) catalyze the final steps (MARIC et al. 1994). Both cysteine proteases (TAKAHASHI et al. 1989; NEEFJES et al. 1992; MIZUOCHI et al. 1994) and cathepsin D (DIMENT 1990; MIZUOCHI et al. 1994) are involved in processing of exogenous antigen. Recently, the cysteine protease cathepsin S has been demonstrated to be essential for efficient Ii proteolysis leading to the formation of sodium dodecyl sulfate (SDS)-stable complexes, i.e., $\alpha\beta$ molecules remaining as dimers in SDS-polyacrylamide gel electrophoresis (PAGE) buffers at room temperature (RIESE et al. 1996). Cathepsin S is an endopeptidase with a broad range of pH activity expressed by B lymphocytes, macrophages, and bone marrow-derived dendritic cells. This suggests that cathepsin S may be active in organelles other than lysosomes and consequently that lysosome compartments are not the sites at which Ii degradation occurs. The specific inhibition of cathepsin S activity leads to the accumulation of the CLIP fragment and prevents antigen presentation.

Ii has sequence similarity with the cystatin family of protease inhibitors and has recently been demonstrated to inhibit the enzymatic activity of cathepsin L and H, whereas cathepsin B was not inhibited (KATUNUMA et al. 1994). In addition, the p41 form of Ii is more resistant to proteolysis than the p33 form (KÄMPGEN et al. 1991; ARUNACHALAM et al. 1994), resulting in a 12-kDa fragment remaining associated with MHC class II for a prolonged period of time (FINESCHI et al. 1995). The peptide fragment encoded by the extra exon in p41Ii has been coisolated with cathepsin L in human kidneys (OGRINC et al. 1993) and found to inhibit the activity of this protease (FINESCHI et al. 1996; BEVEC et al. 1997) as well as the cathepsin L-like enzyme cruzipain (BEVEC et al. 1997). Thus another role of Ii in antigen presentation may be to modulate proteolysis in endosomes. This function of Ii may also explain Ii-mediated endosomal accumulation/retention, as altered proteolytic activity decreases the rate of endocytic flow (NEEFJES et al. 1992; ZACHGO et al. 1992).

3.3 Functions of HLA-DM

The analysis of mutant B cell lines unable to present exogenous antigen (MELLINS et al. 1990) led to the search of other genes mapping to the class II region of the MHC locus (MELLINS et al. 1991; CEMAN et al. 1992, 1994; RIBERDY et al. 1992a; MALNATI et al. 1993). The two genes HLA-DMA and HLA-DMB encode a heterodimer HLA-DM, which has been shown to play a critical regulatory role in MHC class II-restricted antigen presentation (FLING et al. 1994; MORRIS et al. 1994; SLOAN et al. 1995; Fig. 1). Cell lines lacking HLA-DM are defective in the presentation of a number of epitopes derived from intact protein antigen, but not of exogenously supplied peptides, and their class II molecules lack the characteristics of mature, peptide-loaded molecules, e.g., SDS stability and recognition by conformation-specific antibodies. The majority of class II molecules in such cells lack a wild-type repertoire of endogenous peptides; instead, they are associated with the CLIP fragment of Ii (RIBERDY et al. 1992b; SETTE et al. 1992). However, not all class II molecules require HLA-DM for proper peptide loading, and it has been suggested that HLA-DM dependency might be allele and/or species specific (STEBBINS et al. 1995).

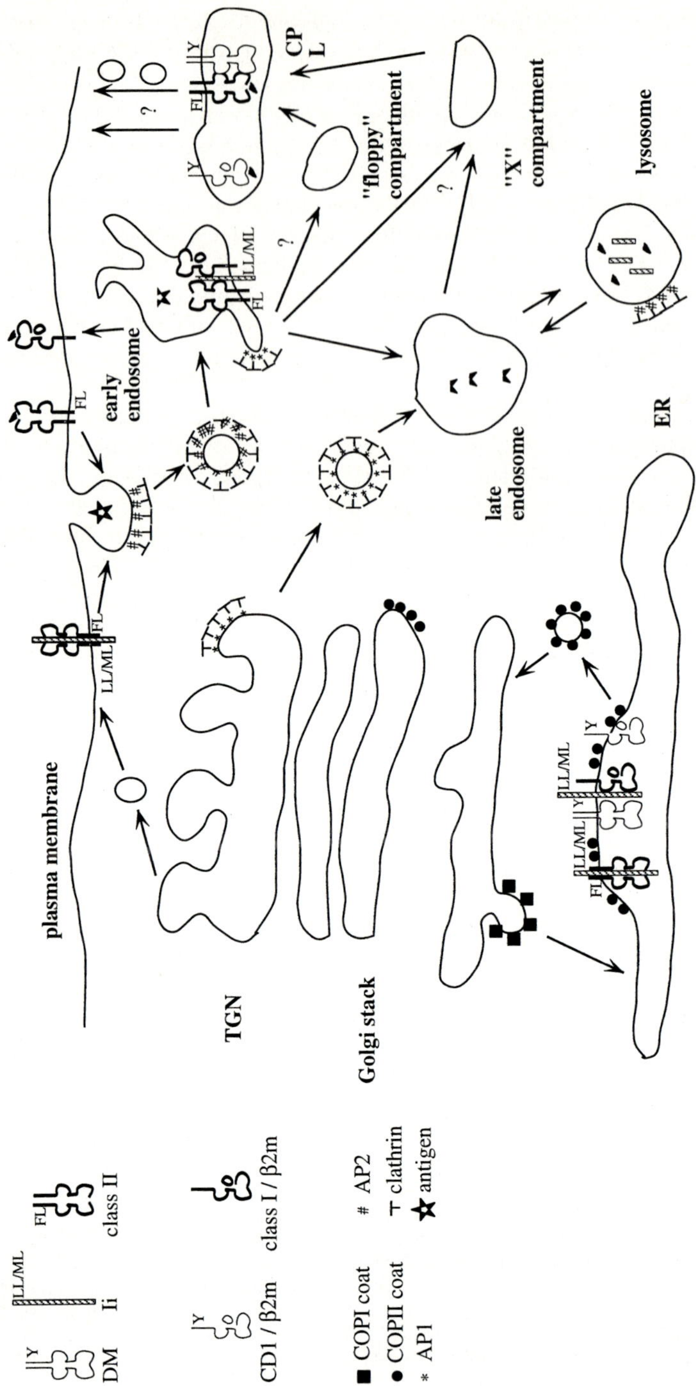
plasma membrane
early endosome
CP L
"floppy" compartment
"X" compartment
lysosome
late endosome
ER
TGN
Golgi stack
DM
Ii
class II
CD1 / β2m
class I / β2m
COPI coat
COPII coat
AP1
AP2
clathrin
antigen

In vitro-isolated HLA-DM molecules enhance peptide loading directly by accelerating the off-rate of CLIP (DENZIN et al. 1995; SHERMAN et al. 1995; SLOAN et al. 1995), suggesting that HLA-DM is necessary for active removal of CLIP in vivo. HLA-DM is expressed at a low steady state level compared to class II molecules (1:23), but in a peptide-loading compartment the ratio is 1:5 (SCHAFER et al. 1996). However, as HLA-DM is able to mediate CLIP removal in an enzyme-like fashion, one HLA-DM molecule may facilitate loading of 3–12 HLA-DR molecules per minute, depending on the HLA-DR alleles and peptides used (VOGT et al. 1996). Based on coprecipitation analysis showing that more HLA-DM precipitates from HLA-DR-CLIP complexes than from HLA-DR complexes loaded with stable peptides, it has been suggested that the latter is an unfavorable substrate for HLA-DM (DENZIN et al. 1996; KROPSHOFER et al. 1996).

Whereas the catalytic effect of HLA-DM involves a transient interaction with MHC class II, prolonged association may be necessary to prevent aggregation of empty class II molecules (DENZIN et al. 1996; KROPSHOFER et al. 1997), thus taking over the role of Ii during $\alpha\beta$ assembly and transport to a peptide-loading compartment. Data suggest that HLA-DM also physically interact with class II molecules during their functional maturation (MONJI et al. 1994), and both HLA-DM (SANDERSON et al. 1996) and its murine equivalent, H-2M (MONJI et al. 1994) have been shown to associate with Ii during synthesis. This suggests that the low pH of a peptide-loading compartment may not be a prerequisite for HLA-DM function and that HLA-DM may also play novel roles during biosynthesis and intracellular transport of $\alpha\beta$-Ii complexes.

The question then arises as to the criteria HLA-DM uses to differentiate between CLIP fragments and peptides destined for stable association with class II molecules.

Fig. 1. Transport routes for proteins involved in the presentation of exogenous antigen. Exogenous antigen are internalized by endocytosis and degraded to peptide antigen, primarily in late endocytic compartments, including phagolysosomes. Class I and CD1 molecules associate with β_2-microglobulin ($\beta 2m$) in the endoplasmic reticulum (*ER*), whereas class II molecules associate with trimeric invariant chain (*Ii*), forming a nonameric complex. These interactions are crucial for subsequent release from ER chaperones such as calnexin and BiP and subsequent transport out of the ER. DM (encoded by HLA-DM) and MHC class I molecules may associate with Ii, but the significance of these interactions remains uncertain. Transport from the ER to the Golgi apparatus is mediated by COPII-coated vesicles. In a post-ER compartment, the p35 Ii isoform may be recognized, possibly by COPI, for ER retrieval. Direct transport from the trans-Golgi network (TGN) to endosomes is mediated via clathrin/AP-1-coated vesicles, whereas possible coats on vesicles destined for the plasma membrane have not yet been characterized. However, molecules on the plasma membrane may be internalized into clathrin/AP-2-coated vesicles, the coat is stripped off, and the vesicles fuse with early endosomes. Here, proteolysis of Ii and internalized antigen is initiated and processed antigenic peptides may then bind to the groove of class II, and possibly also class I molecules. However, efficient exchange of CLIP (the Ii fragment associated with the class II groove) and peptide does not take place until MHC class II reaches the DM-containing compartment for peptide loading (CPL). The "X" compartment is thought to be an intermediate in the transport of class II/CLIP to the CPL, whereas the proposed "floppy" compartment may contain recycling class II molecules that have lost their ligand. Class II–peptide complexes may then be transported to the cell surface by direct fusion of the CPL with the plasma membrane or by some vesicle shuttle mechanism. To our knowledge, nothing is currently known about the route to the cell surface taken by class I molecules and CD1 molecules after acquisition of antigen in endocytic compartments

Recent experiments show that it is the off-rate rather than the affinity of the peptide that determines whether HLA-DM can remove it from the class II groove (KROPSHOFER et al. 1996; WEBER et al. 1996). A dynamic model for HLA-DM function was suggested by KROPSHOFER et al. (1997) whereby HLA-DM executes its function by proof-reading the kinetics of the αβ-peptide association. At low pH, HLA-DM interaction with αβ-peptide complexes will stabilize the open-state conformation of the peptide-binding groove. In this model, low-stability peptides with a higher off-rate than HLA-DM will be released, whereas high-stability peptides will remain bound, leading to HLA-DM dissociation. Thus a single class II molecule may undergo several rounds of peptide editing by HLA-DM until a high-stability peptide is captured.

The efficiency of HLA-DM-mediated editing of the peptide repertoire presented by class II molecules relies on several factors, such as the ratio of MHC class II to HLA-DM, the availability of high- versus low-stability peptides, and the pH of the reaction milieu. Other molecules may also be involved in the modulation of the peptide repertoire. Another MHC class II-like protein, HLA-DO, is expressed in some MHC class II-positive cell types, including dendritic cells and subpopulations of thymus cells (DOUEK et al. 1997). HLA-DO may associate with Ii (DOUEK et al. 1997) and HLA-DM (LILJEDAHL et al. 1996) during intracellular transport and codistributes with HLA-DR (DOUEK et al. 1997) and HLA-DM (LILJEDAHL et al. 1996) in the cell, but so far there is no proof for a role of HLA-DO in antigen presentation. In addition, signal transduction initiated by surface receptors such as immunoglobulins and possibly class II molecules themselves (for a review, see SCHOLL et al. 1994) could modulate peptide loading by several mechanisms. Interestingly, BAROIS et al. (1997) have recently shown that activation of PKC by B cell receptor ligation modulates Ii proteolysis and leads to delayed formation of SDS-stable class II molecules and reduced antigen presentation.

3.4 Intracellular Sites for Peptide Loading

Several laboratories have isolated a type of compartment which constitutes the major site for peptide loading (CPL, compartment for peptide loading) (AMIGORENA et al. 1994; QIU et al. 1994; TULP et al. 1994; WEST et al. 1994; CASTELLINO et al. 1995; ESCOLA et al. 1995). Subcellular fractionation combined with pulse-chase labeling experiments followed by immunoprecipitations demonstrated that the CPL is the first site at which newly synthesized class II molecules and processed peptides encounter one another. CPLs were characterized in B cell lines and splenic B lymphoblasts, melanoma cells and dendritic cells. Percoll gradient centrifugation (QIU et al. 1994; WEST et al. 1994; ESCOLA et al. 1995) or free-flow electrophoresis (AMIGORENA et al. 1994) were among the most powerful methods to demonstrate that the CPL do not have the same characteristics as either early or late endosomes or lysosomes, suggesting that they may be specialized compartments in APC. It has also been reported that entry to this compartment is mediated by distinct transport vesicles, the so-called X-compartment, containing SDS-labile class II molecules still associated with Ii (XU et al. 1995).

At the ultrastructural level the CPL correspond to multivesicular or multilaminar structures enriched with class II molecules and with a diameter of 0.2–0.5 μm. However, the properties of isolated CPL differed depending on fractionation procedures, cell types, and species. In one of the first published observations of endosomal vesicle labeling for MHC class II, multilamellar compartments were described and termed MIIC (MHC class II compartment; PETERS et al. 1991). In humans, MIIC appear to be more closely related to lysosomes than to early endosomes, as they also label for the lysosomal-associated membrane protein (LAMP)-1 and β-hexoaminidase, but are negative for the cation-dependent and cation-independent mannose-6-phosphate receptors which accumulate in late endosomes. However, in a recent study of Epstein-Barr virus (EBV)-transformed human B cells, we found CPL characterized by the colocalization of HLA-DM and class II molecules associated with CLIP or loaded with peptide in mannose-6-phosphate receptor-positive/prelysosomal compartments (STANG et al., submitted). SANDERSON et al. (1994) and ROBBINS et al. (1996) have also shown that MIIC contain HLA-DM molecules. The class II compartments isolated by subcellular fractionation from melanoma cells and B lymphoblastoid cells are also of the MIIC type (TULP et al. 1994; WEST et al. 1994; FERNANDEZ-BORJA et al. 1996). In contrast, in mouse A20 B cells, the corresponding class II-containing vesicle (CIIV) compartment isolated by free-flow electrophoresis by AMIGORENA et al. (1994) differs from MIIC in that it contains the transferrin receptor early endocytic marker and is devoid of lysosomal protein LAMP-1.

In a murine B cell line, ESCOLA et al. (1995, 1996) identified a lysozyme-loading compartment (LLC) with characteristics similar to that of CIIV, i.e., presence of transferrin receptor and absence of lysosomal enzymes. However, ultrastructural studies showed that the morphology of the LLC is somewhat similar to that of the MIIC. It has been proposed that both LLC and CIIV arise from early endosomes (AMIGORENA et al. 1995; ESCOLA et al. 1996). Based on the kinetics of the appearance of bovine serum albumin (BSA)-gold in the compartments, it has been suggested that the multivesicular MIIC are positioned earlier in the pathway than the multilaminar MIIC (RAPOSO et al. 1996). In the CH27 B cell lymphoma, at least two different compartments were defined based on density shift on percoll gradient and formation of "floppy" versus "compact" forms of class II molecules (QIU et al. 1994). The functional significance of these differences and the possible relationship between the various CPL has not yet been established.

MHC class II molecules accumulate in an intracellular compartment that is favorable for peptide loading and, as discussed above, many studies claim that these are unique for APC. However, as they are mainly characterized by morphology and the presence of MHC class II, it is really not possible to make a direct comparison with cells not expressing these molecules. In one study, it was concluded that expression of HLA-DR without Ii is sufficient for the formation of endocytic structures morphologically analogous to MIIC (CALAFAT et al. 1994). In another study it was found that HLA-DR, Ii, and HLA-DM were all required to reconstitute an operational CPL in non-APC (KARLSSON et al. 1994). In conclusion, CPL could be a specialized compartment unique to APC, but it is just as likely that it represents a general part of the endocytic pathway where MHC class II accumulate under certain conditions, perhaps regulated by MHC class II, Ii, and HLA-DM, which are all components of this compartment.

Class II molecules have also been localized to endosomes early in the pathway (PIETERS et al. 1991; HARDING et al. 1992), and several studies have shown that antigens can be processed and bound to MHC class II in early endosomal compartments (McCOY et al. 1993a,b; GAGLIARDI et al. 1994; AMIGORENA et al. 1995; GRIFFIN et al. 1997). Moreover, retrograde traffic between terminal lysosomes and late endosomes (JAHRAUS et al. 1994) could enable binding of terminally processed antigen without entry of class II molecules into terminal lysosomes. Based on the different characteristics of the CPL described and the different degradative requirements for antigen processing reported, we conclude that peptide loading may occur at various steps of the endocytic pathway (ESCOLA et al. 1995; GRIFFIN et al. 1997) depending on the specific degradative capacity of endosomal populations of different cells as well as the vulnerability of disparate antigens to proteolysis.

3.5 Routes Back to the Surface

Once class II molecules have been loaded with high-affinity peptides, the complexes are ready to migrate to the cell surface. Peptide loading promotes a conformational change in class II molecules, converting them to an SDS-resistant compact form (GERMAIN et al. 1991; SADEGH-NASSERI et al. 1991; WETTSTEIN et al. 1991; NEEFJES et al. 1992). The existence of floppy dimers, i.e., SDS-stable class II dimers that have lost their ligand, but retain enhanced denaturation resistance except in the binding domain itself (DORNMAIR et al. 1989; SADEGH-NASSERI et al. 1991; BIKOFF et al. 1993, 1995), argues that high-affinity peptide binding leads to some refolding of the $\alpha\beta$-dimer rather than just gluing the chains together (for a review, see GERMAIN et al. 1996). Surface receptors such as the epidermal growth factor (EGF) receptor are efficiently internalized only after ligand binding (for a review, see SCHLESSINGER et al. 1983), and it is tempting to suggest a similar mechanism for peptide-MHC class II departure from a peptide-loading compartment. Another possibility is that $\alpha\beta$-dimers liberated from Ii follow a possible default route back to the surface.

Transport of complexes to the cell surface could involve a direct fusion of CPL with the plasma membrane or vesicle-mediated membrane traffic between LLC, CIIV, or MIIC and the plasma membrane. In human lymphoblastoid B cells, multivesicular MIIC have the propensity to fuse directly with the plasma membrane and deliver internal vesicles called exosomes, able to stimulate T_H lymphocytes, into the medium. However, this direct fusion was found to be too slow to represent a major pathway for antigen presentation (RAPOSO et al. 1996). This fusion has not yet been described in murine B cells, but LLC membrane fractions prepared for electron microscopy show the presence of budding areas with internal membranes (ESCOLA et al. 1996). The significance of these buds is not clear, but they might generate vesicular compartments containing internal vesicles en route to the cell surface in a microtubule-independent manner. Upon fusion with the plasma membrane, these vesicular compartments will release their content of vesicles to the external medium.

Therefore, they could constitute a potential source of microvesicles with class II molecules exposed at their surface interacting with T lymphocytes.

Transport intermediates between CPL and the cell surface have not been reported, possibly due to a transient existence and lack of markers or other special characteristics. By stably expressing a chimeric molecule with the green fluorescence protein (GFP) attached to the cytoplasmic tail of HLA-DRβ, WUBBOLTS et al. (1996) showed that acidic HLA-DRβ-GFP-containing vesicles moved rapidly to the plasma membrane in the Mel JuSo melanoma cell line, suggesting a direct fusion of the lysosome-like compartments with the plasma membrane. In a recent electron microscopy study of EBV-transformed human B cells, we found small cytoplasmic vesicles that contained only mature class II molecules and no endocytic markers (STANG et al., submitted). These vesicles might in principle be small shuttling vesicles from the CPL to the plasma membrane. In conclusion, there is some evidence that transport from the CPL to the plasma membrane could occur via a direct fusion of multivesicular compartments with the plasma membrane, but this does not rule out the possibility that shuttling vesicles might also be used to transport mature class II molecules to the plasma membrane.

4 Presentation of Exogenous Antigens by Molecules Other Than MHC Class II

4.1 Presentation of Endocytosed Antigens by MHC Class I Molecules

It was earlier believed that a strict division of labor exists between class I and class II molecules; class I molecules bind peptides generated from endogenously synthesized cytoplasmic proteins, whereas class II molecules present peptides derived from internalized degraded material. However, recent data suggest exceptions to this focus on cytoplasmic versus endocytic origin of the antigen (for a review, see GERMAIN 1994; HARDING 1996; WATTS 1997). Indeed, class I molecules are able to present exogenous peptides on certain cell types (ROCK et al. 1990, 1993), but the mechanisms involved remain unclear. In principle, for loading of exogenous peptides on class I molecules, the antigen has to be targeted to the ER, where class I molecules usually bind peptides, or the class I molecules could be targeted to endosomes. Experimental data are now available on both mechanisms.

Studies on phagocytic cells have revealed that peptides derived from antigen-coated latex beads may be presented by class I molecules in a TAP-dependent fashion, indicating that peptide loading takes place in the ER (KOVACSOVICS-BANKOWSKI et al. 1993, 1995). Noncoupled antigens fed to macrophages together with latex beads may also be processed to class I peptides by a variety of cell lines (REIS E SOUSA and GERMAIN 1995), and antigen might obtain access to the cytoplasm by rupture of the macrovesicular membranes (NORBURY et al. 1995; REIS E SOUSA and GERMAIN 1995). A distinct pathway from phagosomes to the ER has also been suggested (PFEIFER et al. 1993), and this may be a potential source of class I peptides.

Class I molecules can associate with Ii in the ER (CERUNDOLO et al. 1992; SUGITA et al. 1995), possibly by interactions in the peptide-binding groove (VIGNA et al. 1996), and MHC class I has been found to colocalize with Ii in endosomes in HeLa cells (SUGITA et al. 1995). Thus it is possible that class I molecules directed to endosomes by Ii could acquire peptide in a way analogous to class II molecules. However, the immunological significance of MHC class I presentation of exogenous antigen is uncertain, but it could be an important mechanism in the defense against pathogens residing in endosomes.

4.2 Presentation of Nonclassical Antigens by Nonclassical Molecules

The CD1 family is composed of members which are distantly related to MHC class I and class II molecules. CD1 molecules are cell surface glycoproteins consisting of a heterodimer composed of an α-chain with a molecular weight of 43 000–49 000 associated with β_2-microglobulin (CALABI et al. 1989). The human CD1 genes are located on chromosome 1 and are not linked to the MHC locus. However, they carry an intron/exon structure similar to that of MHC class I genes and display limited, but significant homology with MHC class I and class II molecules. The four members of the human CD1 family are noncovalently associated with β_2-microglobulin, a characteristic that identified CD1 molecules as class I-related molecules. The members of the CD1 family have been identified in the human, mouse, rat, rabbit, and sheep. They are characterized by both limited polymorphism and distribution (thymus, dendritic cells, intestinal epithelium, GM-CSF-activated monocytes, and lymph nodes). Different subclasses have been identified in several mammalian species and characterized by their specific tissue distribution. For example, CD1d is mainly expressed in enterocytes and B lymphocytes, and CD1a–c in dendritic cells and thymocytes (BENDELAC 1995).

It has become clear that CD1 molecules play a unique role in antigen presentation. The first line of evidence that CD1 molecules act as presenting molecules of non-peptide antigens came from the demonstration that CD1b-restricted *Mycobacterium tuberculosis* T cell clone recognized mycolic acid, a heterogeneous branched and long-chain fatty acid mainly expressed in mycobacteria (BECKMAN et al. 1994). Two other T cell clones were shown to recognize the lipoarabinomannan (LAM) of *M. leprae*, a lipoglycan composed of a hydrophobic lipid-containing phosphatidylinositol group linked to polysaccharide complexes (SIELING et al. 1995). Recently, it was shown that LAM can also be presented in the context of CD1c (BECKMAN et al. 1996). In contrast to class Ib molecules, CD1b and CD1c presentation is TAP-independent and does not require the HLA-DM molecule (PORCELLI et al. 1992; BECKMAN et al. 1996). However, it has been shown that endosomal acidification is required for efficient presentation and that concanamycin A, an inhibitor of Na/H ATPase, blocks CD1b presentation (PORCELLI et al. 1992; SUGITA et al. 1996). These results indicate that non-peptide antigen loading and presentation by CD1b have the characteristics of class II-restricted peptide presentation. Peptides and glycolipids seem to require internalization for interactions with CD1b, which is compatible with the specific intracellular location of CD1b in several cell types (SUGITA et al. 1996).

The intracellular compartment enriched in CD1b corresponds to a prelysosomal/lysosomal compartment with intralaminar and multivesicular structures. It contains lysosomal enzymes, HLA-DM, and MHC class II molecules. This suggests that CD1b molecules are able to reach the peptide-loading compartment. This unique endocytic property of CD1b to be endocytosed and delivered to MIIC compartment is due the presence of the YQNI internalization motif in the CD1b cytoplasmic tail of the molecule (SUGITA et al. 1996; Table 1). These data suggest that glycolipid-CD1b association might occur in CPL and that molecule receptors might exist that target glycolipids to CPL. Conversely, mouse CD1d, mainly expressed in enterocytes, can bind to hydrophobic peptides with features different from peptides bound to MHC class I and class II (CASTANO et al. 1995). In particular, TAP expression is not required.

The T cell-recognizing antigens in the context of CD1 presentation include $CD4^-$,$CD8^-$ $\alpha\beta$ and $\gamma\delta$ T cells, $CD8^+$ $\alpha\beta$ T cells (PORCELLI et al. 1992; CASTANO et al. 1995; SIELING et al. 1995), $CD4^-$,$NK1^-$ T cells, and $CD4^+$,$NK1^+$ T cells (BENDELAC 1995). In addition, many of the CD1-restricted cell lines seem to be autoreactive in that they recognize $CD1^+$ cells without foreign antigens (PORCELLI 1995). The non-peptide antigens identified so far are presented by nonpolymorphic antigen-presenting molecules to a subset of T lymphocytes probably recruited upon infection. These molecules should provide new tools in vaccine development, since they should be recognized in all animal or human targets.

5 Signals for Sorting to the Peptide-Loading Compartments

5.1 General Comments

The cytoplasmic tails of membrane proteins contain the information for sorting to various intracellular destinations. Two main classes of endosomal sorting signals have been identified, one class characterized by an essential tyrosine, e.g., in HLA-DM, and one by a leucine-based motif, e.g., in Ii and the tail of the class II β-chain (Table 1; for a review, see SANDOVAL and BAKKE 1994). There is increasing evidence that both classes of sorting signals interact directly with the adaptor complexes AP-1 and AP-2 for clathrin-dependent sorting from the TGN or the plasma membrane, respectively (GLICKMAN et al. 1989; BOLL et al. 1995; HEILKER et al. 1996). Studies on adaptor complexes binding to tail columns indicate a multivalent attachment of aggregated adaptor complexes (BOLL et al. 1996; MARKS et al. 1997), suggesting a model in which binding to the signals is associated with clustering of the adaptors. In fact, adaptor complexes tend to drive the formation of coats in vitro (BECK et al. 1991), and a requirement for AP clustering is in agreement with the current model in which adaptors are first recruited to the membrane by a docking protein and then associate to form a coat to which the membrane proteins diffuse laterally (for a review, see PEARCE and ROBINSON 1990).

Table 1. Sorting signals of proteins to the endosomal/lysosomal route

Type of signal	Membrane protein	Sorting signal	Destination
Tyrosine	Transferrin receptor	17aa-LS**Y**TR**F**-45aa-tm	Early endosomes/PM
	CI mannose-6-P receptor	tm-24aa-YK**Y**SK**V**-135aa	TGN/late endosomes
	LAMP-1	tm-RKRSHA**GY**QT**I**	Lysosomes
	LAMP-2	tm-KHHHA**GY**EQ**F**	Lysosomes
	CD63	tm-KSIRS**GY**EV**M**	Lysosomes
	CD3γ	tm-15aa-QL**Y**QP**L**-19aa	Lysosomes
	HLA-DMβ	tm-WRRAGHSS**Y**TP**L**P-12aa	CPL
	CD1b	tm-RRRS**Y**QN**I**P	CPL/PM
Leucine	CI mannose-6-P receptor	tm-151aa-DDSDF**LL**HV	TGN/late endosomes
	Invariant chain	M**DDQ**RD**LI**S-22aa-tm-	Endosomes
		9aa-NN**E**QL**PML**-13aa-tm	Endosomes
	Limp II	tm-GQGSTDERAP**LI**RT	Lysosomes
	HLA-DR1 β-chain	tm-RNQKGHSGLQPTG**FL**S	CPL/PM

Sequences are shown for tyrosine- and leucine-based sorting signals that have been demonstrated to be involved in targeting to endosomal compartments, including the compartment for peptide loading. The amino acids critical for sorting are shown in bold. The sorting motifs were obtained from references in the text and databases and are examples that may differ somewhat between various species.
aa, amino acids; *TGN*, trans-Golgi network; *tm*, transmembrane region; *CPL*, compartment for peptide loading; *PM*, plasma membrane; *LAMP*, lysosomal-associated membrane protein; *CI*, cation independent.

A specific interacting subunit of the AP complexes was recently demonstrated for the tyrosine-based signals, as the YXXØ motifs (in a single-letter amino acid code, Y represents tyrosine, X is any amino acid, and Ø represents a bulky, hydrophobic amino acid). are shown to interact with the medium-chain subunits of both AP-1 and AP-2 (Ohno et al. 1995; Rapoport et al. 1997). No such interaction was demonstrated for the leucine class signals in their assays, although there are data indicating that these also mediate binding to the AP complexes (Heilker et al. 1996; Dietrich and Geisler, submitted) and can in particular also bind to the medium chains (Bremnes et al., submitted). Marks et al. (1996) have shown that proteins containing tyrosine-based sorting signals compete for components of the sorting machinery. Using an inducible expression system, we have found a protein (Ii) containing a leucine-based signal to compete with a protein containing a tyrosine-based signal (transferrin receptor) for internalization from the plasma membrane (T.W. Nordeng and O. Bakke, manuscript in preparation), indicating that both types of signals can interact with common components of the sorting machinery. However, others have failed to demonstrate competition between leucine- and tyrosine-based signals (Marks et al. 1996) and even between tyrosine-based signals (Warren et al. 1997), indicating that such interactions also can depend on sequences surrounding the actual signal. Furthermore, phosphatidylinositol 3-phosphate, a product of phosphoinositide 3-kinases (for a review, see De Camilli et al. 1996), is found to bind to AP-2 (Gaidarov et al. 1996) and to actually increase their affinity for tyrosine-based signals (Rapoport et al. 1997), suggesting that the sorting machinery is regulated by signal transduction.

The sorting of the immune molecules to the peptide loading compartment is most likely directed by the various adaptor complexes, i.e., AP-1 in sorting from the Golgi complex and AP-2 in sorting from the plasma membrane. Adaptor complexes have recently also been identified on endocytic membranes – AP-1 on early endosomes (LE BORGNE et al. 1996) and AP-2 on lysosomes (TRAUB et al. 1996) and, under special conditions, also on early endosomes (SEAMAN et al. 1993) – and these may be involved in similar sorting mechanisms to direct the molecules at sorting stations in endosomes and lysosomes and possibly back from the peptide-loading compartment to the plasma membrane (Fig. 1; for a review, see KREIS et al. 1995; ROBINSON 1997). Based on sequence similarity with the characterized adaptors, there are candidate molecular machineries for other sorting stations, such as AP-3 (SIMPSON et al. 1996; DELL'ANGELICA et al. 1997) and a novel clathrin coat identified on early endosomes (STOORVOGEL et al. 1996), and more are likely to appear in the near future. In addition to the adaptors and their interacting molecules, targeting ligands on the transport vesicles and corresponding receptors on the target membrane are necessary similar to other transport steps involving vesicular transport (for a review, see ROBINSON 1997; SCHIMMÖLLER et al. 1997). Finally, the transport has to be powered by motors interacting with the cytoskeleton, the best characterized of which are dynein and kinesin (for a review, see GOODSON et al. 1997).

Altogether, the details of how membrane molecules are recognized and travel within the endosomal pathway still largely remain unclear. The area that has been most thoroughly elucidated over the last few years comprises the actual sorting signals encoded in the cytoplasmic domains of the molecules, whereas the rest of the machinery and how it works is still largely unknown. Table 1 lists sorting signals of molecules that are sorted to the endosomal/lysosomal pathway and found to colocalize with the immune molecules at different intracellular locations.

5.2 Sorting to Endosomes

αβ-Ii nonamers leave the ER in coatamer-coated vesicles and are transported through the Golgi complex, possibly also by COPII-coated transport vesicles (Fig. 1). In a post-ER compartment, a double-arginine motif in the prolonged tail of the p35 Ii isoform is recognized for ER retrieval (SCHUTZE et al. 1994), possibly by COPI in a mechanism resembling the recognition of double-lysine motifs in type I molecules (COSSON et al. 1994; LETOURNEUR et al. 1994). The intracellular route used by MHC class II-Ii to the endosomal compartments is still under debate. The MHC class II-Ii complex could either be sorted directly from the TGN to an endosomal compartment (NEEFJES et al. 1990; PETERS et al. 1991; ODORIZZI et al. 1994; BÉNAROCH et al. 1995; WARMERDAM et al. 1996) or indirectly via the plasma membrane (ROCHE et al. 1993; BREMNES et al. 1994; ODORIZZI et al. 1994; HENNE et al. 1995; Fig. 1). Transport via the plasma membrane/early endosomes is estimated to be rapid, and plasma membrane residence may be less than 1 min (ROCHE et al. 1993; BREMNES et al. 1994); the time occupied in this part of the pathway is therefore short compared to the biosynthetic pathway, which is in the range of 1–3 h for class II molecules to appear on the plasma membrane (NEEFJES et al. 1990). In a recent study, SALAMERO et al. (1996)

found that αβ-Ii expression recruited AP-1 adaptor complexes to Golgi membranes. This recruitment was dependent on the Ii cytoplasmic tail, again suggesting that AP-1 interacts with leucine signals.

On the basis of all these results, we conclude that MHC class II-Ii complexes may reach endosomes by dual pathways, one direct to endosomes and the other via the plasma membrane. Similar, dual endosomal routes have also been reported for other membrane proteins, including the LAMP-1 (CARLSSON et al. 1992), the lysosomal membrane glycoprotein lgp120 (lgp-A) (HARTER et al. 1992), and the mannose-6-phosphate/insulin-like growth factor II receptor (JOHNSON et al. 1992).

Why should there be more than one pathway to the endosomes and peptide-loading compartment? Cointernalization of antigens and MHC class II-Ii complexes in the same endocytic vesicle could obviously ensure that MHC class II molecules and antigenic peptides meet at the correct processing stage of easily degradable antigens. Indeed, antigen processing and peptide association to MHC class II have been demonstrated in early endosomes (MCCOY et al. 1993a,b; GAGLIARDI et al. 1994). On the other hand, direct sorting of MHC class II-Ii complexes to a late endosomal compartment could prevent occupancy of all available class II molecules by peptides from easily degradable antigens. Alternative entry levels of MHC class II-Ii complexes to the endocytic pathway may thus promote the presentation of a broader spectrum of antigenic peptides, regardless of the vulnerability of the endocytosed antigen to protease activity. The multiple sorting signals in the Ii cytoplasmic tail and MHC class II might be involved in the fine-tuning of the intracellular transport of the MHC class II-Ii complex, e.g.,sorting between endosomal populations or retention in a particular endosomal maturation stage.

5.3 Sorting Signals Contained in the Cytoplasmic Tail of Invariant Chain

As discussed below, there are several sorting signals within the MHC class II-Ii complex, and all of these could contribute to the routing and the final destination of the molecules. Ii alone is efficiently sorted to endosomes if released from the ER (BAKKE et al. 1990; LOTTEAU et al. 1990), and the cytoplasmic tail of Ii has two leucine-based motifs; leucine-isoleucine (LI) in positions 7 and 8 and methionine-leucine (ML) in positions 16 and 17. These signals are independently sufficient for endosomal localization of Ii and for rapid internalization of Ii from the plasma membrane ($t_{/2}$, 1 min) (BREMNES et al. 1994). Two-dimensional magnetic resonance imaging (MRI) studies on a peptide corresponding to the cytoplasmic tail of Ii showed that the LI motif is located within a regular α-helix (MOTTA et al. 1995). This prediction was supported by biological data showing that residues neighboring on LI in the helical secondary structure could abolish internalization, whereas residues opposite LI were mutated without influencing sorting, making this combined motif a putative signal patch (MOTTA et al. 1995). Amino-terminal acidic residues were essential for a functional leucine signal, a feature which is also found in similar sorting motifs (POND et al. 1995). Data from ARNESON and MILLER (1995) show furthermore that multimers of the Ii cytoplasmic tail may be required for efficient endosomal sorting of MHC class II-Ii complexes. In a recent MRI study, the cytoplasmic tail of Ii was seen to be able to form trimers in solution (MOTTA et al. 1997), indicating that the cytoplasmic tail also has an intrinsic property to interact. Interestingly, these trimers were not parallel, but arranged in an up-down-up orientation, which might indicate that different trimers can interact with each other. The full cytoplasmic tail of Ii is also involved in endosomal sorting, resulting in retention of endocytosed material and an enlargement of the endosomes (PIETERS et al. 1993; ROMAGNOLI et al. 1993; STANG et al. 1997). This special phenotype is not seen when part of the tail is removed or the residues are mutated (PIETERS et al. 1993; ROMAGNOLI et al. 1993; POND et al. 1995). From the above data, one may conclude that properties other than LI and ML signals also influence Ii trafficking, but the mechanisms remain to be elucidated.

5.4 Sorting Signal(s) in the Cytoplasmic Tail of MHC Class II Molecules

Although it was initially reported that the major sorting information for endocytic transport of αβ-Ii complexes resided within Ii, it has become clear that class II molecules also contain endosomal sorting information (SALAMERO et al. 1990; HUMBERT et al. 1993; SIMONSEN et al. 1993). Class II molecules can be internalized and recycled (REID et al. 1990), and PINET et al. (1995) have shown that the cytoplasmic tails of class II heterodimers are essential for this routing of MHC class II. Recently, ZHONG et al. (1997) reported that the mouse LL sequence in the C terminus of the β-chain of A^K is a sorting signal. We have confirmed this for the corresponding FL motif in HLA-DR1 (A. Simonsen and O. Bakke, unpublished). Both the α- and the β-cytoplasmic tails are necessary for efficient internalization from the plasma membrane (ZHONG et al. 1997) and, as seen below, these signals are also involved in the polarized sorting of class II.

Cells expressing class II molecules in the absence of Ii or together with a truncated Ii lacking the endosomal sorting signals have been shown to present certain antigens efficiently (ANDERSON et al. 1993; NIJENHUIS et al. 1994; PINET et al. 1994; ZHONG et al. 1997). Ii-independent antigen presentation has been demonstrated to be either dependent (HARDING et al. 1989; KAKIUCHI et al. 1990, 1991; ST.-PIERRE et al. 1990; PINET et al. 1994; SWIER et al. 1995; ZHONG et al. 1997) or independent (HARDING et al. 1989; KAKIUCHI et al. 1990; NIJENHUIS et al. 1994; PINET et al. 1994, 1995; ZHONG et al. 1997) of newly synthesized class II molecules, suggesting that recycling class II molecules can reach endosomes containing degraded antigens. The function of MHC class II recycling has been proposed to be the loading of easily degradable antigenic peptides within an early endosomal compartment (PINET et al. 1995). The route taken by such recycling molecules back to the cell surface could possibly involve a transport vehicle resembling the floppy compartment (QIU et al. 1994; Fig. 1). Furthermore, in light of the observation that truncation of the class II β-chain tail results in a decreased fraction of CLIP-containing class II molecules on the surface (SMILEY et al. 1996), it is possible that the sorting signals within class II tails are also involved in the sorting within endosomes.

5.5 A Tyrosine-Based Sorting Signal in the HLA-DM Cytoplasmic Tail

HLA-DM is sorted to the same intracellular location as MHC class II, but is largely absent from the cell surface(KARLSSON et al. 1994; SANDERSON et al. 1994). The cytoplasmic tail of the β-chain contains a typical tyrosine-based signal, and three groups have confirmed that the sorting motif was indeed YTPL (LINDSTEDT et al. 1995; COPIER et al. 1996; MARKS et al. 1996). Interestingly, LINDSTEDT et al. (1995) found that Ii could associate with the mouse equivalent, H2-M, and that the sorting of Ii could replace a deleted H2-M signal. The fact that HLA-DM, like most known lysosomal membrane residents, contains a tyrosine signal, whereas MHC class II and Ii contain leucine signals, indicates that both groups of signals can sort molecules to intracellular CPL.

5.6 Basolateral Sorting Signals Within MHC Class II and Invariant Chain

The plasma membrane of polarized epithelial cells is divided by tight junctions into apical and basolateral domains. Class II molecules have been found at the basolateral surface and in intracellular vesicles in tissue epithelial cells of humans and rodents (HART et al. 1981; SELBY et al. 1981; SKOSKIEWICZ et al. 1985; BLAND et al. 1986; SARLES et al. 1987; VIDAL et al. 1995). The basolateral surface faces the vascular space, where the class II molecules may encounter and present antigen to emigrating T cells. Staining of the apical brush border surface, in addition to basolateral and intracellular staining, has been observed in epithelial cells of the human intestine (SCOTT et al. 1980; HIRATA et al. 1986). Recently, we have studied the sorting of MHC class II and Ii in polarized MDCK II cells (SIMONSEN et al. 1997). Leucine-based sorting motifs were found to be individually sufficient for sorting of Ii to endosomes, and at least fractions of the molecules are transported via the basolateral plasma membrane; Ii is also required for efficient targeting of the class II molecules to the basolateral surface. In addition, a separate novel basolateral signal is located within the ten membrane-proximal residues of the Ii cytoplasmic tail (SIMONSEN et al. 1997). MHC class II is routed to the basolateral domain after the release from Ii; these molecules must therefore also contain basolateral sorting information. In particular, the cytoplasmic tails of the class II molecules were found to be essential for this transport (A. Simonsen et al., unpublished).

The biological significance of several distinct basolateral sorting signals located within the Ii cytoplasmic tail and the class II molecules might relate to an essential function of epithelial cells and other polarized tissue in performing MHC class II-dependent antigen presentation. The various signals might also be involved in fine-tuning of the intracellular transport of the MHC class II-Ii complex, such as sorting between endosomal populations, e.g., to the compartment for peptide loading. As cognate apical and basolateral transport pathways exist in nonpolarized cells (MUSCH et al. 1996; YOSHIMORI et al. 1996), the redundancy of sorting signals may be generally applicable to other cell types. Whether the complexity of basolateral sorting of the MHC class II-Ii complex reflects its immunological importance in polarized cells is still an open question.

6 Concluding Remarks on the Many Roles of Invariant Chain

Although several reports have shown that MHC class II antigen presentation may also function independently of Ii (for a review, see HÄMMERLING and MORENO 1990), data from Ii knockout mice (BIKOFF et al. 1993; VIVILLE et al. 1993; BODMER et al. 1994; ELLIOTT et al. 1994; WONG et al. 1996) and cell lines (STOCKINGER et al. 1989; PETERSON et al. 1990; BERTOLINO et al. 1991; NADIMI et al. 1991; HUMBERT et al. 1993) show that this accessory molecule is indeed essential for class II antigen presentation. The past years of research have provided information about how motifs encoded by all seven exons of Ii contribute to its many functions (for a review, see GERMAIN et al. 1996). Interestingly, evidence is now accumulating for distinct functional roles for the different Ii isoforms. The p41 form of Ii, which can be found in large quantities in dendritic cells and Langerhans cells (KÄMPGEN et al. 1991), may be a more potent stimulator of antigen presentation than the p33 form under certain conditions (PETERSON et al. 1992), possibly by modulating the proteolytic machinery (FINESCHI et al. 1996; BEVEC et al. 1997). However, others have failed to show any effect of p41 Ii in cell lines (STOCKINGER et al. 1989; SERWE et al. 1997) and in p41 knockout mice (TAKAESU et al. 1995, 1997), and the exact role of this Ii isoform is thus still unclear.

Another point of regulation are the two start codons within the human Ii gene; use of the first, and weakest, AUG gives rise to the p35 form, which is found overexpressed in certain cancer cells (VEENSTRA et al. 1993, 1996). p35 Ii is effectively retained in the ER, but those molecules that escape are preferentially transported to endosomes via the plasma membrane, whereas the p33 form takes a more direct route from the TGN to the peptide-loading compartment (WARMERDAM et al. 1996). The effect of this property of p35 Ii has not yet been tested in the context of MHC class II presentation, but it could be speculated that cointernalization of class II molecules and antigen in early endosomes could be beneficial in cases in which the antigen is very sensitive to proteolysis. Table 2 summarizes the reported effects of Ii expressed as different isoforms and at various levels.

In this review, we have described the intracellular transport of molecules involved in the presentation of exogenous antigen mediated by class I and class II molecules and members of the CD1 family. Ii associates with both class I and class II molecules as well as HLA-DO and HLA-DM and is shown to play a central role in antigen presentation by the MHC molecules. In addition, the possibility cannot be ruled out that Ii also influences presentation by CD1 molecules through its interaction with the proteolytic machinery in the endocytic pathway. In addition to its roles in modulating intracellular events related to antigen processing and capture, Ii can also stimulate T_H cells directly by interactions with CD44. It has also recently been demonstrated that Ii is required for B cell maturation and function by an MHC class II-independent

Table 2. Effects of invariant chain (Ii) regulation

	Expression level		Ii isoforms		Other
Phenotype	Low expression, p33	High expression, p33	Alternative AUG (p35)	Alternative splicing (p41)	Glycosylation
Transport	Class II retained in ER	Endosomal retention of class II and ag	ER retention Via the plasma membrane	As p33	n.d.
Proteolysis	Normal	Reduced	Reduced (due to ER retention)	Reduced (inhibits cathepsins and has relatively long half-life)	n.d.
Antigen presentation	Limited repertoire of peptides presented	Enhanced/reduced (depends on the antigen)	Decreased presentation of endogenous peptides	Stimulates	Enhanced by chondroitin sulfate

Currently known effects of expressing different isotypes of invariant chain and expression of invariant chain at different levels. The p35 isoform is a result of initiation of transcription at an upstream start codon of the human invariant chain (Ii) gene, whereas the p41 form arises due to alternative splicing of an additional exon. *n.d.*, not determined; *ag*, antigen; *ER*, endoplasmic reticulum.

mechanism (SHACHAR et al. 1996). Thus, in the future, Ii may turn out to be an important molecule in other processes as well as antigen presentation.

Acknowledgments. We would like to thank Dmitrii Rodionov, Espen Stang, Harald Kropshofer, Anne Vogt, and Norbert Koch for critical reading of this manuscript. We would also like to thank the Norwegian Research Council, Novo Nordisk Foundation, and the Anders Jahre Foundation for financial support and the Norwegian Cancer Society for the fellowship to T.W. Nordeng. This work was supported by EC contract no ERBFMRXCT960069.

References

Accolla RS, Carra G, Guardiola J (1985) Reactivation by a trans-acting factor of human major histocompatibility complex Ia gene expression in interspecies hybrids between an Ia-negative human B-cell variant and an Ia-positive mouse B-cell lymphoma. Proc Natl Acad Sci USA 82:5145–5149

Ahluwalia N, Bergeron JJ, Wada I, Degen E, Williams DB (1992) The p88 molecular chaperone is identical to the endoplasmic reticulum membrane protein, calnexin. J Biol Chem 267:10914–10918

Amigorena S, Drake JR, Webster P, Mellman I (1994) Transient accumulation of new class II MHC molecules in a novel endocytic compartment in B lymphocytes. Nature 369:113–120

Amigorena S, Webster P, Drake J, Newcomb J, Cresswell P, Mellman I (1995) Invariant chain cleavage and peptide loading in major histocompatibility complex class II vesicles. J Exp Med 181:1729–1741

Anderson MS, Miller J (1992) Invariant chain can function as a chaperone protein for class II major histocompatibility complex molecules. Proc Natl Acad Sci USA 89:2282–2286

Anderson KS, Cresswell P (1994) A role for calnexin (IP90) in the assembly of class II MHC molecules. EMBO J 13 (3):675–682

Anderson MS, Swier K, Arneson L, Miller J (1993) Enhanced antigen presentation in the absence of the invariant chain endosomal localization signal. J Exp Med 178:1959–1969

Arneson LS, Miller J (1995) Efficient endosomal localization of major histocompatibility complex class II-invariant chain complexes requires multimerization of the invariant chain targeting sequence. J Cell Biol 129:1217–1228

Arunachalam B, Cresswell P (1995) Molecular requirements for the interaction of class II major histocompatibility complex molecules and invariant chain with calnexin. J Biol Chem 270:2784–2790

Arunachalam B, Lamb CA, Cresswell P (1994) Transport properties of free and MHC class II-associated oligomers containing different isoforms of human invariant chain. Int Immunol 6:439–451

Bakke O, Dobberstein B (1990) MHC class II-associated invariant chain contains a sorting signal for endosomal compartments. Cell 63:707–716

Bangia N, Watts TH (1995) Evidence for invariant chain 85–101 (CLIP) binding in the antigen binding site of MHC class II molecules. Int Immunol 7:1585–1591

Barois N, Forquet F, Davoust J (1997) Selective modulation of the major histocompatibility complex class II antigen presentation pathway following B cell receptor ligation and protein kinase C activation. J Biol Chem 272:3641–3647

Beck KA, Keen JH (1991) Self-association of the plasma membrane-associated clathrin assembly protein AP-2. J Biol Chem 266:4437–4441

Beckman EM, Porcelli SA, Morita CT, Behar SM, Furlong ST, Brenner MB (1994) Recognition of a lipid antigen by CD1-restricted alpha beta+ T cells. Nature 372:691–694

Beckman EM, Melian A, Behar SM, Sieling PA, Chatterjee D, Furlong ST, Matsumoto R, Rosat JP, Modlin RL, Porcelli SA (1996) CD1c restricts responses of mycobacteria-specific T cells. Evidence for antigen presentation by a second member of the human CD1 family. J Immunol 157:2795–2803

Bénaroch P, Yilla M, Raposo G, Ito K, Miwa K, Geuze HJ, Ploegh HL (1995) How MHC class II molecules reach the endocytic pathway. EMBO J 14:37–49

Bendelac A (1995) CD1: presenting unusual antigens to unusual T lymphocytes. Science 269:185–186

Berg T, Gjoen T, Bakke O (1995) Physiological functions of endosomal proteolysis. Biochem J 307:313–326

Bertolino P, Forquet F, Pont S, Koch N, Gerlier D, Rabourdin-Combe C (1991) Correlation between invariant chain expression level and capability to present antigen to MHC class II-restricted T cells. Int Immunol 3:435–443

Bertolino P, Staschewski M, Trescol-Biémont M-C, Freisewinkel IM, Schenck K, Chrétien I, Forquet F, Gerlier D, Rabourdin-Combe C, Koch N (1995) Deletion of a C-terminal sequence of the class II-associated invariant chain abrogates invariant chains oligomer formation and class II antigen presentation. J Immunol 154:5620–5629

Bevec T, Stoka V, Pungercic G, Cazzulo JJ, Turk V (1997) A fragment of the major histocompatibility complex class II-associated p41 invariant chain inhibits cruzipain, the major cysteine proteinase from Trypanosoma cruzi. FEBS Lett 401:259–261

Bijlmakers ME, Benaroch P, Ploegh HL (1994) Mapping functional regions in the lumenal domain of the class II-associated invariant chain. J Exp Med 180:623–629

Bikoff EK, Huang LY, Episkopou V, van Meerwijk J, Germain RN, Robertson EJ (1993) Defective major histocompatibility complex class II assembly, transport, peptide acquisition, and CD4+ T cell selection in mice lacking invariant chain expression. J Exp Med 177:1699–1712

Bikoff EK, Germain RN, Robertson EJ (1995) Allelic differences affecting invariant chain dependency of MHC class II subunit assembly. Immunity 2:301–310

Bland PW, Warren LG (1986) Antigen presentation by epithelial cells of the rat small intestine. II. Selective induction of suppressor T cells. Immunology 58:9–14

Blum JS, Cresswell P (1988) Role for intracellular proteases in the processing and transport of class II HLA antigens. Proc Natl Acad Sci USA 85:3975–3979

Bodmer H, Viville S, Benoist C, Mathis D (1994) Diversity of endogenous epitopes bound to MHC class II molecules limited by invariant chain. Science 263:1284–1286

Boll W, Gallusser A, Kirchhausen T (1995) Role of the regulatory domain of the EGF-receptor cytoplasmic tail in selective binding of the clathrin-associated complex AP-2. Curr Biol 5:1168–1178

Boll W, Ohno H, Zhou SY, Rapoport I, Cantley LC, Bonifacino JS, Kirchhausen T (1996) Sequence requirements for the recognition of tyrosine-based endocytic signals by clathrin AP-2 complexes. EMBO J 15:5789–5795

Bonnerot C, Marks MS, Cosson P, Robertson EJ, Bikoff EK, Germain RN, Bonifacino JS (1994) Association with BiP and aggregation of class II MHC molecules synthesized in the absence of invariant chain. EMBO J 13:934–944

Bowman EJ, Siebers A, Altendorf K (1988) Bafilomycins: a class of inhibitors of membrane ATPases from microorganisms, animal cells, and plant cells. Proc Natl Acad Sci USA 85:7972–7976

Bremnes B, Madsen T, Gedde-Dahl M, Bakke O (1994) An LI and ML motif in the cytoplasmic tail of the MHC-associated invariant chain mediate rapid internalization. J Cell Sci 107:2021–2032

Bremmes T, Lauvrak V, Lindqvist B, Bakke O (1997) A region from the medium adaptor subunit (μ) recognises leucine- and tyrosine based sorting signals (submitted).

Brown AM, Barr CL, Ting JP (1991) Sequences homologous to class II MHC W, X, and Y elements mediate constitutive and IFN-gamma-induced expression of human class II-associated invariant chain gene. J Immunol 146:3183–3189

Brown AM, Wright KL, Ting JP-Y (1993) Human major histocompatibility complex class II-associated invariant chain gene promoter. Functional analysis and in vivo protein/DNA interactions of constitutive and IFN-gamma-induced expression. J Biol Chem 268:26328–26333

Busch R, Cloutier I, Sékaly RP, Hämmerling GJ (1996) Invariant chain protects class II histocompatibility antigens from binding intact polypeptides in the endoplasmic reticulum. EMBO J 15:418–428

Buus S (1994) MHC restricted antigen presentation and T cell recognition. Dan Med Bull 41:345–358

Calabi F, Belt KT, Yu CY, Bradbury A, Mandy WJ, Milstein C (1989) The rabbit CD1 and the evolutionary conservation of the CD1 gene family. Immunogenetics 30:370–377

Calafat J, Nijenhuis M, Janssen H, Tulp A, Dusseljee S, Wubbolts R, Neefjes J (1994) Major histocompatibility complex class II molecules induce the formation of endocytic MIIC-like structures. J Cell Biol 126:967–977

Carlsson SR, Fukuda M (1992) The lysosomal membrane glycoprotein lamp-1 is transported to lysosomes by two alternative pathways. Arch Biochem Biophys 296:630–639

Castano AR, Tangri S, Miller JE, Holcombe HR, Jackson MR, Huse WD, Kronenberg M, Peterson PA (1995) Peptide binding and presentation by mouse CD1. Science 269:223–226

Castellino F, Germain RN (1995) Extensive trafficking of MHC class II-invariant chain complexes in the endocytic pathway and appearance of peptide-loaded class II in multiple compartments. Immunity 2:73–88

Ceman S, Rudersdorf R, Long EO, Demars R (1992) MHC class II deletion mutant expresses normal levels of transgene encoded class II molecules that have abnormal conformation and impaired antigen presentation ability. J Immunol 149:754–761

Ceman S, Petersen JW, Pinet V, Demars R (1994) Gene required for normal MHC class II expression and function is localized to approximately 45 kb of DNA in the class II region of the MHC. J Immunol 152:2865–2873

Cerundolo V, Elliott T, Elvin J, Bastin J, Townsend A (1992) Association of the human invariant chain with H-2 Db class I molecules. Eur J Immunol 22:2243–2248

Chapman RE, Munro S (1994) Retrieval of TGN proteins from the cell surface requires endosomal acidification. EMBO J 13:2305–2312

Chervonsky A, Sant AJ (1995) In the absence of major histocompatibility complex class II molecules, invariant chain is translocated to late endocytic compartments by autophagy. Eur J Immunol 25:911–918

Claesson-Welsh L, Peterson PA (1985) Implications of the invariant gamma-chain on the intracellular transport of class II histocompatibility antigens. J Immunol 135:3551–3557

Clague MJ, Urbe S, Aniento F, Gruenberg J (1994) Vacuolar ATPase activity is required for endosomal carrier vesicle formation. J Biol Chem 269:21–24

Collins DS, Unanue ER, Harding CV (1991) Reduction of disulfide bonds within lysosomes is a key step in antigen processing. J Immunol 147:4054–4059

Collins T, Korman AJ, Wake CT, Boss JM, Kappes DJ, Fiers W, Ault KA, Gimbrone MA, Jr, Strominger JL, Pober JS (1984) Immune interferon activates multiple class II major histocompatibility complex genes and the associated invariant chain gene in human endothelial cells and dermal fibroblasts. Proc Natl Acad Sci USA 81:4917–4921

Copier J, Kleijmeer MJ, Ponnambalam S, Oorschot V, Potter P, Trowsdale J, Kelly A (1996) Targeting signal and subcellular compartments involved in the intracellular trafficking of HLA-DMB. J Immunol 157:1017–1027

Cosson P, Bonifacino JS (1992) Role of transmembrane domain interactions in the assembly of class II MHC molecules. Science 258:659–662

Cosson P, Letourneur F (1994) Coatomer interaction with di-lysine endoplasmic reticulum retention motifs. Science 263:1629–1631

Daibata M, Xu M, Humphreys RE, Reyes VE (1994) More efficient peptide binding to MHC class II molecules during cathepsin B digestion of Ii than after Ii release. Mol Immunol 31:255–260

De Camilli P, Emr SD, McPherson PS, Novick P (1996) Phosphoinositides as regulators in membrane traffic. Science 271:1533–1539

Degen E, Williams DB (1991) Participation of a novel 88-kD protein in the biogenesis of murine class I histocompatibility molecules. J Cell Biol 112:1099–1115

Dell'Angelica EC, Ohno H, Ooi CE, Rabinovich E, Roche KW, Bonifacino JS (1997) AP-3: An adaptor-like protein complex with ubiquitous expression. EMBO J 16:917–928

Denzin LK, Cresswell P (1995) HLA-DM induces CLIP dissociation from MHC class II $\alpha\beta$ dimers and facilitates peptide loading. Cell 82:155-165

Denzin LK, Hammond C, Cresswell P (1996) HLA-DM interactions with intermediates in HLA-DR maturation and a role for HLA-DM in stabilizing empty HLA-DR molecules. J Exp Med 184:2153-2165

Dietrich J, Kastrup J, Nielsen BL, Odum N, Geisler C (1997) Regulation and function of the CD3 gamma DxxxLL motif: a binding site for adaptor protein-1 and adaptor protein-2 in vitro. J Cell Biol 138:271–281

Diment S (1990) Different roles for thiol and aspartyl proteases in antigen presentation of ovalbumin. J Immunol 145:417–422

Dodi AI, Brett S, Nordeng TW, Sidhu S, Batchelor RJ, Lombardi G, Bakke O, Lechler RI (1994) The invariant chain inhibits presentation of endogenous antigens by a human fibroblast cell line. Eur J Immunol 24:1632–1639

Dornmair K, Rothenhausler B, McConnell HM (1989) Structural intermediates in the reactions of antigenic peptides with MHC molecules. Cold Spring Harb Symp Quant Biol 54 Pt 1:409–416

Douek DC, Altmann DM (1997) HLA-DO is an intracellular class II molecule with distinctive thymic expression. Int Immunol 9:355–364

Eades AM, Litfin M, Rahmsdorf HJ (1990) The IFN-gamma response of the murine invariant chain gene is mediated by a complex enhancer that includes several MHC class II consensus elements. J Immunol 144:4399–4409

Elliott EA, Drake JR, Amigorena S, Elsemore J, Webster P, Mellman I, Flavell RA (1994) The invariant chain is required for intracellular transport and function of major histocompatibility complex class II molecules. J Exp Med 179:681–694

Ericson ML, Sundström M, Sansom DM, Charron DJ (1994) Mutually exclusive binding of peptide and invariant chain to major histocompatibility complex class II antigens. J Biol Chem 269:26531–26538

Escola J-M, Grivel J-C, Chavrier P, Gorvel J-P (1995) Different endocytic compartments are involved in the tight association of class II molecules with processed hen egg lysozyme and ribonuclease A in B cells. J Cell Sci 108:2337–2345

Escola JM, Deleuil F, Stang E, Boretto J, Chavrier P, Gorvel JP (1996) Characterization of a lysozyme major histocompatibility complex class II molecule-loading compartment as a specialized recycling endosome in murine B lymphocytes. J Biol Chem 271:27360–27365

Fernandez-Borja M, Verwoerd D, Sanderson F, Aerts H, Trowsdale J, Tulp A, Neefjes J (1996) HLA-DM and MHC class II molecules co-distribute with peptidase-containing lysosomal subcompartments. Int Immunol 8:625–640

Fineschi B, Arneson LS, Naujokas MF, Miller J (1995) Proteolysis of major histocompatibility complex class II-associated invariant chain is regulated by the alternatively spliced gene product, p41. Proc Natl Acad Sci USA 92:10257–10261

Fineschi B, Sakaguchi K, Appella E, Miller J (1996) The proteolytic environment involved in MHC class II-restricted antigen presentation can be modulated by the p41 form of invariant chain. J Immunol 157:3211–3215

Fling SP, Arp B, Pious D (1994) HLA-DMA and -DMB genes are both required for MHC class II/peptide complex formation in antigen-presenting cells. Nature 368:554–558

Freisewinkel IM, Schenck K, Koch N (1993) The segment of invariant chain that is critical for association with major histocompatibility complex class II molecules contains the sequence of a peptide eluted from class II polypeptides. Proc Natl Acad Sci USA 90:9703–9706

Gagliardi M-C, Nisini R, Benvenuto R, De Petrillo G, Michel M-L, Barnaba V (1994) Soluble transferrin mediates targeting of hepatitis B envelope antigen to transferrin receptor and its presentation by activated T cells. Eur J Immunol 24:1372-1376

Gaidarov I, Chen Q, Falck JR, Reddy KK, Keen JH (1996) A functional phosphatidylinositol 3,4,5-trisphosphate/phosphoinositide binding domain in the clathrin adaptor AP-2 α subunit – implications for the endocytic pathway. J Biol Chem 271:20922–20929

Galvin K, Krishna S, Ponchel F, Frohlich M, Cummings DE, Carlson R, Wands JR, Isselbacher KJ, Pillai S, Ozturk M (1992) The major histocompatibility complex class I antigen-binding protein p88 is the product of the calnexin gene. Proc Natl Acad Sci USA 89:8452–8456

Gautam AM, Pearson C, Quinn V, McDevitt HO, Milburn PJ (1995) Binding of an invariant-chain peptide, CLIP, to I-A major histocompatibility complex class II molecules. Proc Natl Acad Sci USA 92:335–339

Gedde-Dahl M, Freisewinkel I, Staschewski M, Schenck K, Koch N, Bakke O (1997) Exon 6 is essential for invariant chain trimerization and induction of large endosomal structures. J Biol Chem 272:8281–8287

Germain RN (1994) MHC-dependent antigen processing and peptide presentation: providing ligands for T lymphocyte activation. Cell 76:287–299

Germain RN, Hendrix LR (1991) MHC class II structure, occupancy and surface expression determined by post-endoplasmic reticulum antigen binding. Nature 353:134–139

Germain RN, Margulies DH (1993) The biochemistry and cell biology of antigen processing and presentation. Annu Rev Immunol 11:403–450

Germain RN, Castellino F, Han R, Reis ES, Romagnoli P, Sadegh-Nasseri S, Zhong GM (1996) Processing and presentation of endocytically acquired protein antigens by MHC class II and class I molecules. Immunol Rev 151:5–30

Ghosh P, Amaya M, Mellins E, Wiley DC (1995) The structure of an intermediate in class II MHC maturation: CLIP bound to HLA-DR3. Nature 378:457–462

Glickman JN, Conibear E, Pearse BM (1989) Specificity of binding of clathrin adaptors to signals on the mannose-6-phosphate/insulin-like growth factor II receptor. EMBO J 8:1041–1047

Goodson HV, Valetti C, Kreis TE (1997) Motors and membrane traffic. Curr Opin Cell Biol 9:18–28

Gorvel J-P, Escola J-M, Stang E, Bakke O (1995) Invariant chain induces a delayed transport from early to late endosomes. J Biol Chem 270:2741–2746

Griffin JP, Chu R, Harding CV (1997) Early endosomes and a late endocytic compartment generate different peptide-class II MHC complexes via distinct processing mechanisms. J Immunol 158:1523–1532

Guo TL, Mudzinski SP, Lawrence DA (1996) Regulation of HLA-DR and invariant chain expression by human peripheral blood mononuclear cells with lead, interferon-gamma, or interleukin-4. Cell Immunol 171:1–9

Hämmerling GJ, Moreno J (1990) The function of the invariant chain in antigen presentation by MHC class II molecules.. Immunol Today 11:337–340

Hammond C, Braakman I, Helenius A (1994) Role of N-linked oligosaccharide recognition, glucose trimming, and calnexin in glycoprotein folding and quality control. Proc Natl Acad Sci USA 91:913–917

Hampl J, Gradehandt G, Kalbacher H, Rude E (1992) In vitro processing of insulin for recognition by murine T cells results in the generation of A chains with free CysSH. J Immunol 148:2664–2671

Harding CV (1996) Class I MHC presentation of exogenous antigens. J Clin Immunol 16:90–96

Harding CV, Unanue ER (1989) Antigen processing and intracellular Ia. Possible roles of endocytosis and protein synthesis in Ia function. J Immunol 142:12–19

Harding CV, Unanue ER (1990) Low-temperature inhibition of antigen processing and iron uptake from transferrin: deficits in endosome functions at 18 degrees C. Eur J Immunol 20:323–329

Harding CV, Geuze HJ (1992) Class II MHC molecules are present in macrophage lysosomes and phagolysosomes that function in the phagocytic processing of Listeria monocytogenes for presentation to T cells. J Cell Biol 119:531–542

Harding CV, Collins DS, Kanagawa O, Unanue ER (1991a) Liposome-encapsulated antigens engender lysosomal processing for class II MHC presentation and cytosolic processing for class I presentation. J Immunol 147:2860–2863

Harding CV, Collins DS, Slot JW, Geuze HJ, Unanue ER (1991b) Liposome-encapsulated antigens are processed in lysosomes, recycled, and presented to T cells. Cell 64:393–401

Hart DN, Fuggle SV, Williams KA, Fabre JW, Ting A, Morris PJ (1981) Localization of HLA-ABC and DR antigens in human kidney. Transplantation 31:428–433

Harter C, Mellman I (1992) Transport of the lysosomal membrane glycoprotein lgp120 (lgp-A) to lysosomes does not require appearance on the plasma membrane. J Cell Biol 117:311–325

Heilker R, Manning-Krieg U, Zuber JF, Spiess M (1996) In vitro binding of clathrin adaptors to sorting signals correlates with endocytosis and basolateral sorting. EMBO J 15:2893–2899

Henne C, Schwenk F, Koch N, Möller P (1995) Surface expression of the invariant chain (CD74) is independent of concomitant expression of major histocompatibility complex class II antigens. Immunology 84:177–182

Hirata I, Austin LL, Blackwell WH, Weber JR, Dobbins WO (1986) Immunoelectron microscopic localization of HLA-DR antigen in control small intestine and colon and in inflammatory bowel disease. Dig Dis Sci 31:1317–1330

Hochstenbach F, David V, Watkins S, Brenner MB (1992) Endoplasmic reticulum resident protein of 90 kilodaltons associates with the T- and B-cell antigen receptors and major histocompatibility complex antigens during their assembly. Proc Natl Acad Sci USA 89:4734–4738

Humbert M, Bertolino P, Forquet F, Rabourdin-Combe C, Gerlier D, Davoust J, Salamero J (1993) Major histocompatibility complex class II-restricted presentation of secreted and endoplasmic reticulum resident antigens requires the invariant chains and is sensitive to lysosomotropic agents. Eur J Immunol 23:3167–3172

Jahraus A, Storrie B, Griffiths G, Desjardins M (1994) Evidence for retrograde traffic between terminal lysosomes and the prelysosomal/late endosome compartment. J Cell Sci 107:145–157

Jensen PE (1991) Enhanced binding of peptide antigen to purified class II major histocompatibility glycoproteins at acidic pH. J Exp Med 174:1111–1120

Johnson KF, Kornfeld S (1992) The cytoplasmic tail of the mannose 6-phosphate/insulin-like growth factor-II receptor has two signals for lysosomal enzyme sorting in the Golgi. J Cell Biol 119:249–257

Kageyama T, Yonezawa S, Ichinose M, Miki K, Moriyama A (1996) Potential sites for processing of the human invariant chain by cathepsins D and E. Biochem Biophys Res Commun 223:549–553

Kakiuchi T, Watanabe M, Hozumi N, Nariuchi H (1990) Differential sensitivity of specific and nonspecific antigen-presentation by B cells to a protein synthesis inhibitor. J Immunol 145:1653–1658

Kakiuchi T, Takatsuki A, Watanabe M, Nariuchi H (1991) Inhibition by brefeldin A of the specific B cell antigen presentation to MHC class II-restricted T cells. J Immunol 147:3289–3295

Kämpgen E, Koch N, Koch F, Stoger P, Heufler C, Schuler G, Romani N (1991) Class II major histocompatibility complex molecules of murine dendritic cells: synthesis, sialylation of invariant chain, and antigen processing capacity are down-regulated upon culture. Proc Natl Acad Sci USA 88:3014–3018

Karlsson L, Péléraux A, Lindstedt R, Liljedahl M, Peterson PA (1994) Reconstitution of an operational MHC class II compartment in nonantigen-presenting cells. Science 266:1569–1573

Katunuma N, Kakegawa H, Matsunaga Y, Saibara T (1994) Immunological significances of invariant chain from the aspect of its structural homology with the cystatin family. FEBS Lett 349:265–269

Kaufman JF, Auffray C, Korman AJ, Shackelford DA, Strominger J (1984) The class II molecules of the human and murine major histocompatibility complex. Cell 36:1–13

Kjær-Nielsen L, Perera JD, Boyd LF, Margulies DH, McCluskey J (1990) The extracellular domains of MHC class II molecules determine their processing requirements for antigen presentation. J Immunol 144:2915–2924

Klagge I, Kopp U, Koch N (1997) Granulocyte-machrophage colony-stimulating factor elevates invariant chain expression in immature myelomonocytic cell lines. Immunology 91:114–120

Klionsky DJ, Nelson H, Nelson N (1992) Compartment acidification is required for efficient sorting of proteins to the vacuole in Saccharomyces cerevisiae. J Biol Chem 267:3416–3422

Koch N (1988) Posttranslational modifications of the Ia-associated invariant protein p41 after gene transfer. Biochemistry 27:4097–4102

Kolk DP, Floyd-Smith G (1992) The HXY box regulatory element modulates expression of the murine IA antigen-associated invariant chain in L fibroblasts. DNA Cell Biol 11:745–754

Kolk DP, Floyd-Smith G (1993) Induction of the murine class-II antigen-associated invariant chain by TNF-alpha is controlled by an NF-kappa B-like element. Gene 126:179–185

Kovacsovics-Bankowski M, Rock KL (1995) A phagosome-to-cytosol pathway for exogenous antigens presented on MHC class I molecules. Science 267:243–246

Kovacsovics-Bankowski M, Clark K, Benacerraf B, Rock KL (1993) Efficient major histocompatibility complex class I presentation of exogenous antigen upon phagocytosis by macrophages. Proc Natl Acad Sci USA 90:4942–4946

Kreis TE, Lowe M, Pepperkok R (1995) COPs regulating membrane traffic. Annu Rev Cell Biol 11:677–706

Kropshofer H, Vogt AB, Hämmerling GJ (1995a) Structural features of the invariant chain fragment CLIP controlling rapid release from HLA-DR molecules and inhibition of peptide binding. Proc Natl Acad Sci USA 92:8313–8317

Kropshofer H, Vogt AB, Stern LJ, Hämmerling GJ (1995b) Self-release of CLIP in peptide loading of HLA-DR molecules. Science 270:1357–1359

Kropshofer H, Vogt AB, Moldenhauer G, Hammer J, Blum JS, Hämmerling GJ (1996) Editing of the HLA-DR-peptide repertoire by HLA-DM. EMBO J 15:6144–6154

Kropshofer H, Hämmerling GJ, Vogt AB (1997) How HLA-DM edits the MHC class II peptide repertoire: survival of the fittest? Immunol Today 18:77–82

Kvist S, Wiman K, Claesson L, Peterson PA, Dobberstein B (1982) Membrane insertion and oligomeric assembly of HLA-DR histocompatibility antigens. Cell 29:61–69

Lamb CA, Cresswell P (1992) Assembly and transport properties of invariant chain trimers and HLA-DR-invariant chain complexes. J Immunol 148:3478–3482

Lamb CA, Yewdell JW, Bennink JR, Cresswell P (1991) Invariant chain targets HLA class II molecules to acidic endosomes containing internalized influenza virus. Proc Natl Acad Sci USA 88:5998–6002

Layet C, Germain RN (1991) Invariant chain promotes egress of poorly expressed, haplotype-mismatched class II major histocompatibility complex A alpha A beta dimers from the endoplasmic reticulum/cis-Golgi compartment. Proc Natl Acad Sci USA 88:2346–2350

Le Borgne R, Griffiths G, Hoflack B (1996) Mannose 6-phosphate receptors and ADP-ribosylation factors cooperate for high affinity interaction of the AP-1 Golgi assembly proteins with membranes. J Biol Chem 271:2162–2170

Lee Cz, McConnell HM (1995) A general model of invariant chain association with class II major histocompatibility complex proteins. Proc Natl Acad Sci USA 92:8269–8273

Lenardo MJ, Baltimore D (1989) NF-kappa B: a pleiotropic mediator of inducible and tissue-specific gene control. Cell 58:227–229

Letourneur F, Gaynor EC, Hennecke S, Demolliere C, Duden R, Emr SD, Riezman H, Cosson P (1994) Coatomer is essential for retrieval of dilysine-tagged proteins to the endoplasmic reticulum. Cell 79:1199–1207

Liang MN, Beeson C, Mason K, McConnell HM (1995) Kinetics of the reactions between the invariant chain (85–99) peptide and proteins of the murine class II MHC. Int Immunol 7:1397–1404

Liang MN, Lee C, Xia Y, McConnell HM (1996) Molecular modeling and design of invariant chain peptides with altered dissociation kinetics from class II MHC. Biochemistry 35:14734–14742

Liljedahl M, Kuwana T, Fung-Leung WP, Jackson M, Peterson PA, Karlsson L (1996) HLA-DO is a lysosomal resident which requires association with HLA-DM for efficient intracellular transport. EMBO J 15:4817–4824

Lindstedt R, Liljedahl M, Peleraux A, Peterson PA, Karlsson L (1995) The MHC class II molecule H2-M is targeted to an endosomal compartment by a tyrosine-based targeting motif. Immunity 3:561–572

Long EO (1985) In search of a function for the invariant chain associated with Ia antigens. Surv Immunol Res 4:27–34

Long EO, Mach B, Accolla RS (1984) Ia-negative B-cell variants reveal a coordinate regulation in the transcription of the HLA class II gene family. Immunogenetics 19:349–353

Long EO, LaVaute T, Pinet V, Jaraquemada D (1994) Invariant chain prevents the HLA-DR-restricted presentation of a cytosolic peptide. J Immunol 153:1487–1494

Lotteau V, Teyton L, Peleraux A, Nilsson T, Karlsson L, Schmid SL, Quaranta V, Peterson PA (1990) Intracellular transport of class II MHC molecules directed by invariant chain. Nature 348:600–605

Maffei A, Perfetto C, Ombra N, Del Pozzo G, Guardiola J (1989) Transcriptional and post-transcriptional regulation of human MHC class II genes require the synthesis of short-lived proteins. J Immunol 142:3657–3661

Malcherek G, Gnau V, Jung G, Rammensee H-G, Melms A (1995) Supermotifs enable natural invariant chain-derived peptides to interact with many major histocompatibility complex-class II molecules. J Exp Med 181:527–536

Malnati MS, Ceman S, Weston M, Demars R, Long EO (1993) Presentation of cytosolic antigen by HLA-DR requires a function encoded in the class II region of the MHC. J Immunol 151:6751–6756

Maric MA, Taylor MD, Blum JS (1994) Endosomal aspartic proteinases are required for invariant-chain processing. Proc Natl Acad Sci USA 91:2171–2175

Marks MS, Blum JS, Cresswell P (1990) Invariant chain trimers are sequestered in the rough endoplasmic reticulum in the absence of association with HLA class II antigens. J Cell Biol 111:839–855

Marks MS, Germain RN, Bonifacino JS (1995) Transient aggregation of major histocompatibility complex class II chains during assembly in normal spleen cells. J Biol Chem 270:10475–10481

Marks MS, Woodruff L, Ohno H, Bonifacino JS (1996) Protein targeting by tyrosine- and di-leucine-based signals: evidence for distinct saturable components. J Cell Biol 135:341–354

Marks MS, Ohno H, Kirchhausen T, Bonifacino SJ (1997) Protein sorting by tyrosine-based signals: adapting to the Ys and wherefores. Trends Cell Biol 7:124–128

McCabe MJ Jr, Lawrence DA (1990) The heavy metal lead exhibits B cell-stimulatory factor activity by enhancing B cell Ia expression and differentiation. J Immunol 145:671–677

McCabe MJ Jr, Dias JA, Lawrence DA (1991) Lead influences translational or posttranslational regulation of Ia expression and increases invariant chain expression in mouse B cells. J Biochem Toxicol 6:269–276

McCoy KL, Gainey D, Inman JK, Stutzman R (1993a) Antigen presentation by B lymphoma cells: requirements for processing of exogenous antigen internalized through transferrin receptors. J Immunol 151:4583–4594

McCoy KL, Noone M, Inman JK, Stutzman R (1993b) Exogenous antigens internalized through transferrin receptors activate CD4+ T cells. J Immunol 150:1691–1704

Mellins E, Smith L, Arp B, Cotner T, Celis E, Pious D (1990) Defective processing and presentation of exogenous antigens in mutants with normal HLA class II genes. Nature 343:71–74

Mellins E, Kempin S, Smith L, Monji T, Pious D (1991) A gene required for class II-restricted antigen presentation maps to the major histocompatibility complex. J Exp Med 174:1607–1615

Mizuochi T, Yee S-T, Kasai M, Kakiuchi T, Muno D, Kominami E (1994) Both cathepsin B and cathepsin D are necessary for processing of ovalbumin as well as for degradation of class II MHC invariant chain. Immunol Lett 43:189–193

Momburg F, Koch N, Moller P, Moldenhauer G, Butcher GW, Hammerling GJ (1986) Differential expression of Ia and Ia-associated invariant chain in mouse tissues after in vivo treatment with IFN-gamma. J Immunol 136:940–948

Monji T, McCormack AL, Yates JR, Pious D (1994) Invariant-cognate peptide exchange restores class II dimer stability in HLA-DM mutants. J Immunol 153:4468–4477

Morkowski S, Goldrath AW, Eastman S, Ramachandra L, Freed DC, Whiteley P, Rudensky AY (1995) T cell recognition of major histocompatibility complex class II complexes with invariant chain processing intermediates. J Exp Med 182:1403–1413

Morris P, Shaman J, Attaya M, Amaya M, Goodman S, Bergman C, Monaco JJ, Mellins E (1994) An essential role for HLA-DM in antigen presentation by class II major histocompatibility molecules. Nature 368:551–554

Motta A, Bremnes B, Morelli MAC, Frank RW, Saviano G, Bakke O (1995) Structure-activity relationship of the leucine-based sorting motifs in the cytosolic tail of the major histocompatibility complex-associated invariant chain. J Biol Chem 270:27165–27171

Motta A, Amodeo P, Fucile P, Castiglione Morelli MA, Bremmes B, Bakke O (1997) A new triple stranded α-helical bundle in solution: the assembly of the cytosolic tail of MHC-associated invariant chain Structure (in press).

Musch A, Xu HX, Shields D, Rodriguez-Boulan E (1996) Transport of vesicular stomatitis virus G protein to the cell surface is signal mediated in polarized and nonpolarized cells. J Cell Biol 133:543–558

Nadimi F, Moreno J, Momburg F, Heuser A, Fuchs S, Adorini L, Hammerling GJ (1991) Antigen presentation of hen egg-white lysozyme but not of ribonuclease A is augmented by the major histocompatibility complex class II-associated invariant chain. Eur J Immunol 21:1255–1263

Naujokas MF, Morin M, Anderson MS, Peterson M, Miller J (1993) The chondroitin sulfate form of invariant chain can enhance stimulation of T cell responses through interaction with CD44. Cell 74:257–268

Neefjes JJ, Ploegh HL (1992) Inhibition of endosomal proteolytic activity by leupeptin blocks surface expression of MHC class II molecules and their conversion to SDS resistance alpha beta heterodimers in endosomes. EMBO J 11:411–416

Neefjes JJ, Momburg F (1993) Cell biology of antigen presentation. Curr Opin Immunol 5:27–34

Neefjes JJ, Stollorz V, Peters PJ, Geuze HJ, Ploegh HL (1990) The biosynthetic pathway of MHC class II but not class I molecules intersects the endocytic route. Cell 61:171–183

Newcomb JR, Cresswell P (1993) Structural analysis of proteolytic products of MHC class II-invariant chain complexes generated in vivo. J Immunol 151:4153-4163

Newcomb JR, Carboy-Newcomb C, Cresswell P (1996) Trimeric interactions of the invariant chain and its association with major histocompatibility complex class II αβ dimers. J Biol Chem 271:24249–24256

Nguyen QV, Humphreys RE (1989) Time course of intracellular associations, processing, and cleavages of Ii forms and class II major histocompatibility complex molecules. J Biol Chem 264:1631–1637

Nijenhuis M, Calafat J, Kuijpers KC, Janssen H, de Haas M, Nordeng TW, Bakke O, Neefjes JJ (1994) Targeting major histocompatibility complex class II molecules to the cell surface by invariant chain allows antigen presentation upon recycling. Eur J Immunol 24:873–883

Noelle RJ, Kuziel WA, Maliszewski CR, McAdams E, Vitetta ES, Tucker PW (1986) Regulation of the expression of multiple class II genes in murine B cells by B cell stimulatory factor-1 (BSF-1). J Immunol 137:1718–1723

Norbury CC, Hewlett LJ, Prescott AR, Shastri N, Watts C (1995) Class I MHC presentation of exogenous soluble antigen via macropinocytosis in bone marrow macrophages. Immunity 3: 783–791

Odorizzi CG, Trowbridge IS, Xue L, Hopkins CR, Davis CD, Collawn JF (1994) Sorting signals in the MHC class II invariant chain cytoplasmic tail and transmembrane region determine trafficking to an endocytic processing compartment. J Cell Biol 126:317–330

Ogrinc T, Dolenc I, Ritonja A, Turk V (1993) Purification of the complex of cathepsin L and the MHC class II-associated invariant chain fragment from human kidney. FEBS Lett 336:555–559

Ohno H, Stewart J, Fournier MC, Bosshart H, Rhee I, Miyatake S, Saito T, Gallusser A, Kirchhausen T, Bonifacino JS (1995) Interaction of tyrosine-based sorting signals with clathrin-associated proteins. Science 269:1872–1875

O'Sullivan DM, Noonan D, Quaranta V (1987) Four Ia invariant chain forms derive from a single gene by alternate splicing and alternate initiation of transcription/translation. J Exp Med 166:444–460

Ou WJ, Cameron PH, Thomas DY, Bergeron JJ (1993) Association of folding intermediates of glycoproteins with calnexin during protein maturation. Nature 364:771–776

Pearse BM, Robinson MS (1990) Clathrin, adaptors, and sorting. Biochem J 6:151–171

Pessara U, Koch N (1990) Tumor necrosis factor alpha regulates expression of the major histocompatibility complex class II-associated invariant chain by binding of an NF-kappa B-like factor to a promoter element. Mol Cell Biol 10:4146–4154

Peters PJ, Neefjes JJ, Oorschot V, Ploegh HL, Geuze HJ (1991) Segregation of MHC class II molecules from MHC class I molecules in the Golgi complex for transport to lysosomal compartments. Nature 349:669–676

Peterson M, Miller J (1990) Invariant chain influences the immunological recognition of MHC class II molecules. Nature 345:172–174

Peterson M, Miller J (1992) Antigen presentation enhanced by the alternatively spliced invariant chain gene product p41. Nature 357:596–598

Pfeifer JD, Wick MJ, Roberts RL, Findlay K, Normark SJ, Harding CV (1993) Phagocytic processing of bacterial antigens for class I MHC presentation to T cells. Nature 361:359–362

Pieters J, Horstmann H, Bakke O, Griffiths G, Lipp J (1991) Intracellular transport and localization of major histocompatibility complex class II molecules and associated invariant chain. J Cell Biol 115:1213–1223

Pieters J, Bakke O, Dobberstein B (1993) The MHC class II associated invariant chain contains two endosomal targeting siganls within its cytoplasmic tail. J Cell Sci 106:831–846

Pinet V, Malnati MS, Long EO (1994) Two processing pathways for the MHC class II-restricted presentation of exogenous influenza virus antigen. J Immunol 152:4852–4860

Pinet V, Vergelli M, Martin R, Bakke O, Long EO (1995) Antigen presentation mediated by recycling of surface HLA-DR molecules. Nature 375:603–606

Polla BS, Poljak A, Ohara J, Paul WE, Glimcher LH (1986) Regulation of class II gene expression: analysis in B cell stimulatory factor 1-inducible murine pre-B cell lines. J Immunol 137:3332–3337

Pond L, Kuhn LA, Teyton L, Schutze MP, Tainer JA, Jackson MR, Peterson PA (1995) A role for acidic residues in di-leucine motif-based targeting to the endocytic pathway. J Biol Chem 270:19989–19997

Porcelli SA (1995) The CD1 family: a third lineage of antigen-presenting molecules. Adv Immunol 59:1–98

Porcelli S, Morita CT, Brenner MB (1992) CD1b restricts the response of human CD4-8-T lymphocytes to a microbial antigen. Nature 360:593–597

Pure E, Inaba K, Crowley MT, Tardelli L, Witmer-Pack MD, Ruberti G, Fathman G, Steinman RM (1990) Antigen processing by epidermal Langerhans cells correlates with the level of biosynthesis of major histocompatibility complex class II molecules and expression of invariant chain. J Exp Med 172:1459–1469

Qiu Y, Xu X, Wandinger-Ness A, Dalke DP, Pierce SK (1994) Separation of subcellular compartments containing distinct functional forms of MHC class II. J Cell Biol 125:595–605

Quaranta V, Majdic O, Stingl G, Liszka K, Honigsmann H, Knapp W (1984) A human Ia cytoplasmic determinant located on multiple forms of invariant chain (gamma, gamma 2, gamma 3). J Immunol 132:1900–1905

Rahmsdorf HJ, Koch N, Mallick U, Herrlich P (1983) Regulation of MHC class II invariant chain expression: induction of synthesis in human and murine plasmocytoma cells by arresting replication. EMBO J 2:811–816

Rahmsdorf HJ, Harth N, Eades AM, Litfin M, Steinmetz M, Forni L, Herrlich P (1986) Interferon-gamma, mitomycin C, and cycloheximide as regulatory agents of MHC class II-associated invariant chain expression. J Immunol 136:2293–2299

Rapoport I, Miyazaki M, Boll W, Duckworth B, Cantley LC, Shoelson S, Kirchhausen T (1997) Regulatory interactions in the recognition of endocytic sorting signals by AP-2 complexes. EMBO J 16:2240–2250

Raposo G, Nijman HW, Stoorvogel W, Liejendekker R, Harding CV, Melief CJ, Geuze HJ (1996) B lymphocytes secrete antigen-presenting vesicles. J Exp Med 183:1161–1172

Reid PA, Watts C (1990) Cycling of cell-surface MHC glycoproteins through primaquine-sensitive intracellular compartments. Nature 346:655–657

Reis e Sousa C, Germain RN (1995) Major histocompatibility complex class I presentation of peptides derived from soluble exogenous antigen by a subset of cells engaged in phagocytosis. J Exp Med 182:841–851

Riberdy JM, Cresswell P (1992a) The antigen-processing mutant T2 suggests a role for MHC-linked genes in class II antigen presentation [published erratum appears in J Immunol 1992 Aug 1;149(3):1113]. J Immunol 148:2586–2590

Riberdy JM, Newcomb JR, Surman MJ, Barbosa JA, Cresswell P (1992b) HLA-DR molecules from an antigen-processing mutant cell line are associated with invariant chain peptides. Nature 360:474–477

Riese RJ, Wolf PR, Bromme D, Natkin LR, Villadangos JA, Ploegh HL, Chapman HA (1996) Essential role for cathepsin S in MHC class II-associated invariant chain processing and peptide loading. Immunity 4:357–366

Robbins NF, Hammond C, Denzin LK, Pan M, Cresswell P (1996) Trafficking of major histocompatibility complex class II molecules through intracellular compartments containing HLA-DM. Hum Immunol 45:13–23

Robinson MS (1997) Coats and vesicle budding. Trends Cell Biol 7:99–102

Roche PA, Cresswell P (1990) Invariant chain association with HLA-DR molecules inhibits immunogenic peptide binding. Nature 345:615–618

Roche PA, Cresswell P (1991a) Proteolysis of the class II-associated invariant chain generates a peptide binding site in intracellular HLA-DR molecules. Proc Natl Acad Sci USA 88:3150–3154

Roche PA, Marks MS, Cresswell P (1991b) Formation of a nine-subunit complex by HLA class II glycoproteins and the invariant chain. Nature 354:392–394

Roche PA, Teletski CL, Stang E, Bakke O, Long EO (1993) Cell surface HLA-DR-invariant chain complexes are targeted to endosomes by rapid internalization. Proc Natl Acad Sci USA 90:8581–8585

Rock KL, Gamble S, Rothstein L (1990) Presentation of exogenous antigen with class I major histocompatibility complex molecules. Science 249:918–921

Rock KL, Rothstein L, Gamble S, Fleischacker C (1993) Characterization of antigen-presenting cells that present exogenous antigens in association with class I MHC molecules. J Immunol 150:438–446

Romagnoli P, Germain RN (1994) The CLIP region of invariant chain plays a critical role in regulating major histocompatibility complex class II folding, transport, and peptide occupancy. J Exp Med 180:1107–1113

Romagnoli P, Layet C, Yewdell J, Bakke O, Germain RN (1993) Relationship between invariant chain expression and major histocompatibility complex class II transport into early and late endocytic compartments. J Exp Med 177:583–596

Sadegh-Nasseri S, Germain RN (1991) A role for peptide in determining MHC class II structure. Nature 353:167–170

Salamero J, Humbert M, Cosson P, Davoust J (1990) Mouse B lymphocyte specific endocytosis and recycling of MHC class II molecules. EMBO J 9:3489–3496

Salamero J, Le Borgne R, Saudrais C, Goud B, Hoflack B (1996) Expression of major histocompatibility complex class II molecules in HeLa cells promotes the recruitment of AP-1 Golgi-specific assembly proteins on Golgi membranes. J Biol Chem 271:30318–30321

Sanderson F, Kleijmeer MJ, Kelly A, Verwoerd D, Tulp A, Neefjes JJ, Geuze HJ, Trowsdale J (1994) Accumulation of HLA-DM, a regulator of antigen presentation, in MHC class II compartments. Science 266:1566–1569

Sanderson F, Thomas C, Neefjes J, Trowsdale J (1996) Association between HLA-DM and HLA-DR in vivo. Immunity 4:87–96

Sandoval IV, Bakke O (1994) Targeting of membrane proteins to endosomes and lysosomes. Trends Cell Biol 4:292–298

Sant AJ, Cullen SE, Giacoletto KS, Schwartz BD (1985a) Invariant chain is the core protein of the Ia-associated chondroitin sulfate proteoglycan. J Exp Med 162:1916–1934

Sant AJ, Cullen SE, Schwartz BD (1985b) Biosynthetic relationships of the chondroitin sulfate proteoglycan with Ia and invariant chain glycoproteins. J Immunol 135:416–422

Sarles J, Gorvel JP, Olive D, Maroux S, Mawas C, Giraud F (1987) Subcellular localization of class I (A,B,C) and class II (DR and DQ) MHC antigens in jejunal epithelium of children with coeliac disease. J Pediatr Gastroenterol Nutr 6:51–56

Schafer PH, Green JM, Malapati S, Gu L, Pierce SK (1996) HLA-DM is present in one-fifth the amount of HLA-DR in the class II peptide-loading compartment where it associates with leupeptin-induced peptide (LIP)-HLA-DR complexes. J Immunol 157:5487965495

Schaiff WT, Hruska KA,Jr., Bono C, Shuman S, Schwartz BD (1991) Invariant chain influences post-translational processing of HLA-DR molecules. J Immunol 147:603–608

Schimmöller F, Itin C, Pfeffer S (1997) Vesicle traffic: get your coat. Curr Biol 7:R235-R237

Schlessinger J, Schreiber AB, Levi A, Lax I, Libermann T, Yarden Y (1983) Regulation of cell proliferation by epidermal growth factor. CRC Crit Rev Biochem 14:93–111

Scholl PR, Geha RS (1994) MHC class II signaling in B-cell activation. Immunol Today 15:418–422

Schreiber KL, Bell MP, Huntoon CJ, Rajagopalan S, Brenner MB, McKean DJ (1994) Class II histocompatibility molecules associate with calnexin during assembly in the endoplasmic reticulum. Int Immunol 6:101–111

Schutze M-P, Peterson PA, Jackson MR (1994) An N-terminal double-arginine motif maintains type II membrane proteins in the endoplasmic reticulum. EMBO J 13:1696–1705

Scott H, Solheim BG, Brandtzaeg P, Thorsby E (1980) HLA-DR-like antigens in the epithelium of the human small intestine. Scand J Immunol 12:77–82

Seaman MN, Ball CL, Robinson MS (1993) Targeting and mistargeting of plasma membrane adaptors in vitro. J Cell Biol 123:1093–1105

Selby WS, Janossy G, Goldstein G, Jewell DP (1981) T lymphocyte subsets in human intestinal mucosa: the distribution and relationship to MHC-derived antigens. Clin Exp Immunol 44:453–458

Serwe M, Reuter G, Sponaas A, Koch S, Koch N (1997) Both invariant chain isoforms Ii31 and Ii41 promote class II antigen presentation. Int Immunol 9:983–991

Sette A, Ceman S, Kubo RT, Sakaguchi K, Appella E, Hunt DF, Davis TA, Michel H, Shabanowitz J, Rudersdorf R et al (1992) Invariant chain peptides in most HLA-DR molecules of an antigen-processing mutant. Science 258:1801–1804

Sette A, Southwood S, Miller J, Appella E (1995) Binding of major histocompatibility complex class II to the invariant chain-derived peptide, CLIP, is regulated by allelic polymorphism in class II. J Exp Med 181:677–683

Shachar I, Flavell RA (1996) Requirement for invariant chain in B cell maturation and function. Science 274:106–108

Sherman MA, Weber DA, Jensen PE (1995) DM enhances peptide binding to class II MHC by release of invariant chain-derived peptide. Immunity 3:197–205

Shih N-Y, Floyd-Smith G (1995) Invariant chain (CD74) gene regulation: enhanced expression associated with activation of protein kinase Cδ in a murine B lymphoma cell line. Mol Immunol 32:643-650

Sieling PA, Chatterjee D, Porcelli SA, Prigozy TI, Mazzaccaro RJ, Soriano T, Bloom BR, Brenner MB, Kronenberg M, Brennan PJ et al (1995) CD1-restricted T cell recognition of microbial lipoglycan antigens. Science 269:227–230

Simonis S, Miller J, Cullen SE (1989) The role of the Ia-invariant chain complex in the posttranslational processing and transport of Ia and invariant chain glycoproteins. J Immunol 143:3619–3625

Simonsen A, Momburg F, Drexler J, Hämmerling JH, Bakke O (1993) Intracellular distribution of the MHC class II molecules and the associated invariant chain (Ii) in different cell lines. Int Immunol 5, No. 8:903–917

Simonsen A, Stang E, Bremnes B, Roe M, Prydz K, Bakke O (1997) Sorting of MHC class II molecules and the associated invariant chain (Ii) in polarized MDCK cells. J Cell Sci 110:597–609

Simpson F, Bright NA, West MA, Newman LS, Darnell RB, Robinson MS (1996) A novel adaptor-related protein complex. J Cell Biol 133:749–760

Skoskiewicz MJ, Colvin RB, Schneeberger EE, Russell PS (1985) Widespread and selective induction of major histocompatibility complex-determined antigens in vivo by gamma interferon. J Exp Med 162:1645–1664

Sloan VS, Cameron P, Porter G, Gammon M, Amaya M, Mellins E, Zaller DM (1995) Mediation by HLA-DM of dissociation of peptides from HLA-DR. Nature 375:802-806

Smiley ST, Rudensky AY, Glimcher LH, Grusby MJ (1996) Truncation of the class II β-chain cytoplasmic domain influences the level of class II invariant chain-derived peptide complexes. Proc Natl Acad Sci USA 93:241–244

St.-Pierre Y, Watts TH (1990) MHC class II-restricted presentation of native protein antigen by B cells is inhibitable by cycloheximide and brefeldin A. J Immunol 145:812–818

Stang E, Bakke O (1997) MHC class II associated invariant chain induced enlarged endosomal structures. A morphological study. Exp Cell Res 235:79–82

Stebbins CC, Loss GE Jr, Elias CG, Chervonsky A, Sant AJ (1995) The requirement for DM in class II-restricted antigen presentation and SDS-stable dimer formation is allele and species dependent. J Exp Med 181:223–234

Stockinger B, Pessara U, Lin RH, Habicht J, Grez M, Koch N (1989) A role of Ia-associated invariant chains in antigen processing and presentation. Cell 56:683–689

Stoorvogel W, Oorschot V, Geuze HJ (1996) A novel class of clathrin-coated vesicles budding from endosomes. J Cell Biol 132:21–33

Stössel H, Koch F, Kämpgen E, Stöger P, Lenz A, Heufler C, Romani N, Schuler G (1990) Disappearance of certain acidic organelles (endosomes and Langerhans cell granules) accompanies loss of antigen processing capacity upon culture of epidermal Langerhans cells. J Exp Med 172:1471–1482

Stang E, Guerra CB, Amaya M, Paterson Y, Bakke O, Mellins ED (1997) DR/CLIP and DR/peptide complexes co-localize in pre-lysosomes in human B lymphoblastoid cells. Exp Cell Res 235:79–92

Strubin M, Berte C, Mach B (1986) Alternative splicing and alternative initiation of translation explain the four forms of the Ia antigen-associated invariant chain. EMBO J 5:3483–3488

Sugita M, Brenner MB (1995) Association of the invariant chain with major histocompatibility complex class I molecules directs trafficking to endocytic compartments. J Biol Chem 270:1443–1448

Sugita M, Jackman RM, Van Donselaar E, Behar SM, Rogers RA, Peters PJ, Brenner MB, Porcelli SA (1996) Cytoplasmic tail-dependent localization of CD1b antigen-presenting molecules to MIICs. Science 273:349–352

Sung E, Jones PP (1981) The invariant chain of murine Ia antigens: its glycosylation, abundance and subcellular localization. Mol Immunol 18:899–913

Swier K, Miller J (1995) Invariant chain-independent antigen presentation depends primarily upon the pool of newly synthesized MHC class II molecules. J Immunol 155:1851–1861

Takaesu NT, Lower JA, Robertson EJ, Bikoff EK (1995) Major histocompatibility class II peptide occupancy, antigen presentation, and CD4+ T cell function in mice lacking the p41 isoform of invariant chain. Immunity 3:385–396

Takaesu NT, Lower JA, Yelon D, Robertson EJ, Bikoff EK (1997) In vivo functions mediated by the p41 isoform of the MHC class II-associated invariant chain. J Immunol 158:187–199

Takahashi H, Cease KB, Berzofsky JA (1989) Identification of proteases that process distinct epitopes on the same protein. J Immunol 142:2221–2229

Tartakoff AM (1983) Perturbation of the structure and function of the Golgi complex by monovalent carboxylic ionophores. Methods Enzymol 98:47–59

Thorens B, Vassalli P (1986) Chloroquine and ammonium chloride prevent terminal glycosylation of immunoglobulins in plasma cells without affecting secretion. Nature 321:618–620

Traub LM, Bannykh SI, Rodel JE, Aridor M, Balch WE, Kornfeld S (1996) AP-2-containing clathrin coats assemble on mature lysosomes. J Cell Biol 135:1801–1814

Tulp A, Verwoerd D, Dobberstein B, Ploegh HL, Pieters J (1994) Isolation and characterization of the intracellular MHC class II compartment. Nature 369:120–126

Veenstra H, Jacobs P, Dowdle EB (1993) Processing of HLA-class II invariant chain and expression of the p35 form is different in malignant and transformed cells. Blood 82:2494-2500

Veenstra H, Jacobs P, Dowdle EB (1996) Abnormal association between invariant chain and HLA class II α and β chains in chronic lymphocytic leukemia. Cell Immunol 171:68–73

Vidal K, Grosjean I, Kaiserlian D (1995) Antigen presentation by a mouse duodenal epithelial cell line (MODE-K). Adv Exp Med Biol 371A:225–228

Vidard L, Rock KL, Benacerraf B (1992) Diversity in MHC class II ovalbumin T cell epitopes generated by distinct proteases. J Immunol 149:498–504

Vigna JL, Smith KD, Lutz CT (1996) Invariant chain association with MHC class I – preference for HLA class I β2-microglobulin heterodimers, specificity, and influence of the MHC peptide-binding groove. J Immunol 157:4503-4510

Viville S, Neefjes J, Lotteau V, Dierich A, Lemeur M, Ploegh H, Benoist C, Mathis D (1993) Mice lacking the MHC class II-associated invariant chain. Cell 72:635–648

Vogt AB, Stern LJ, Amshoff C, Dobberstein B, Hämmerling GJ, Kropshofer H (1995) Interference of distinct invariant chain regions with superantigen contact area and antigenic peptide binding groove of HLA-DR. J Immunol 155:4757–4765

Vogt AB, Kropshofer H, Moldenhauer G, Hämmerling GJ (1996) Kinetic analysis of peptide loading onto HLA-DR molecules mediated by HLA-DM. Proc Natl Acad Sci USA 93:9724–9729

Volc-Platzer B, Majdic O, Knapp W, Wolff K, Hinterberger W, Lechner K, Stingl G (1984) Evidence of HLA-DR antigen biosynthesis by human keratinocytes in disease. J Exp Med 159:1784–1789

Warmerdam PAM, Long EO, Roche PA (1996) Isoforms of the invariant chain regulate transport of MHC class II molecules to antigen processing compartments. J Cell Biol 133:281–291

Warren RA, Green FA, Enns CA (1997) Saturation of the endocytic pathway for the transferrin receptor does not affect the endocytosis of the epidermal growth factor receptor. J Biol Chem 272:2116–2121

Watts C (1997) Capture and processing of exogenous antigens for presentation on MHC molecules. Annu Rev Immunol 15:821–850

Weber DA, Evavold BD, Jensen PE (1996) Enhanced dissociation of HLA-DR-bound peptides in the presence of HLA-DM. Science 274:618–620

Weenink SM, Milburn PJ, Gautam AM (1997) A continuous central motif of invariant chain peptides, CLIP, is essential for binding to various I-A MHC class II molecules. Int Immunol 9:317–325

West MA, Lucocq JM, Watts C (1994) Antigen processing and class II MHC peptide-loading compartments in human B-lymphoblastoid cells. Nature 369:147–151

Wettstein DA, Boniface JJ, Reay PA, Schild H, Davis MM (1991) Expression of a class II major histocompatibility complex (MHC) heterodimer in a lipid-linked form with enhanced peptide/soluble MHC complex formation at low pH. J Exp Med 174:219–228

Wong P, Rudensky AY (1996) Phenotype and function of CD4+ T cells in mice lacking invariant chain. J Immunol 156:2133–2142

Woo JT, Shinohara C, Sakai K, Hasumi K, Endo A (1992) Inhibition of the acidification of endosomes and lysosomes by the antibiotic concanamycin B in macrophage J774. Eur J Biochem 207:383–389

Wright KL, Moore TL, Vilen BJ, Brown AM, Ting JPY (1995) Major histocompatibility complex class II-associated invariant chain gene expression is up-regulated by cooperative interactions of Sp1 and NF-Y. J Biol Chem 270:20978–20986

Wu SH, Gorski J (1996) The MHC class II-associated invariant chain-derived peptide clip binds to the peptide-binding groove of class II molecules. Mol Immunol 33:371–377

Wubbolts R, Fernandez-Borja M, Oomen L, Verwoerd D, Janssen H, Calafat J, Tulp A, Dusseljee S, Neefjes J (1996) Direct vesicular transport of MHC class II molecules from lysosomal structures to the cell surface. J Cell Biol 135:611–622

Xu M, Capraro GA, Daibata M, Reyes VE, Humphreys RE (1994) Cathepsin B cleavage and release of invariant chain from MHC class II molecules follow a staged pattern. Mol Immunol 31:723–731

Xu XX, Song WX, Cho H, Qiu Y, Pierce SK (1995) Intracellular transport of invariant chain MHC class II complexes to the peptide-loading compartment. J Immunol 155:2984–2992

Yilla M, Tan A, Ito K, Miwa K, Ploegh HL (1993) Involvement of the vacuolar H(+)-ATPases in the secretory pathway of HepG2 cells. J Biol Chem 268:19092–19100

Yoshimori T, Yamamoto A, Moriyama Y, Futai M, Tashiro Y (1991) Bafilomycin A1, a specific inhibitor of vacuolar-type H(+)-ATPase, inhibits acidification and protein degradation in lysosomes of cultured cells. J Biol Chem 266:17707–17712

Yoshimori T, Keller P, Roth MG, Simons K (1996) Different biosynthetic transport routes to the plasma membrane in BHK and CHO cells. J Cell Biol 133:247–256

Zachgo S, Dobberstein B, Griffiths G (1992) A block in degradation of MHC class II-associated invariant chain correlates with a reduction in transport from endosome carrier vesicles to the prelysosome compartment. J Cell Sci 103:811–822

Zeuzem S, Feick P, Zimmermann P, Haase W, Kahn RA, Schulz I (1992) Intravesicular acidification correlates with binding of ADP-ribosylation factor to microsomal membranes. Proc Natl Acad Sci USA 89:6619–6623

Zhong GM, Castellino F, Romagnoli P, Germain RN (1996) Evidence that binding site occupancy is necessary and sufficient for effective major histocompatibility complex (MHC) class II transport through the secretory pathway redefines the primary function of class II-associated invariant chain peptides (CLIP). J Exp Med 184:2061–2066

Zhong GM, Romagnoli P, Germain RN (1997) Related leucine-based cytoplasmic targeting signals in invariant chain and major histocompatibility complex class II molecules control endocytic presentation of distinct determinants in a single protein. J Exp Med 185:429–438

Zhu L, Jones PP (1990) Transcriptional control of the invariant chain gene involves promoter and enhancer elements common to and distinct from major histocompatibility complex class II genes. Mol Cell Biol 10:3906–3916

Biologic Consequences of Defective Major Histocompatibility Complex Class II Presentation

MARTHA M. EIBL and HERMANN M. WOLF

1 Introduction to MHC Class II – Structure and Function

Major histocompatibility (MHC) class II molecules are heterodimeric transmembrane glycoproteins that are constitutively expressed on certain cells of the immune system, where they play an important role in the development and function of the immune system. Three different MHC class II isotypes, HLA-DR, -DQ and -DP, have been described in humans, each encoded by distinct α-chain and β-chain genes. A high degree of allelic polymorphism of the MHC class II genes is responsible for the genetic diversity of the MHC class II system. The function of MHC class II molecules that was first understood was their role as transplantation antigens (BACH and VAN ROOD 1976). The crucial role of MHC class II in the initiation and regulation of the immune response was later suggested by the discovery that MHC class II genes are responsible for the genetic basis of high and low immunological responsiveness to antigen (immune response genes) (BENACERRAF 1981). Since then, the pivotal function of MHC class II molecules in the context of antigen presentation to CD4$^+$ T cells has been unraveled in a number of important studies (e.g. BABBITT et al. 1985; SELLE at al. 1987).

Institute of Immunology, University of Vienna, Borschkegasse 8A, 1090 Vienna, Austria

Following their assembly inside the cell, MHC class II heterodimers are transported to a specific endocytic compartment, where their peptide-binding groove (located within the polymorphic part of the molecule) is loaded with peptides derived from processed exogenous protein antigens (CRESSWELL 1994a). The peptide-MHC class II complex is then transported to the cell surface, where it is recognised by the T cell receptor (TCR), leading to MHC class II allele-restricted and antigenic peptide-specific activation of $CD4^+$ T cells. Professional antigen-presenting cells (APC) such as cells of the monocyte/macrophage lineage, B cells or dendritic cells (e.g. Langerhans cells) express MHC class II molecules constitutively, although MHC class II expression on these cells can be significantly increased following activation. MHC class II molecules are essential for cognate interaction of B cells with CD4-expressing T helper cells. These helper T cells, activated by APC presenting the antigenic peptides in the context of MHC class II, interact with B cells in T-dependent immunoglobulin production and isotype switching. De novo MHC class II expression is inducible in a variety of otherwise MHC class II-negative cell types such as T cells, fibroblasts or endothelial cells upon activation with a variety of stimuli such as cytokines, thereby enabling these cells to present antigenic peptides to $CD4^+$ T cells under physiological or certain immunopathological conditions.

In addition to antigen presentation, MHC class II molecules mediate superantigen-induced T cell activation. Superantigens are microbial products such as bacterial toxins, e.g. staphylococcal enterotoxins (SE), toxic shock syndrome toxin (TSST)-1, that bind to MHC class II molecules without a requirement for antigen processing and induce T cell activation upon binding to the TCR β-chain at a site outside the antigen-binding groove (HERMAN et al. 1991). MHC class II molecules themselves are also capable of eliciting signals following interaction with ligands such as superantigens, leading to the activation of an MHC class II-positive cell such as the B cell or the monocyte (SCHOLL and GEHA 1994). Furthermore, expression of MHC class II molecules on cells within the thymus is critical for the development and maturation of T cells of the $CD4^+$ lineage.

2 MHC Class II-Deficient Mice

Patients with primary immunodeficiency diseases provided opportunities to study the consequences of the complete absence of certain components of the immune system. These "experiments of nature" showed immunological defects indicative for the role of the absent or defective structure in immune function. During the last few years, however, a new and powerful genetic approach has made in-depth studies possible that have contributed significantly to our understanding of the role that individual gene products such as lymphokines or cell surface molecules play in the development and function of the immune system. By combining the techniques of homologous recombination in embryonic stem cells with blastocyst injection, genetically modified mouse strains can be generated that show targeted mutations in specific genes (CAPECCHI 1989; FUNG-LEUNG and MAK 1992). This genetic approach has been utilised to generate mice that are deficient in MHC class II expression. In the mouse, two isotypes of MHC class II molecules are expressed, I-A and I-E, each composed of a specific α- and β-chain. The requirement for surface expression of MHC class II molecules is the formation of a pair consisting of one α- and one β-chain. Most of the embryonic stem cell lines used for gene targeting are derived from mouse strains of the H-2b haplotype, which already has a deletion in the I-Eα gene, resulting in the inability to produce I-Eα protein and to express I-E on the cell surface (MATHIS et al. 1983). Targeted disruption of either the I-Aα or the I-Aβ gene resulted in the inability to produce I-Aα or I-Aβ protein, respectively, and in H-2b embryonic stem cells already negative for I-E expression led to an MHC class II-deficient phenotype (COSGROVE et al. 1991; GRUSBY et al. 1991; KONTGEN et al. 1993).

More recently, mice deficient in MHC class II expression have been generated by targeted disruption of the MHC class II transactivator (CIITA) gene (CHANG et al. 1996). CIITA is a *trans*-acting regulatory factor that is involved in the induction of the coordinate expression of MHC class II genes and is defective in one type of patients with MHC class II deficiency (REITH et al. 1995). CIITA-deficient mice do not express conventional MHC class II molecules on the surface of splenic B cells, dendritic cells or peritoneal resident macrophages. However, a subset of thymic epithelial cells express MHC class II molecules. The consequences of the CIITA-defect in the mouse resemble the findings observed in patients with a CIITA defect (e.g. BCH; see MANNHALTER et al. 1991). The numbers of mature $CD4^+$ T cells in the periphery are reduced in CIITA-deficient mice, and T-dependent antigen responses are impaired.

In addition to mouse strains deficient in MHC class II expression, MHC class I-deficient mice have been generated by gene targeting and disruption of the β_2-microglobulin gene (KOLLER et al. 1990; ZIJLSTRA et al. 1990; RAULET 1994). Mice bearing a mutated β_2-microglobulin gene were mated with MHC class II-deficient mice, and subsequent intercrossing of offsprings heterozygous for both mutations generated animals homozygous for mutations at both the β_2-microglobulin and the MHC class II loci, resulting in deficient expression of both MHC class I and class II antigens (GRUSBY et al. 1993).

3 Invariant Chain-Deficient Mice

The invariant chain is a nonpolymorphic type II transmembrane protein that plays an important role in the intracellular transport and peptide loading of MHC class II molecules (CRESSWELL 1994b). The invariant chain associates with MHC class II molecules in the endoplasmatic reticulum (ER), inhibits their aggregation with other proteins, facilitates α- to β-dimer assembly and protects the unoccupied peptide-binding groove from binding of peptides encountered in the ER or during transit from the ER to the endocytic, exogenous antigenic peptide-containing compartment. It enhances the egress of MHC class II molecules from the ER and participates in the transport through the Golgi complex to the endosomal compartments. Somewhere within the endosome, the invariant chain is proteolytically degraded and dissociates from the MHC class II dimer, followed by binding of short peptides derived from proteolytic cleavage of endocytosed foreign and self-protein antigens. Peptide-MHC class II complexes are then transported to the cell surface.

Studies in invariant chain-deficient mice generated by targeted disruption of the invariant chain gene have demonstrated that the appearance of peptide-loaded MHC class II molecules at the cell surface depends critically on the invariant chain. Peptides bound in the groove of MHC class II molecules are needed for MHC complexes to achieve the proper confirmation and full stability at the cell surface, and loading of exogenous peptides onto class II molecules is inefficient in the absence of invariant chain (BUSCH et al. 1995). In mice lacking the invariant chain, most class II molecules are held up in the ER, and the small amounts that eventually exit bypass the endosomal compartments and appear at the surface in reduced numbers and in an atypical form (BIKOFF et al. 1993; VIVILLE et al. 1993; ELLIOTT et al. 1994). Sodium dodecyl sulphate polyacrylamide gel electrophoresis (SDS-PAGE) analysis shows that the class II molecules that are able to reach the cell surface in invariant chain-deficient mice lack the compact conformation typically associated with stable peptide binding, suggesting that they may be empty or only occupied by losely bound peptides. However, the altered class II molecules expressed in reduced amounts on cells from invariant chain-deficient mice can bind peptides added exogenously and thereby convert to the compact form (BIKOFF et al. 1993, 1995; VIVILLE et al. 1993; ELLIOTT et al. 1994), showing that the structure of MHC class II molecules is principally intact. As a result of the defective MHC class II transport to the cell surface, APC from invariant chain-deficient mice showed a strongly reduced ability to present several exogenously supplied proteins such as HEL or OVA, while they could efficiently present the corresponding optimal peptides. Efficient presentation of proteins in association with MHC class II was only possible at high antigen concentrations (BIKOFF et al. 1993; VIVILLE et al. 1993).

4 MHC Class II Deficiency – A Regulatory Defect in MHC Class II Gene Transcription

In view of the importance of MHC class II molecules for the development and function of the immune system, it is evident that abnormalities in MHC class II expression are likely to be associated with severe immunodeficiency (KLEIN et al. 1993) or may play a role in the pathological T cell activation that can be observed in certain autoimmune diseases (ACHA ORBEA and MCDEVITT 1990).

MHC class II deficiency is an autosomal recessive disease characterised by cellular and humoral immunodeficiency. A lack of surface expression of MHC class II molecules leads to a severe impairment in antigen presentation, resulting in defective T cell responses to recall antigen and defective T-dependent antibody responses (GRISCELLI et al. 1993). The majority of patients have hypogammaglobulinaemia and an impaired production of specific antibodies following infection or after vaccination

(SMITH et al. 1988; GRISCELLI et al. 1993; KLEIN et al. 1993). Within the first year of life, the patients present with severe failure to thrive, and deterioration of the clinical picture occurs due to severe bacterial, fungal and viral infections. The disease is associated with high mortality at an early age, and bone marrow transplantation (BMT) is considered to be the only possible cure for the patients (GRISCELLI et al. 1993; KLEIN et al. 1993). Engraftment of bone marrow-derived MHC class II-positive APC of donor origin results in the correction of the immunodeficiency; however, expression of MHC class II in non-haematologic cells most likely remains defective, although this has not been directly demonstrated by studies in patients after BMT up to now. The possible long-term clinical consequences of such a residual defect are unclear.

In patients with MHC class II deficiency, constitutive and inducible cell membrane expression of MHC class II molecules is defective, while expression of the invariant chain is normal, and MHC class I expression is normal or only slightly decreased. In all patients in whom there is a defect in the coordinate expression of all MHC class II genes, especially if the defect has been characterised and affects coordinate MHC class II gene expression, e.g. CIITA or RFX-5, constitutive MHC class II expression is absent on all APC. The defect can be detected by the lack of HLA-DR expression on the surface of peripheral blood mononuclear cells such as B lymphocytes and monocytes (GRISCELLI et al. 1993). In addition, inducible expression of MHC class II molecules is defective, such as in activated T cells and in fibroblasts stimulated with interferon (IFN)-γ (DE PREVAL et al. 1988). The latter finding indicates that defective MHC class II expression in the patients is not limited to bone marrow-derived cell types. Preliminary reports have indicated that other constitutively MHC class II-positive cells, such as Langerhans cells of the skin, show an abnormal MHC class II expression in the patients (GRISCELLI et al. 1993), but definite studies have not been published. In very preliminary statements, a defective expression of MHC class II has also been reported for non-haematologic cells other than fibroblasts, such as gastrointestinal epithelial cells and endothelial cells (GRISCELLI et al. 1993).

Several observations provided evidence that MHC class II deficiency is caused by a regulatory defect at the transcriptional level. The defect in MHC class II expression was observed at both the mRNA and protein level and affects the expression of more than one of the MHC class II genes, HLA-DR, -DP, -DQ, α- and β-chains. IFN-γ is known to coordinately upregulate the expression of all MHC class II genes at the transcriptional level, but in general fails to induce normal MHC class II mRNA expression in patients with MHC class II deficiency. In addition, family studies showed that the disease segregates independently from the MHC class II gene loci (DE PREVAL et al. 1985), further indicating that the genetic defect is not located in the MHC class II genes themselves but rather involves *trans*-acting regulatory factors essential for MHC class II gene transcription. It is thus important to emphasise that the defects in the MHC class II gene-targeted mice and in the MHC class II-deficient patients are principally different.

Further progress in the molecular characterisation of the defect in MHC class II deficiency was achieved by the discovery that patients from different families could be ascribed to different groups of genetic defects, indicating that defects in different *trans*-acting regulatory factors can lead to an MHC class II-negative phenotype in humans. Somatic cell hybridisation experiments employing Epstein-Barr virus (EBV)-transformed B cell lines from different patients and in vitro-generated MHC class II-deficient mutant cell lines showed that the majority of patients with MHC class II deficiency fall into three distinct complementation groups, A–C (HUME and LEE 1989; BENICHOU and STROMINGER 1991; SEIDL et al. 1992; LISOWSKA-GROSPIERRE et al. 1994). One experimentally generated MHC class II-negative mutant cell line defines a fourth complementation group D to which no patient has yet been ascribed (GLADSTONE and PIOUS 1978). In addition, recent reports have described three patients from two unrelated families with new phenotypes of MHC class II deficiency characterised by a residual expression of certain MHC class II genes (HAUBER et al. 1995; PEIJNENBURG et al. 1995). Both families were shown to represent new and different complementation groups (PEIJNENBURG et al. 1995; DOUHAN et al. 1996; B. Lisowska-Grospierre, personal communication), so that currently patients with MHC class II deficiency can be subdivided into at least five different complementation groups.

The expression of the MHC class II genes is regulated mainly at the transcriptional level (MACH et al. 1996). A highly conserved promoter-proximal region is located upstream of the transcription initiation site and is sufficient to mediate both constitutive and inducible expression of MHC class II genes (BENOIST and MATHIS 1989; GLIMCHER and KARA 1992; TING and BALDWIN 1993). This promoter-proximal region contains sequence motifs called S, X, X2 and Y boxes. Transient transfection experiments (BENOIST and MATHIS 1989; GLIMCHER and KARA 1992; TING and BALDWIN 1993), studies in transgenic MHC class II-deficienct mice (DORN et al. 1987; VAN EWIJK et al. 1988) and in vitro transcription experiments (HUME and LEE 1990; DURAND et al. 1994) have led to the concept that the MHC class II promoter behaves as a single functional unit in which the different boxes all contribute to optimal promoter activity by specifically binding different nuclear proteins in a coordinated manner.

Two types of molecular defects affecting different *trans*-acting promoter-binding proteins have been discovered in patients with MHC class II deficiency (REITH et al. 1995; MACH et al. 1996). RFX is a complex of nuclear proteins binding to the X box of MHC class II promoter regions and consists of at least two subunits, p75 and p36. By in vitro binding assays, a defect in the binding of RFX to the X box was detected in patients with MHC class II deficiency belonging to complementation groups B and C, while in patients from complementation group A binding of all factors including RFX to the MHC promoter region was normal (REITH et al. 1988; STIMAC et al. 1991; HERRERO SANCHEZ et al. 1992; HASEGAWA et al. 1993; DURAND et al. 1994). In accordance, analysis of MHC class II promoter occupation in vivo showed unoccupied promoter regions in RFX-deficient cells, while in vivo promoter occupation was normal in cells where normal in vitro binding of *trans*-regulatory factors such as RFX to the MHC class II promoter regions was detected.

A genetic complementation approach based on the transfection of mammalian cDNA expression libraries into MHC class II-deficient patient cell lines to functionally complement the defect led to the identification of the genetic defects in patients belonging to complementation groups A and C (HUME and LEE 1989; BENICHOU and STROMINGER 1991; STEIMLE et al. 1993, 1995; STEIMLE and MACH 1995). Transfection of a 4.5-kb cDNA called CIITA into all cell lines belonging to complementation group A restored normal levels of HLA-DR, -DQ and -DP (STEIMLE et al. 1993). Subsequently, mutations in the CIITA gene (gene deletions and point mutations) were detected in patient cell lines from complementation group A, and all mutated CIITA alleles were shown to be inactive in transfection experiments, thereby confirming that the defect in MHC class II expression observed in complementation group A is caused by mutations in the CIITA gene (STEIMLE et al. 1993).

The same genetic complementation approach was applied to investigate the defect in patient cell lines of complementation group C. A 3.4-kb cDNA was isolated that was capable of restoring normal MHC class II expression (STEIMLE et al. 1995) and by sequence analysis was shown to encode a new family member of the X box-binding proteins RFX1 to 4, which was therefore named RFX5. Patients from complementation group C showed mutations in RFX5, confirming that gene defects in RFX5 are responsible for the deficient MHC class II phenotype in these patients. RFX5 is the p75 subunit of the RFX complex found to display defective binding activity in complementation group C (REITH et al. 1988; DURAND et al. 1994; STEIMLE et al. 1995).

Genetic defects leading to defective RFX binding are most likely heterogeneous, since not only patients with a defect in RFX5, but also cell lines belonging to other complementation groups display defective RFX binding activity (REITH et al. 1988; STIMAC et al. 1991; HERRERO SANCHEZ et al. 1992; HASEGAWA et al. 1993; CHIN et al. 1994; DURAND et al. 1994). In these patients, the defect or defects could affect mutations in p36, the smaller subunit of RFX, in another yet unidentified RFX subunit, or in a *trans*-acting factor regulating binding activity of RFX.

Recently, twin brothers with MHC class II deficiency belonging to a new, previously unrecognised complementation group were described and were characterised by an isotype-, chain- and cell-specific residual expression of certain MHC class II genes (HAUBER et al. 1995; DOUHAN et al. 1996). In addition, these patients can be distinguished from previously described patients by the relatively benign clinical symptomatology and by the unexpected finding that T cell and antibody responses could be induced following vaccination, at least against some antigens such as tetanus toxoid (WOLF et al. 1995). Elucidation of the molecular defects affecting these and other patients with residual expression of some MHC class II genes (PEIJNENBURG et al. 1995) could provide additional insights in dyscoordinate and cell-specific levels of the regulation of MHC class II expression. The regulatory defect observed in these families, at a level affecting only certain MHC class II genes and probably certain cell types more than others, is in marked contrast to the situation observed in patients with a defect in RFX5 or CIITA, defects that clearly affect the basic and coordinate regulation of MHC class II gene expression.

5 Biological Consequences of MHC Class II Deficiencies

5.1 Role of MHC Class II in $CD4^+$ T Cell Development

The capacity of the immune system to discriminate between self and non-self is largely defined by the T cell repertoire. Mature T cells exclusively respond to foreign antigen presented on self MHC molecules, a phenomenon called MHC restriction of antigen-induced T cell activation. Two mechanisms are responsible for the development of a T cell repertoire that is self-tolerant and self-MHC restricted, namely positive and negative selection, both processes which T cells undergo during thymic development. Negative selection eliminates T cells that express TCR with potential specificity for self determinants by clonal deletion, while positive selection selects T cells that are capable of recognising foreign antigens in the context of self-MHC. In addition to TCR-dependent thymic selection mechanisms that regulate the appropriate T cell repertoire, signals have to be delivered to immature $CD4^+CD8^+$ double-positive thymocytes to promote selective commitment to either the $CD4^+$ or the $CD8^+$ single-positive lineage. Although the exact mechanisms regulating the intrathymic development of $CD4^+$ or $CD8^+$ single-positive T cells are still under debate (GUIDOS 1996), interaction of TCR and CD4 as a co-receptor with MHC class II molecules expressed on cells within the thymus is considered to be critical for the development and maturation of the $CD4^+$ single-positive T cell lineage, and interaction of TCR and CD8 as a co-receptor with MHC class I antigens is important for selection of $CD8^+$ single-positive T cells.

The first set of experimental evidence in support of the importance of MHC class II in the development of $CD4^+$ T cells came from treatment of mice with antibodies against MHC class II (KRUISBEEK et al. 1985a). Long-term treatment of newborn mice with anti-MHC class II had a profound effect on T cell development (KRUISBEEK et al. 1983) and resulted in the absence of mature T cells of the $CD4^+CD8^-$ T cell lineage (KRUISBEEK et al. 1985a). Mice treated with anti-Ia in the neonatal period had markedly reduced Ia antigen expression in thymic and splenic tissues (FULTZ et al. 1982; KRUISBEEK et al. 1983) and showed an abrogation of self-class II-restricted and non-self class II-specific allo-T cell proliferation (KRUISBEEK et al. 1983, 1985b). Exogenous interleukin (IL)-2 restored the defect, indicating that the reason for the impaired alloresponse is a malfunction in the IL-2-producing cell population known to be mainly associated with the CD4 subset. Spleen cells from anti-MHC class II-treated mice had a significantly reduced IL-2 production as compared to normal mice, and the frequency of IL-2-producing cells was also markedly decreased. Antigen-presenting function is practically abrogated in these animals. The defective T cell-proliferative response correlated with defective thymic APC function (KRUISBEEK et al. 1985a). Further investigations in mice treated with anti-Ia antibody revealed that $CD4^+$ T cells ($LYT2^-L3T4^+$ T cells) were virtually absent in the spleen and in the thymus, while the other T cell subsets ($LYT2^+L3T4^+$ cortical immature T cells, $LYT2^+L3T4^-$ functional CTL precursors and $LYT2^-L3T4^-$ precursor thymocytes) were present in normal numbers.

Further experimental evidence that MHC class II molecules are critical for the development of $CD4^+$ T cells by positive selection comes from the analysis of transgenic mice that express rearranged TCR of defined specificity and MHC restriction (THE et al. 1988; BENOIST and MATHIS 1989; BERG et al. 1989; BILL and PALMER 1989; SCOTT et al. 1989). The most compelling evidence for the role of MHC class II in T cell development comes from mice made MHC class II deficient by gene targeting. These mice show an almost complete absence of mature $CD4^+CD8^-$ T cells in the thymus and the periphery (COSGROVE et al. 1991; GRUSBY et al. 1991; KONTGEN et al. 1993). In invariant chain-deficient mice, defective MHC class II-dependent selection of $CD4^+$ T cells by thymic epithelial cells could also be observed and leads to a reduction in the numbers of peripheral $CD4^+$ T cells to approximately 25% of normal and to alterations in the T cell repertoire due to the inability of normal presentation of self-peptides in association with MHC class II during intrathymic T cell development (TOURNE et al. 1995; WONG and RUDENSKY 1996).

Two models tried to explain how immature $CD4^+CD8^+$ (double-positive) T cells become committed to either the CD4 or the CD8 lineage (GRUSBY and GLIMCHER 1995). According to the stochastic model, $CD4^+CD8^+$ cells randomly down-regulate either one of the co-receptor molecules; transitional cells expressing a TCR and CD4 or CD8 bind to MHC class II or I, respectively, and undergo positive selection. The second model suggests that ligation of the TCR and either CD4 or CD8 instructs the cell to down-regulate the other unused co-receptor (instructional model). There is experimental evidence for both the stochastic (GUIDOS et al. 1990; CROMPTON et al. 1992) and the instructional hypothesis (BORGULYA et al. 1991; ROBEY et al. 1991). It also appears feasible that the instructional model could apply in the differentiation for $CD4^+$ and the stochastic for $CD8^+$ T cells. The fact that CD4 transitional cells ($CD4^+CD8^{lo}$) can be detected in MHC class II-deficient mice (CHAN et al. 1993; CRUMP et al. 1993; GRUSBY and GLIMCHER 1995) that show a marked reduction in the numbers of mature $CD4^+$ single-positive T cells, and that $CD4^{lo}CD8^+$ cells (CD8 transitional cells) are detectable in β_2-microglobulin-deficient mice that lack mature $CD8^+$ single-positive T cells would favour, but does not prove the stochastic hypothesis.

In normal mice, CD4 transitional cells that display a TCR restricted for MHC class II undergo positive selection, while those cells that display a class I-restricted TCR are not selected and die. In MHC class II-deficient mice, however, $CD4^+$ transitional cells become arrested in their development due to the lack of positive selection. Transitional cells in MHC class II-deficient mice have down-regulated the expression of both the recombination activating gene (RAG) and the terminal deoxynucleotidyl transferase (TdT) gene while expressing the activation marker CD69 (CHAN et al. 1993; CRUMP et al. 1993), indicating that $CD4^+$ transitional cells had received an activation signal in an MHC class II-deficient environment, possibly through MHC class I. Consistent with this finding, animals defective in the expression of both MHC class I and II do have CD4 and CD8 transitional cells, but these cells do not overexpress the TCR and the CD69 structure to a comparably high extent as observed on transitional cells in the single deficient animals (CHAN et al. 1993; CRUMP et al. 1993).

It has thus been speculated that a combination of stochastic and instructional processes may subsequently be operational. The MHC-independent random down-regulation of the co-receptor molecules is followed by up-regulation of TCR due to an MHC-derived signal independent of CD4 or CD8 lineage commitment; on these transitional cells, expression of TCR and co-receptor molecule with the same MHC restriction then leads to positive selection on this MHC antigen.

The requirement for MHC class II on radioresistant thymic nursing cells for the development of mature CD4$^+$ T cells is well agreed upon (BEVAN 1977; FINK and BEVAN 1978; ZINKERNAGEL et al. 1978a,b), while the question whether bone marrow-derived cells are also capable of this function has been controversial (LONGO and SCHWARTZ 1980; LONGO and DAVIS 1983; LONGO et al. 1985). Information is provided by experimental and clinical results. Irradiated MHC class II-deficient mice were reconstituted with T cell-depleted bone marrow from control animals and irradiated control mice were treated with bone marrow from MHC class II-deficient animals. The former did not develop CD4$^+$ cells when reconstituted with bone marrow from control donor mice, while mature CD8$^+$ cells could be detected. Normal mice reconstituted with MHC class II-deficient bone marrow were able to develop both CD4$^+$ and CD8$^+$ cells. These findings indicate that thymic class II-expressing cells are required for normal CD4$^+$ T cell development and that bone marrow-derived cells cannot take over as substitutes. The effect of MHC molecules on shaping the T cell repertoire has been addressed by several investigators using transgenic mice (KISIELOW et al. 1988; THE et al. 1988; BILL and PALMER 1989; SCOTT et al. 1989). It was demonstrated in studies performed in mice expressing transgenic TCR or MHC class II structures that positive selection of T cells requires interaction of the TCR with thymic major histocompatibility antigens (THE et al. 1988; SCOTT et al. 1989) and that positive selection of CD4$^+$ cells is mediated by MHC class II-positive cells in the thymic cortex (BILL and PALMER 1989). Further investigations extended this notion and demonstrated that positive selection in thymocyte differentiation is a rather late event occurring at the stage of maturation of cortical double-positive cells to single-positive T cells (BENOIST and MATHIS 1989). If MHC class II is only expressed in the thymic medulla, positive selection will be impaired. Furthermore, TCR–MHC interaction in the thymus is also responsible for the increased export of T cells into the periphery (BERG et al. 1989).

The findings obtained in MHC class II-deficient mice were confirmed and extended by results in MHC class II-deficient patients. Thymocytes obtained from the thymus of an MHC class II-negative patient contained reduced numbers of CD4$^+$CD8$^-$ T cells when compared to thymocytes derived from normal, MHC class II-expressing thymus. Furthermore, only one third of the patient's CD4$^+$CD8$^-$ thymocytes co-expressed the CD3 antigen, and the level of CD3 expression was lower than in control thymocytes. While double-positive CD4$^+$CD8$^+$ thymocytes seemed to be normal, the population of CD4$^-$CD8$^+$ thymocytes was significantly increased. Despite the absence of peripheral MHC class II expression, CD4$^+$CD8$^-$ T cells could be detected in the periphery of patients with MHC class II deficiency, albeit in significantly reduced numbers, and the majority of the CD4$^+$CD8$^-$ T cells co-expressed the CD45RO marker. These CD4$^+$ cells could have developed via MHC class I antigens or novel structures with MHC-like function (BACCALA et al. 1991; KELLY et al 1991; CHAN et al. 1993). Restrictions in the use of available TCR V-gene family pool were not observed, but some deviation in the usage of V-gene family members was seen. (EGGERMOND et al. 1993).

The question of TCR V-gene segment usage in $CD4^+CD8^-$ T cells of bare lymphocyte syndrome (BLS) patients has also been addressed in another study using TCR V-region-specific monoclonal antibodies and a semi-quantitative polymerase chain reaction (PCR) technique for Vα and Vβ gene region families (LAMBERT et al. 1992). These studies indicated that some of the Vα gene segments were used less frequently in the $CD4^+$ T cell population of patients with BLS, and TCR $V\alpha_{12}$ transcripts were greatly diminished in $CD4^+$ and $CD8^+$ cells of a patient, indicating a skewing in the usage frequency of some of the Vα gene segments (LAMBERT et al. 1992). More recently, the same group described that human T cell repertoire development in the absence of MHC class II expression results in a circulating $CD4^+CD8^-$ T cell population bearing TCR with changes in the physico-chemical properties of complementarity-determining region 3 (HENWOOD et al. 1996).

The impact of an MHC class II-negative thymic environment on lymphokine profiles of activated $CD4^+$ T cells has been examined in one patient, and the results indicate that $CD4^+CD8^-$ T cells matured in an MHC class II-deficient environment display lymphokine transcription patterns comparable to CD4 cells of normal, healthy individuals. (LAMBERT et al. 1993).

Patients with MHC class II deficiency and a defect of coordinated expression of all MHC class II structures have low numbers of $CD4^+$ T cells in the periphery even after many years of successful BMT and even if clinical symptoms of increased susceptibility to infections are absent. In a recent study, patients were analysed up to 10 years after BMT (KLEIN et al. 1995), and $CD4^+$ cells ranged between 100 and 650 per μl, certainly far below the age-related normal range. The patients were clinically healthy, the bone marrow applied for reconstitution was histo-identical in some and haplo-identical in others, and this obviously had little bearing on the CD4 lymphopaenia observed. Whether these low numbers of $CD4^+$ cells represent a special subset of the CD4 population or a subset with restricted specificity is not known. Taking information from humans and gene-targeted mice together, the results suggest that thymic cortical elements expressing MHC class II and shown to be radioresistant are crucial in the development and differentiation of $CD4^+$ T cells and cannot be replaced by cells of the bone marrow.

5.2 MHC Class II and B Cell Function

Even though direct evidence is scarce, it is the assumption that MHC class II expression might be necessary for normal B cell development and/or function. Immature pre-B cells in the bone marrow express MHC class II molecules (TARLINTON 1993), and antibodies against MHC class II inhibit differentiation of B cell precursors into mature B cells in vitro. It came as a surprise that the phenotype and function of B lymphocytes in MHC class II-deficient animals was normal (GRUSBY and GLIMCHER 1995). Flow cytometric analysis of pro-B, pre-B and mature B cells did not reveal detectable differences in numbers between normal and MHC class

II-deficient mice. Furthermore, markers detecting expression in mature B cell phenotypes provided comparable results. Unexpectedly, even though lymph nodes of MHC class II-deficient animals have abnormally low numbers of germinal centres, B cells are still able to terminally differentiate into plasma cells and secrete immunoglobulins of all isotypes, although circulating IgM is slightly increased and levels of IgG_1 are reduced.

The production of specific antibodies has been analysed in MHC class II-deficient mice immunised with trinitrophenyl-lipopolysaccaride (TNP-LPS), TNP-Ficoll or TNP-ovalbumin. TNP-specific antibodies predominantly, but not exclusively of the IgM isotype have been produced in equal or higher concentrations than in controls in the TNP-LPS- and TNP-Ficoll-immunised animals (MARKOWITZ et al. 1993), while TNP antibodies have not been elicited in TNP-ovalbumin-immunised animals. These results indicate that MHC class II-deficient animals are capable of immunoglobulin production and of switching from IgM to other isotypes, but that the B cells are incapable of producing antibodies to T-dependent antigens due to an impairment of cognate T cell-B cell interaction. Allogeneic T cells were able to induce B cells to proliferate and secrete immunoglobulin after pre-activation with immobilised CD3, indicating that B cells are capable of T cell-B cell interaction when the requirement for antigen recognition is bypassed.

Results from studies on B cell function in patients with MHC class II deficiency are not conclusive. While most of the patients are hypogammaglobulinaemic, the decrease in serum immunoglobulin levels is not uniform (KLEIN et al. 1993). Most patients with a defect in the coordinate expression of MHC class II genes show a severe impairment in the production of specific antibodies following infection or vaccination (SMITH et al. 1988; KLEIN et al. 1993). In vitro studies of B cell function have been performed and indicate that polyclonal activation of Ig production by pokeweed mitogen (PWM), staphlococcus aureus cowan (SAC) or anti-IgM antibody may be impaired (RIJKERS et al. 1987; CLEMENT et al. 1988). Two patients (KEN and KER) in whom MHC class II expression was absent when mononuclear cells were examined by flow cytometry, but who did express mRNA for DR-α, DQ-α and DP-β and had low amounts of expression of mRNA of the complementary chains in adherent cells as well, did produce antibodies, e.g. against tetanus toxoid, within the normal range following vaccination. MHC class II-deficient patients successfully treated by BMT do produce antibodies even if these are chimeras with recipient B cells and donor T cells (KLEIN et al. 1995).

5.3 Recognition of MHC Class I and II in MHC Deficiency

Before discussing allorecognition in MHC class I- and MHC class II-deficient mice and patients with BLS, the mechanisms involved in alloreactivity need to be analysed.

When skin completely lacking MHC class II is transplanted into normal allogeneic recipients, it is rejected with kinetics comparable to allografts from donors expressing MHC class II. When MHC class II-negative skin allografts are transplanted with only a few minor antigenic disparities between donor and recipient, survival of these grafts is significantly prolonged when the recipient is depleted of $CD4^+$ cells. These results indicate that $CD4^+$ cells are instrumental in allograft rejection even when MHC class II is not expressed on the graft (AUCHINCLOSS et al. 1993; GRUSBY and GLIMCHER 1995). When allograft recipient mice are depleted of $CD8^+$ T cells to eliminate recognition of MHC class I molecules, rapid rejection of grafts, even of MHC class II-deficient skin grafts, still occurs (LEE et al. 1994). These results indicate that rejection of allografts requires the presence of both $CD4^+$ and $CD8^+$ T cells.

The question of alloreactivity has been addressed in mice lacking MHC class I and class II molecules (GRUSBY et al. 1993). These mice were obtained by mating β2-microglobulin-deficient with class II-deficient animals and are depleted of $CD4^+$ and $CD8^+$ T cells in peripheral lymphoid organs, while the B cell and natural killer (NK) cell compartment is increased. These mice did not respond with T cell activation or antibody production to T cell-dependent antigens, but mounted specific antibody responses to T-independent antigens. The proliferative response of spleen cells of these MHC-deficient mice in the mixed lymphocyte reaction was hardly above background and could be completely abolished when the responding cell population was treated with CD4 monoclonal antibody plus complement, suggesting that a few $CD4^+$ cells were present in these animals and these cells provided the response. When spleen cells of MHC-deficient animals were used as stimulators, the response of normal spleen cells to these stimuli was minimal and again dependent on CD4. It came as a surprise, therefore, that allogeneic skin grafts in these MHC-deficient animals have been rejected in a similar time course as in non-MHC-deficient controls (13 vs. 9.5 days) and skin grafts from the MHC-deficient mice as donors were rejected as rapidly as from mice expressing MHC normally. Skin graft rejection in MHC-deficient mice is dependent on both $CD4^+$ and $CD8^+$ T cells (GRUSBY et al. 1993). The role of $CD4^+$ cells is controversial. They have been discussed as helper cells (GRUSBY et al. 1993) and as effector cells (DIERICH et al. 1993). It is beyond doubt, however, that cells from MHC-deficient mice are well recognised as foreign, results which are not surprising in view of the fact that discrimination between self and non-self is a crucial and archaic requirement in evolution and development.

Information on allorecognition in MHC class II-deficient humans is incomplete and preliminary. In one of our patients extensively studied, T cells, B cells and NK cells were present in the circulation in normal numbers. However, the total number of $CD4^+$ cells was substantially decreased. Only 10% (369 cells per mm^3) of the patient's circulating lymphocytes were $CD4^+$ as compared to 30%–40% (1000–1800 cells per mm^3) in healthy controls (MANNHALTER et al. 1994). $CD8^+$ cells were increased (60% of lymphocytes, 2128 cells per mm^3). All subsets of mononuclear cells (monocytes, B cells, activated T cells) were shown to lack MHC class II (MANNHALTER et al. 1991), as were EBV-transformed B cells (MANNHALTER et al. 1994) and gut epithelial cells (T. RADASZKIEWICZ, unpublished observation). MHC class I expression was normal (MANNHALTER et al. 1991). When a T-enriched cell population from the patient's peripheral blood was co-cultured with irradiated mononuclear cells from an MHC-incompatible volunteer, an alloresponse could be induced. This alloresponse was substantially reduced compared to the response obtained by T cells from the patient's histo-identical healthy brother stimulated with the same donor cell population. While the alloantigen-induced proliferation was inhibited by up to 95% by an antibody to MHC class II in the patient's brother, the inhibition did not exceed 10%–30% when the patient's T cells were tested (MANNHALTER et al. 1994). Cells of the MHC class II-deficient patient could also be primed for cytotoxicity following allogeneic stimulation. After stimulation with the cell line JY, the cytotoxicity expressed by the patient's cells was 14.6% specific lysis at an effector target ratio of 40:1, as compared to 37%±16% obtained with a panel of healthy control subjects (n=25) (MANNHALTER et al. 1991).

6 Immune Response, Susceptibility to Infection and Autoimmunity

6.1 MHC Class II-Deficient Mice

The importance of histocompatibility antigens in the immune response was discovered long before the exact role of these structures in immune recognition and effector function was determined. $CD4^+$ T cells need the antigen presented by MHC class II for clonal expansion. Thus antigen-specific T cell proliferation will be highly reduced or absent in gene-targeted mice completely lacking MHC class II and in MHC class II-deficient patients with a defect in coordinate expression of all HLA-D genes. As a consequence of defective MHC-TCR interaction, production of both IL-2 and certain other cytokines by $CD4^+$ cells is reduced upon antigen stimulation. Due to the defect in antigen-specific T cell activation and clonal expansion, the ability to respond to T cell-dependent antigens is highly reduced to absent.

When the MHC class II defect is incomplete, some of these functions may be preserved. Mice with a defect in invariant chain expression and humans with an incomplete defect of HLA-D gene expression will be able to mount CD4-dependent antigen-specific responses correlating to the extent of functional $CD4^+$ cells present. T-helper cell-dependent antibody responses were produced following infection or immunisation in invariant chain-deficient mice (BATTEGAY et al. 1996). Invariant chain-deficient mice mounted surprisingly efficient IgG responses against vesicular stomatitis virus (VSV) antigens, although presentation of VSV antigen to a specific T-helper cell hybridoma was reduced by a factor of more than 300; it has been reported that the efficiency of antigen presentation in invariant chain-deficient cells apparently depends upon the particular T cell hybridoma used for the experiment (VIVILLE et al. 1993). T-helper cell-independent cytotoxic T cell and B cell responses were normal in the absence of MHC class II or invariant chain. In contrast, a T-helper cell-dependent class switch from IgM to IgG was absent in class II-deficient mice, but was surprisingly efficient in invariant chain-deficient mice. Locally high intra- and/or extracellular protein concentrations, as caused by viral infections and destruction of infected cells, may be sufficient to overcome the Ii deficiency.

Recent studies performed in mice with absent invariant chain due to gene targeting provided further interesting information. Certain strains of mice are susceptible and others resistant to infection with the parasite *Leishmania major*, and resistance is dependent on $CD4^+$ cell function. The requirement for INF-γ in the control of the organisms in resistant strains has been clearly established, thus the T-helper-1 (TH1)-type response is crucial. *Leishmania* replicates productively in endolysosomal-like compartments of macrophages containing MHC class II molecules (ANTOINE et al. 1991; RUSSELL et al. 1992). Infection of macrophages reduced their capacity for MHC class II-dependent antigen presentation (FRUTH et al. 1993; PRINA et al. 1993), while trafficking of MHC class II to the cell surface was unimpaired. Since the role of the invariant chain is the protection of the peptide cleft of newly synthesised MHC class II molecules during transit from the ER (CRESSWELL 1996), it came as a surprise that resistance to *L. major*, known to be dependent on CD4 function, was unimpaired in invariant chain-deficient mice (BROWN et al. 1997), even though differentiation and/or function of $CD4^+$ cells is highly dependent on MHC class II and known to be impaired in mice lacking the invariant chain. Further studies in these mice revealed that APC are perfectly normal in their function of presenting *L. major* peptide antigens while showing a substantial impairment in the presentation of entire organisms, antigens that have to be processed before peptides can be expressed in the antigen cleft of MHC class II on the cell surface. It could also be demonstrated that bone marrow-derived APC from invariant chain-deficient animals were capable of inducing IL-2 release when pulsed with peptide antigens, whereas they were clearly deficient in this function when *L. major* live promastigotes or live amastigotes were used. Interestingly, INF-γ production was unimpaired in lymph node cells of invariant chain-deficient mice infected with *L. major* upon stimulation with *L. major* antigen, while proliferation and IL-2 production were significantly reduced. Taken together, these results indicate that protection against *L. major* infection known to be dependent on INF-γ production was functional in invariant chain-deficient mice, while IL-2 production and proliferation were strongly reduced. These results further suggest that, under certain conditions, an impairment in antigen presentation might affect IL-2 production and proliferation and leave INF-γ production intact. This could be explained by quantitative differences in T cell activation for IL-2 and INF-γ secretion or by two different pathways of activation with different requirements at the induction step. In another system in lymphocytes of patients with common variable immunodeficiency, a primary human immunodeficiency disease, we observed that IL-2 production and proliferation were impaired upon TCR stimulation, while the expression of an early activation marker, CD69, was not (V.Thon et al. in press).

In the interpretation of these results, we considered that one of two different pathways could be defective or an amplification pathway essential for IL-2 gene transcription could be impaired. If we continue to think along these lines, we realise that individual activation pathways in T cells might already require different signals at the level of induction. Other studies (SLOAN-LANCASTER and ALLEN 1996) could be further indications of this possibility. Experimental infection with lymphocytic choriomeningitis virus and with influenza virus in MHC class II-deficient mice (BODMER et al. 1993; LAUFER et al. 1993) indicated that specific antibodies were not induced, while a cytotoxic T lymphocyte (CTL) response was detectable within the normal range, proving that, in addition to the known alloresponse, a CTL response can also be induced against infected cells in the absence of MHC class II expression. MHC class II-deficient mice were able to clear influenza virus from their lung after infection, and clearance was definitely mediated by T cells, as antibodies against the respective viral antigens could not be detected at any time (BODMER et al. 1993).

In view of recent evidence, conditions involving an impairment of T cell activation but a functional B cell compartment may lead to autoimmune conditions. This is likely to be due to the fact that T cells are involved in keeping the delicate balance between the recognition of non-self and the control of autoreactivity. Gene-targeted mice with functional defects of T cell activation (e.g. IL-2 deficiency or defective TCR) and defects in T cell-regulatory function (e.g. IL-10 deficiency) develop inflammatory bowel disease (IBD) as one of their main clinical problems. IBD is also seen in MHC class II-deficient mice. MHC class II-deficient mice develop histological lesions resembling ulcerative colitis. Thus, irrespective of the exact nature of the underlying defect, impaired MHC class II-T cell interaction may favour the development of IBD. Mice that have B cells but lack functionally active T cells are likely candidates for this condition, while mice with severe combined immune deficiency lacking both B and T cells, such as RAG gene-targeted mice, do not develop disease. Thus it appears likely that impaired function of T cells and regulatory abnormalities of B cells must be present and play a role in the pathogenesis.

The possibility that autoimmune disease arises as a consequence of a dysregulated response to infection has been discussed. Circumstantial evidence suggests that the IBD observed is not due to a specific pathogen. The fact that the gene-targeted animals do develop IBD even when housed under specific pathogen-free conditions and that a similar pathology has been observed in gene-targeted animal colonies in different parts of the world, while combined (both B and T) cell-deficient mice housed under similar conditions do not develop the disease, points to no or insignificant contribution of infection to this pathology.

MHC class II-deficient mice are well suited to help to define the role of individual immune mechanisms in the pathophysiology of certain autoimmune diseases. In these animal models, the pathogenetic significance of autoantibodies and/or T cells, e.g. CTL, as well as requirements for T and B cells and/or T cells and non-specific effector cells, such as NK cells or macrophages, can be characterised. One of the models which has addressed the question was the MRL/LPR gene-targeted mouse. These animals develop lymphadenopathy due to the expansion of $CD4^-CD8^-$ T cells and systemic lupus with autoantibodies against nucleic acids, glomerulonephritis and arthritis (THEOFILOPOULOS and DIXON 1981). MHC class II-deficient MRL/LPR gene-targeted mice failed to develop high levels of IgM and IgG autoantibodies specific for nucleic acids and did not develop autoimmune nephritis, while the development of lymphadenopathy due to the expansion of $CD4^-CD8^-$ T cells did occur (JEVNIKAR et al. 1994). These experiments certainly clarified the contribution of different lymphocyte subsets in the individual features of this model.

Myasthenia gravis is an autoimmune disease known to be caused by autoantibodies against the nicotinic acetylcholine receptor on nerve junctions. Results of studies in MHC class II-deficient mice confirmed the hypothesis of autoantibodies as the only pathogenetic mechanism in this disease, as MHC class II-deficient mice did not develop myasthenia gravis upon immunisation with purified acetylcholine receptor (KAUL et al. 1994).

6.2 MHC Class II-Deficient Patients

Patients with a defect in coordinate expression of MHC class II failed to respond with antibody production to ubiquitous and/or vaccination antigens. The clinical symptomatology in these patients is dominated by the well-known problems of antibody-deficient individuals with recurrent and/or chronic sinopulmonary infections and/or gastrointestinal disease. In vitro tests reveal an impairment in antigen presentation leading to absent antigen-induced clonal expansion as demonstrated by absent proliferative responses to ubiquitous and vaccination recall antigens (MANNHALTER et al. 1991; GRISCELLI et al. 1993; KLEIN et al. 1993).

We specifically addressed this question in one of our patients (MANNHALTER et al. 1991). In this patient, the absolute numbers of T cells and B cells were within the normal range, and mitogen-induced and alloantigen-induced lymphocyte proliferation was normal or close to normal. The alloactivated T cells of the patient did not express MHC class II molecules. However, proliferation and expression of other activation markers, e.g. CD25, were within the normal range. The complete lack of MHC class II on the patient's cells led to severe deficiencies of MHC class II-mediated immune functions such as antigen presentation. In good agreement with others (GRISCELLI et al. 1993; KOVATS et al. 1994), an absence of T cell responses to recall antigens, e.g. tetanus toxoid, tick-borne meningoencephalitis virus antigen or purified protein derivative (PPD) was found. Since a histo-identical sibling was available for study, so-called "criss-cross" experiments could be performed, i.e. T cells from the patient could be cultured in the presence of antigen and autologous accessory cells or in the presence of antigen and accessory cells derived from the healthy histo-identical brother and vice versa (brother's T cells cultured with accessory cells from the brother or the patient and antigen). The results of these studies demonstrated that an antigen-specific proliferative response could not be demonstrated when cells from the patient were used either as APC or as responders (in all combinations employed) (MANNHALTER et al. 1991). Interestingly, other monocyte functions, such as the capacity to release IL-1, were normal. An explanation for the absent response of the patient's T cells to antigen presented by his brother's monocytes can be found in the hypothesis that the patient's T cells have never been properly primed to recall antigens.

The HLA class II-like genes DMA and DMB have been implicated in antigen presentation and are also reduced or absent in patients with BLS. Fusion of different complementation types of BLS cells known to restore the expression of class II genes also restores the expression of DM gene product and the ability of antigen presentation. It therefore appears likely that expression of MHC class II is functionally linked to the expression of products of the DM gene and that a coregulation of the two pathways is usually operational. Whether this is always the case will have to be re-examined in the light of more recent results showing that exceptions in coordinate regulation of all MHC class II genes do occur (HAUBER et al. 1995, 1996).

The function of antigen presentation is obviously linked to the three-dimensional structure of MHC class II expressed in the cell membrane. When BLS-APC are transfected with MHC class II structural genes, the function of antigen presentation is not restored (KOVATS et al. 1994). However, when MHC expression is achieved by either fusion of cells from two patients belonging to different complementation groups or when the genetically absent or mutated *trans*-activating factor is transfected, MHC class II expression will be connected with the functional restoration of antigen presentation.

In two MHC class II-deficient patients (HLA-identical twins), the defect was limited to the expression of one of the chains each of HLA-DR, -DQ and -DP (HAUBER et al. 1995; WOLF et al. 1995). These patients had no detectable MHC class II on the surface of their mononuclear cells, and molecular characterisation revealed an HLA-DRα^+, -DRβ^-, -DQα^+, -DQβ^-, -DPα^-, -DPβ^+, Ii$^+$ phenotype. While flow cytometric analysis failed to detect MHC class II protein on the surface of EBV B cells (and this finding was confirmed by immunoblotting), HLA-DRβ protein could be detected in lysates of the adherent mononuclear cell fraction (HAUBER et al. 1995). These results indicated a cell-specific difference of MHC class II expression in these patients. Results of investigations looking for antigen-specific response revealed that these patients were able to mount a humoral response to tetanus toxoid after vaccination and that the patients' mononuclear cells did respond by proliferation upon tetanus toxoid stimulation (WOLF et al. 1995). In contrast to these findings, EBV-transformed cells were unable to present antigen to an HLA-DR-compatible tetanus toxoid-specific cell line (HAUBER et al. 1995). The antigen-specific response by the patients' mononuclear cells could be explained by the notion that very low numbers of MHC–peptide complexes are sufficient to stimulate T cells. Obviously, biological assessment is more sensitive than detection by flow cytometry.

In addition to the role in antigen presentation, MHC class II molecules are involved in T cell activation by bacterial superantigens such as SE. SE bind to MHC class II molecules on accessory cells without a requirement for antigen processing, and cross-linking of TCR by interaction of MHC-bound SE with a site on the V region of the TCR β-chain outside of the antigen-binding groove induces T cell activation, cytokine release and T cell proliferation. The results depicted in Table 1 were obtained in three patients from two unrelated families and show that, depending on the residual expression of MHC class II molecules on the surface of cells within the mononuclear cell fraction, even if expression is too low to be detectable by flow cytometry, T cells from MHC class II-deficient patients will be able to respond to SE stimulation. Twin brothers from one family (KEN and KER) who show a residual expression of HLA-DR protein (α- and β-chain) in their mononuclear cell fraction, albeit at very low levels, and have the capacity to mount a T cell response following tetanus vaccination (WOLF et al. 1995) showed a normal proliferative response to SE stimulation (Table 1). In contrast, T cells from patient D.Y., who showed a defective expression of all MHC class II genes in resting mononuclear cells (HAUBER et al. 1996), were unable to respond to SE, although the response to mitogenic stimulation (PHA) was normal. This patient was also unable to mount a T cell response to recall antigen (HORNEFF et al. 1994).

Susceptibility to viral infection in MHC class II-deficient patients depends upon the nature of the infecting agent. Most of these patients are known to suffer from severe viral infections caused by enterovirus, adenovirus and cytomegalovirus (CMV). In contrast, neither viruses causing childhood diseases nor live virus vaccines caused life-threatening problems. The same applies for influenza. Knowing that lung

Table 1. T cell proliferative response to superantigen (SE-A, SE-E) in three patients with MHC class II deficiency from two unrelated families

	[^{3}H]Thymidine incorporation (dpm)			
	SE-A	SE-E	PHA	Medium
Controls (*n*=5)	87667 ± 39514	87683 ± 35470	39178 ± 15785	469 ± 219
Patients				
KEN	146466 ± 14937	125534 ± 9754	n.t.	453 ± 105
KER	119954 ± 2962	86254 ± 9701	n.t.	332 ± 69
D.Y.	422 ± 41	380 ± 73	33676 ± 1026	465 ± 115

Peripheral blood mononuclear cells (1×10^5 per well and 200 μl) were incubated for 5 days in 96-well microtitre plates in the presence of bacterial superantigens (SE-A and -E; final concentration, 10 ng/ml), mitogen (phytohaemagglutinin, PHA; final dilution, 1:1250) or medium alone (RPMI-1640 medium; Flow Laboratories, Irvine, England) supplemented with 10% pooled, heat-inactivated (30 min, 56°C) human AB serum, penicillin (100 IU/ml), streptomycin (100 μg/ml) and L-glutamine (2 m*M*). T cell proliferative responses were assessed by measuring [^{3}H]thymidine incorporation. Results are given as dpm: mean±SD of triplicate cultures (patients) or mean±SD of the average values of triplicate cultures obtained in four unrelated healthy controls and the mother of KEN and KER (controls). *dpm*, disintegrations per minute

clearance of influenza will proceed in the complete absence of antibodies in MHC class II-deficient mice, it appears likely that similar mechanisms are functional in MHC class II-deficient patients as well.

One of our patients and several other MHC class II-deficient patients survived Bacille Calmette-Guérin (BCG) vaccination, which was frequently performed in Europe until a decade ago. Normal cytotoxic T cell functions, the capacity to respond to antigen presented in the context of MHC class I and normal NK activity may have been relevant in this context (MANNHALTER et al. 1991).

One of the most prominent clinical features in patients with MHC class II deficiency is their failure to thrive. Gastrointestinal symptoms such as malabsorption, diarrhoea or simply the inability to gain weight are present in the great majority of the patients observed, and the true reasons are unknown. The detection of enterovirus infection in the majority of these patients might certainly be one of the reasons for this clinical condition, but cannot be taken as the only explanation. It is well known that MHC class II is expressed on the luminal surface of the gastrointestinal mucosa, but the role of MHC class II in this location is not well understood. It has been claimed that the MHC class II observed in this location might come from migratory cells embedding below or in between mucosal cells. The morphology and uniformity of MHC class II expression, however, certainly does not support this hypothesis. MHC class II-deficient patients reconstituted by BMT could help to clarify this unsolved problem. The function of MHC class II in this location might be involved in local defence against viral, bacterial and/or parasitic infections. It might also be related to the binding, elimination and/or neutralisation of superantigens or, more generally, of toxins generated in the intestinal lumen. Interestingly, after BMT, patients do not have the gastrointestinal symptoms described, and children suffering from gastrointestinal diseases or who were underweight and underdeveloped clearly catch up after BMT.

Added in Proof. After submission of the manuscript, a genetic defect in RFXAP, a 36 kD subunit of the RFX complex, was identified in a mutant cell line belonging to complementation group D (DURAND et al. 1997). Subsequently, mutations in the RFXAP gene were described in several patients with MHC class II deficiency and these patients were also classified as belonging to complementation group D (VILLARD et al. 1997).

References

Acha Orbea H, McDevitt HO (1990) The role of class II molecules in development of insulin-dependent diabetes mellitus in mice, rats and humans. Curr Top Microbiol Immunol 156:103–19

Antoine J-C,Jouanne C, Lang T, Prina E, de Chastellier C, Frehel C (1991) Localization of major histocompatibility complex class II molecules in phagolysosomes of murine macrophages infected with Leishmania amazonensis. Infect Immunol 59:764–775

Auchincloss HJ, Lee R, Shea S, Markowitz JS, Grusby MJ, Glimcher LH (1993) The role of "indirect" recognition in initiating rejection of skin grafts from major histocompatibility complex class II-deficient mice. Proc Natl Acad Sci USA 90:3373–3377

Babbitt BP, Allen PM, Matsueda G, Haber E, Unanue ER (1985) Binding of immunogenic peptides to Ia histocompatibility molecules. Nature 317:359–361

Baccala R, Kono DH, Walker S, Balderas RS, Theofilopoulos AN (1991) Genomically imposed and somatically modified human thymocyte V beta gene repertoires. Proc Natl Acad Sci USA 88:2908–2912

Bach FH, van Rood JJ (1976) The major histocompatibility complex – genetics and biology. New Engl J Med 295:806–813, 872–878, 927–936

Battegay M, Bachmann MF, Burhkart C, Viville S, Benoist C, Mathis D, Hengartner H, Zinkernagel RM (1996) Antiviral immune responses of mice lacking MHC class II or its associated invariant chain. Cell Immunol 167:115–121

Benacerraf B (1981) Role of MHC gene products in immune regulation. Science 212:1229–1238

Benichou B, Strominger JL (1991) Class II antigen-negative patient and mutant cell lines represent at least three, and probably four, distinct genetic defects defined by complementation analysis. Proc Natl Acad Sci USA 88:4285–4288

Benoist C, Mathis D (1989) Positive selection of the T cell repertoire: where and when does it occur. Cell 58:1027–1033

Benoist C, Mathis D (1990) Regulation of major histocompatibility complex class-II genes: X, Y and other letters of the alphabet. Annu Rev Immunol 8:681–715

Berg LJ, Pullen AM, Fazekas de St. Groth B, Mathis D, Benoist C, Davis MM (1989) Antigen/MHC-specific T cells are preferentially exported from the thymus in the presence of their MHC ligand. Cell 58:1035–1046

Bevan MJ (1977) In a radiation chimaera, host H-2 antigens determine immune responsiveness of donor cytotoxic cells. Nature 269:417–418

Bikoff EK, Huang L-Y, Episkopou V, van Meerwijk J, Germain RN, Robertson EJ (1993) Defective major histocompatibility complex class II assembly, transport, peptide acquisition, and CD4+ T cell selection in mice lacking invariant chain expression. J Exp Med 177:1699–1712

Bikoff EK, Germain RN, Robertson EJ (1995) Allelic differences affecting invariant chain dependency of MHC class II subunit assembly. Immunity 2:301–310

Bill J, Palmer E (1989) Positive selection of CD4+ T cells mediated by MHC class II-bearing stromal cell in the thymic cortex. Nature 341:649–651

Bodmer H, Obert G, Chan S, Benoist C, Mathis D (1993) Environmental modulation of the autonomy of cytotoxic T lymphocytes. Eur J Immunol 23:1649–1654

Borgulya P, Kishi H, Müller U, Kirberg J, von Boehmer H (1991) Development of the CD4 and CD8 lineage of T cells: instruction versus selection. EMBO J 10:913–918

Brown DR, Swier K, Moskowitz NH, Naujokas MF, Locksley RM, Reiner SL (1997) T helper subset differentiation in the absence of invariant chain. J Exp Med 185:31–41

Busch R, Vturina IY, Drexler J, Momburg F, Hammerling J (1995) Poor loading of major histocompatibility complex class II molecules with endogenously synthesized short peptides in the absence of invariant chain. Eur J Immunol 25:48–53

Capecchi MR (1989) Altering the genome by homologous recombination. Science 244:1288–1292

Chan SH, Cosgrove D, Waltzinger C, Benoist C, Mathis D (1993) Another view of the selective model of thymocyte selection. Cell 73:225–236

Chang CH, Guerder S, Hong SC, van Ewijk W, Flavell RA (1996) Mice lacking the MHC class II transactivator (CIITA) show tissue-specific impairment in MHC class II expression. Immunity 4(2):167–178

Chin K, Mao C, Skinner C, Riley JL, Wright KL, Moreno CS, Stark GR, Boss JM, Ting JPY (1994) Molecular analysis of G1B and G3A IFN-gamma mutants reveals that defects in CIITA or RFX result in defective class II MHC and Ii gene induction. Immunity 1:687–697

Clement LT, Plaeger-Marshall S, Haas A, Saxon A, Martin AM (1988) Bare lymphocyte syndrome: consequences of absent class II major histocompatibility antigen expression for B lymphocyte differentiation and function. J Clin Invest 81:669–675

Cosgrove D, Gray D, Dierich A, Kaufman J, Lemeur M, Benoist C, Mathis D (1991) Mice lacking MHC class II molecules. Cell 66:1051–1066

Cresswell P (1994a) Antigen presentation. Getting peptides into MHC class II molecules. Curr Biol 4:541–543

Cresswell P (1994b) Assembly, transport, and function of MHC class II molecules. Annu Rev Immunol 12:259–293

Cresswell P (1996) Invariant chain structure and class II function. Cell 84:505–507

Crompton T, Pircher H, MacDonald HR (1992) CD4+8- thymocytes bearing major histocompatibility complex class I-restricted T cell receptors: evidence for homeostatic control of early stages of CD4/CD8 lineage development. J Exp Med 176:903–907

Crump AL, Grusby MJ, Glimcher LH, Cantor H (1993) Thymocyte development in major histocompatibility complex-deficient mice: evidence for stochastic commmitment to the CD4 and CD8 lineages. Proc Natl Acad Sci USA 90:10739–10743

De Preval C, Lisowska-Grospierre B, Loche M, Griscelli C, Mach B (1985) A trans-acting class II regulatory gene unlinked to the MHC controls expression of HLA class II genes. Nature 318:291–293

De Preval C, Hadam MR, Mach B (1988) Regulation of genes for HLA class II antigens in cell lines from patients with severe combined immunodeficiency. New Engl J Med 318:1295–1300

Dierich A, Chan SH, Benoist C, Mathis D (1993) Graft rejection by T cells not restricted by conventional major histocompatibility complex molecules. Eur J Immunol 23:2725–2728

Dorn A, Durand B, Marfing C, Le Meur M, Benoist C, Mathis D (1987) Conserved major histocompatibility complex class II boxes – X and Y – are transcriptional control elements and specifically bind nuclear proteins. Proc Natl Acad Sci USA 84:6249–6253

Douhan J III, Hauber I, Eibl MM, Glimcher LH (1996) Genetic evidence for a new type of major histocompatibility complex class II combined immunodeficiency characterized by a dyscoordinate regulation of HLA-D alpha and beta chains. J Exp Med 183:1063–1069

Durand B, Kobr M, Reith W, Mach B (1994) Functional complementation of MHC class II regulatory mutants by the purified X box binding protein RFX. Mol Cell Biol 14:6839–6847

Durand B, Sperisen P, Emery P, Barras E, Zufferey M, Mach B, Reith W (1997) RFXAP, a novel subunit of the RFX DNA binding complex is mutated in MHC class II deficiency. EMBO J 16:1045–1055

Eggermond MCJA, Rijkers GT, Kuis W, Zegers BJM, van den Elsen PJ (1993) T cell development in a major histocompatibility complex class II-deficient patient. Eur J Immunol 23:2585–2591

Elliott EA, Drake JR, Amigorena S, Elsemore J, Webster P, Mellman I, Flavell RA (1994) The invariant chain is required for intracellular transport and function of major histocompatibility complex class II molecules. J Exp Med 179:681–694

Fink PJ, Bevan MJ (1978) H-2 antigens of the thymus determine lymphocyte specificity. J Exp Med 148:766–775

Fruth U, Solioz N, Louis JA (1993) Leishmania major interferes with antigen presentation by infected macrophages. J Immunol 150:1857–1864

Fultz MJ, Scher I, Finkelman FD, Kincade P, Mond JJ (1982) Neonatal suppression with anti-Ia antibody. I. Suppression of murine B cell development. J Immunol 129:992–995

Fung-Leung WP, Mak TW (1992) Embryonic stem cells and homologous recombination. Curr Opin Immunol 4:187–194

Gladstone P, Pious D (1978) Stable variants affecting B cell alloantigens in human lymphoid cells. Nature 271:459–461

Glimcher LH, Kara CJ (1992) Sequences and factors: a guide to MHC class-II transcription. Annu Rev Immunol 10:13–49

Griscelli C, Lisowska-Grospierre B, Mach B (1993) Combined immunodeficiency with defective expression in MHC class II genes. In: Rosen FS, Seligmann M (eds) Immunodeficiencies. Harwood Academic, Chur, Switzerland, pp 141–154

Grusby MJ, Glimcher LH (1995) Immune response in MHC class II-deficient mice. Annu Rev Immunol 13:417–435

Grusby MJ, Johnson RS, Papaioannou VE, Glimcher LH (1991) Depletion of CD4+ T cells in major histocompatibility complex class II-deficient mice. Science 253:1417–1420

Grusby MJ, Auchincloss H, Lee R, Johnson RS, Spencer JP, Zulstra M, Jaenisch R, Papaioannou VE, Glimcher LH (1993) Mice lacking major histocompatibility complex class I and class II molecules. Proc Natl Acad Sci USA 90:3913–3917

Guidos CJ (1996) Positive selection of CD4+ and CD8+ T cells. Curr Opin Immunol 8:225–232

Guidos CJ, Danska JS, Fathman CG, Weissman IL (1990) T cell receptor-mediated selection of autoreactive T lymphocyte precursors occurs after commitment to the CD4 or CD8 lineages. J Exp Med 172:835–845

Hasegawa SL, Riley JL, Sloan JH III, Boss JM (1993) Protease treatment of nuclear extracts distinguishes between class II MHC X1 box DNA-binding proteins in wild-type and class II-deficient B cells. J Immunol 150:1781–1793

Hauber I, Gulle H, Wolf HM, Maris M, Eggenbauer H, Eibl MM (1995) Molecular characterization of major histocompatibility complex class II gene expression and demonstration of antigen-specific T cell response indicate a new phenotype in class II-deficient patients. J Exp Med 181:1411–1423

Hauber I, Wolf HM, Eibl MM, Wahn V (1996) More on MHC class II deficiency. New Engl J Med 335:977–978

Henwood J, van Eggermond MCJA, van Boxel-Dezaire AHH, Schipper R, den Hoedt M, Peijnenburg A, Sanal TM, Ersoy F, Rijkers GT, Zegers BJM, Vossen JM, van Tol MJD, van den Elsen PJ (1996) Human T cell repertoire generation in the absence of MHC class II expression results in a circulating CD4+CD8- population with altered physicochemical properties of complementarity-determining region 3. J Immunol 156:895–906

Herman A, Kappler JW, Marrack P, Pullen AM (1991) Superantigens: mechanisms of T cell stimulation and role in immune responses. Annu Rev Immunol 9:745–772

Herrero Sanchez C, Reith W, Silacci P, Mach B (1992) The DNA-binding defect observed in major histocompatibility complex class II regulatory mutants concerns only one member of a family of complexes binding to the X boxes of class II promoters. Mol Cell Biol 12:4076–4083

Horneff G, Seitz RC, Stephan V, Wahn V (1994) Autoimmune haemolytic anaemia in a child with MHC class II deficiency. Arch Dis Child 71:339–342

Hume CR, Lee JS (1989) Congenital immunodeficiencies associated with absence of HLA class II antigens on lymphocytes result from distinct mutations in trans-acting factors. Hum Immunol 26:288–309

Hume CR, Lee JS (1990) Functional analysis of cis-linked regulatory sequences in the HLA DRA promoter by transcription in vitro. Tissue Antigens 36:108–115

Jevnikar AM, Grusby MJ, Glimcher LH (1994) Prevention of nephritis in major histocompatibility complex class-II-deficient Mrl-lpr mice. J Exp Med 179:1137–1143

Kaul R, Shenoy M, Goluszko E, Christadoss P (1994) Major histocompatibility complex class II gene disruption prevents experimental autoimmune myastthenia gravis. J Immunol 152:3152–3157

Kelly AP, Monaco JJ, Cho SG, Trowsdale J (1991) A new human HLA class II-related locus, DM. Nature 353:571–573

Kisielow P, Teh HS, Blüthmann H, von Boehmer H (1988) Positive selection of antigen-specific T cells in thymus by restricting molecules. Nature 335:730–733

Klein C, Lisowska-Grospierre B, LeDeist F, Fischer A, Griscelli C (1993) Major histocompatibility complex class II deficiency: clinical manifestations, immunologic features, and outcome. J Pediatr 123:921–928

Klein C, Cavazzana-Calvo M, Le Deist F, Jabado N, Benkerrou M, Blanche S, Lisowska-Grospierre B, Griscelli C, Fischer A (1995) Bone marrow transplantation in major histocompatibility complex class II deficiency: a single-center study of 19 patients. Blood 85:580–587

Koller BH, Marrack P, Kappler JW, Smithies O (1990) Normal development of mice deficient in beta 2M, MHC class I proteins, and CD8+ T cells. Science 248:1227–1230

Kontgen F, Suss G, Stewart C, Steinmetz M, Bluethmann H (1993) Targeted disruption of the MHC class II Aa gene in C57BL/6 mice. Int Immunol 5:957–964

Kovats S, Drover S, Marshall WH, Freed D, Whiteley PE, Nepom GT, Blum JS (1994) Coordinate defects in human histocompatibility leukocyte antigen class II expression and antigen presentation in bare lymphocyte syndrome. J Exp Med 179:2017–2022

Kruisbeek AM, Fultz MJ, Sharrow SO, Singer A, Mond JJ (1983) Early development of the T cell repertoire. In vivo treatment of neonatal mice with anti-Ia antibodies interferes with differentiation of I-restricted T cells but not K/D restricted T cells. J Exp Med 157:1932–1946

Kruisbeek AM, Mond JJ, Fowlkes BJ, Carmen JA, Bridges S, Longo DL (1985a) Absence of the Lyt-2-,L3T4+ lineage of T cells in mice treated neonatally with anti I-A correlates with absence of intrathymic I-A-bearing antigen-presenting cell function. J Exp Med 161:1029–1047

Kruisbeek AM; Bridges S; Carmen J; Longo DL; Mond JJ (1985b) In vivo treatment of neonatal mice with anti-I-A antibodies interferes with the development of the class I, class II, and Mls-reactive proliferating T cell subset. J Immunol 134:3597–3604

Lambert M, van Eggermond M, Mascart F, Dupont E, van den Elsen P (1992) TCR Vα- and Vβ-gene segment use in T-cell subcultures derived from a type-III bare lymphocyte syndrome patient deficient in MHC class-II expression. Dev Immunol 2:227–236

Lambert M, van Eggermond M, van den Elsen P (1993) Lymphokine gene transcription in CD4+CD8- T cells of a type III bare lymphocyte syndrome patient. Hum Immunol 37:143–148

Laufer TM, von Herrath MG, Grusby MJ, Oldstone MB, Glimcher LH (1993) Autoimmune diabetes can be induced in transgenic major histocompatibility complex class II-deficient mice. J Exp Med 178:589–596

Lee RS, Grusby MJ, Glimcher LH, Winn HJ, Auchincloss HJ (1994) Indirect recognition by helper cells can induce donor-specific cytotoxic T lymphocytes in vivo. J Exp Med 179:865–872

Lisowska-Grospierre B, Fondaneche MC, Rols MP, Griscelli C, Fischer A (1994) Two complementation groups account for most cases of inherited MHC class II deficiency. Hum Mol Genet 3:953–958

Longo DL, Davis ML (1983) Early appearance of donor-type antigen-presenting cells in the thymuses of 1200 R radiation-induced bone marrow chimeras correlates with self-recognition of donor I region gene products. J Immunol 130:2525–2527

Longo DL, Schwartz RH (1980) T-cell specificity for H-2 and Ir gene phenotype correlates with the phenotype of thymic antigen-presenting cells. Nature 287:44–46

Longo DL, Kruisbeek AM, Davis ML, Matis LA (1985) Bone marrow-derived thymic antigen-presenting cells determine self-recognition of Ia-restricted T lymphocytes. Proc Natl Acad Sci USA 82:5900–5904

Mach B, Steimle V, Martinez-Soria E, Reith W (1996) Regulation of MHC class II genes: lessons from a disease. Annu Rev Immunol 14:301–331

Mannhalter JW, Wolf HM,Gadner H, Potschka M, Eibl MM (1991) Cell-mediated immune functions in a patient with MHC class II deficiency. Immunol Invest 20:151–167

Mannhalter JW, Wolf HM, Hauber I, Miricka M, Gadner H, Eibl MM (1994) T cell differentiation and generation of the antigen-specific T cell repertoire in man: observations in MHC class II deficiency. Clin Exp Immunol 97:392–395

Markowitz JS, Rogers PR, Grusby MJ, Parker DC, Glimcher LH (1993) B lymphocyte development and activation independent of MHC class II expression. J Immunol 150:1223–1233

Mathis DJ, Benoist C, Williams V, Kanter M, McDevitt HO (1983) Several mechanisms can account for defective E alpha genen expression in different mouse haplotypes. Proc Natl Acad Sci USA 80:273–277

Peijnenburg A, Godthelp B, van Boxel-Dezaire A, van den Elsen PJ (1995) Definition of a novel complementation group in MHC class II deficiency. Immunogenetics 41:287–294

Prina E, Jouanne C, Lao S, Szabo A, Guillet J-G, Antoine J-C (1993) Antigen presentation capacity of murine macrophages infected with Leishmania amazonensis amastigotes. J Immunol 151:2050–2061

Raulet DH (1994) MHC class I-deficient mice. Adv Immunol 55:381–421

Reith W, Satola S, Herrero Sanchez C, Amaldi I, Lisowska-Grospierre B, Griscelli C, Hadam MR, Mach B (1988) Congenital immunodeficiency with a regulatory defect in MHC class II gene expression lacks a specific HLA-DR promoter binding protein, RF-X. Cell 53:897–906

Reith W, Steimle V, Mach B (1995) Molecular defects in the bare lymphocyte syndrome and regulation of MHC class II genes. Immunol Today 16:539–546

Rijkers GT, Roord JJ, Koning F, Kuis W, Zegers BJM (1987) Phenotypical and functional analysis of B lymphocytes of two siblings with combined immunodeficiency and defective expression of major histocompatibility complex (MHC) class II antigens on mononuclear cells. J Clin Immunol 7:98–106

Robey EA, Fowlkes BJ, Gordon JW, Kioussis D, von Boehmer H, Ramsdell F, Axel R (1991) Thymic selection in CD8 transgenic mice supports an instructive model for commitment to a CD4 or CD8 lineage. Cell 64:99–107

Russell DG, Xu S, Chakraborty P (1992) Intracellular trafficking and the parasitophorous vacuole of Leishmania mexicana-infected macrophages. J Cell Sci 103:1193–1210

Scholl PR, Geha RS (1994) MHC class II signaling in B cell activation. Immunol Today 15:418–422

Scott B, Blüthmann H, Teh HS, von Boehmer H (1989) The generation of mature T cells requires interaction of the αβ T-cell receptor with major histocompatibility antignes. Nature 338:591–593

Seidl C, Saraiya C, Osterweil Z, Fu YP, Lee JS (1992) Genetic complexity of regulatory mutants defective for HLA class II gene expression. J Immunol 148: 1576–1584

Selle A, Buus S, Colon S, Smith JA, Miles C, Grey HM (1987) Structural characteristics of an antigen required for its interaction with Ia and recognition by T cells. Nature 328:395–399

Sloan-Lancaster J, Allen PM (1996) Altered peptide ligand-induced partial T cell activation: molecular mechanisms and role in T cell biology. Annu Rev Immunol 14:1–27

Smith CIE, Bremard-Oury C, le Deist F, Griscelli C, Aucouturier P, Moller G, Hammarström L, Preud'homme JL, Weening RS (1988) The antibody spectrum in individuals with defective expression of HLA class II and the LFA-1 glycoprotein family genes. Clin Exp Immunol 74:449–453

Steimle V, Mach B (1995) Complementation cloning of mammalian transcriptional regulators: the example of MHC class II gene regulation. Curr Opin Genet Dev 5:646–651

Steimle V, Otten LA, Zufferey M, Mach B (1993) Complementation cloning of an MHC class II transactivator mutated in hereditary MHC class II deficiency (or bare lymphocyte syndrome). Cell 75:135–146

Steimle V, Durand B, Barras E, Zufferey M, Hadam MR, Mach B, Reith W (1995) A novel DNA binding regulatory factor is mutated in primary MHC class II deficiency (bare lymphocyte syndrome). Genes Dev 9:1021–1032

Stimac E, Urieli-Shoval S, Kempin S, Pious D (1991) Defective HLA DRA X box binding in the class II transactive transcription factor mutant 6.1.6 and in cell lines from class II immunodeficient patients. J Imunol 146:4398–4405

Tarlinton D (1993) Direct demonstration of MHC class II surface expression on murine pre-B cells. Int Immunol 5:1629–1635

The HS, Kisielow P, Scott B, Kishi H, Uematsu Y, Bluethmann H, von Boehmer H (1988) Thymic major histocompatibility complex antigens and the αβ T-cell receptor determine the CD4/CD8 phenotype of T cells. Nature 335:229–233

Theofilopoulos AN, Dixon FJ (1981) Etiopathogenesis of murine SLE. Immunol Rev 55:179–216

Ting JPY, Baldwin A (1993) Regulation of MHC gene expression. Curr Opin Immunol 5:8–16

Tourne S, Nakano N, Viville S, Benoist C, Mathis D (1995) The influence of invariant chain on the positive selection of single T cell receptor specificities. Eur J Immunol 25:1851–1856

Van Ewijk W, Ron Y, Monaco J, Kappler J, Marrack P, Le Meaur M, Gerlinger P, Durand B, Benoist C, Mathis D (1988) Compartmentalization of MHC class II gene expression in transgenic mice. Cell 53:357–370

Villard J, Lisowska-Grospierre B, van den Elsen, Fischer A, Reith W, Mach B (1997) Mutation of RFXAP, a regulator of MHC class II genes, in primary MHC class II deficiency. New Engl J Med 337:748–753

Viville S, Neefjes J, Lotteau V, Dierich A, Lemeur M, Ploegh H, Benoist C, Mathis D (1993) Mice lacking the MHC class II-associated invariant chain. Cell 72:635–648

Wolf HM, Hauber I, Gulle H, Thon V, Eggenbauer H, Fischer MB, Fiala S, Eibl MM (1995) Twin boys with major histocompatibility complex class II deficiency but inducible immune responses. New Engl J Med 332:86–90

Wong P, Rudensky AY (1996) Phenotype and function of CD4+ T cells in mice lacking invariant chain. J Immunol 156:2133–2142

Zijlstra M, Bix M, Simister NE, Loring JM, Raulet DH, Jaenisch R (1990) Beta 2-microglobulin deficient mice lack CD4-8+ cytolytic T cells. Nature 344:742–746

Zinkernagel RM, Callahan GN, Althage A, Cooper S, Klein PA, Klein J (1978a) On the thymus in the differentiation of "H-2 self-recognition" by T cells: evidence for dual recognition? J Exp Med 147:882–896

Zinkernagel RM, Callahan GN, Klein J, Dennert G (1978b) Cytotoxic T cells learn specificity for self H-2 during differentiation in the thymus. Nature 271:251–253

Subject Index

Printing: Saladruck, Berlin
Binding: Buchbinderei Lüderitz & Bauer, Berlin

Current Topics in Microbiology and Immunology

Volumes published since 1989 (and still available)

Vol. 191: **ter Meulen, Volker; Billeter, Martin A. (Eds.):** Measles Virus. 1995. 23 figs. IX, 196 pp. ISBN 3-540-57389-5

Vol. 192: **Dangl, Jeffrey L. (Ed.):** Bacterial Pathogenesis of Plants and Animals. 1994. 41 figs. IX, 343 pp. ISBN 3-540-57391-7

Vol. 193: **Chen, Irvin S. Y.; Koprowski, Hilary; Srinivasan, Alagarsamy; Vogt, Peter K. (Eds.):** Transacting Functions of Human Retroviruses. 1995. 49 figs. IX, 240 pp. ISBN 3-540-57901-X

Vol. 194: **Potter, Michael; Melchers, Fritz (Eds.):** Mechanisms in B-cell Neoplasia. 1995. 152 figs. XXV, 458 pp. ISBN 3-540-58447-1

Vol. 195: **Montecucco, Cesare (Ed.):** Clostridial Neurotoxins. 1995. 28 figs. XI., 278 pp. ISBN 3-540-58452-8

Vol. 196: **Koprowski, Hilary; Maeda, Hiroshi (Eds.):** The Role of Nitric Oxide in Physiology and Pathophysiology. 1995. 21 figs. IX, 90 pp. ISBN 3-540-58214-2

Vol. 197: **Meyer, Peter (Ed.):** Gene Silencing in Higher Plants and Related Phenomena in Other Eukaryotes. 1995. 17 figs. IX, 232 pp. ISBN 3-540-58236-3

Vol. 198: **Griffiths, Gillian M.; Tschopp, Jürg (Eds.):** Pathways for Cytolysis. 1995. 45 figs. IX, 224 pp. ISBN 3-540-58725-X

Vol. 199/I: **Doerfler, Walter; Böhm, Petra (Eds.):** The Molecular Repertoire of Adenoviruses I. 1995. 51 figs. XIII, 280 pp. ISBN 3-540-58828-0

Vol. 199/II: **Doerfler, Walter; Böhm, Petra (Eds.):** The Molecular Repertoire of Adenoviruses II. 1995. 36 figs. XIII, 278 pp. ISBN 3-540-58829-9

Vol. 199/III: **Doerfler, Walter; Böhm, Petra (Eds.):** The Molecular Repertoire of Adenoviruses III. 1995. 51 figs. XIII, 310 pp. ISBN 3-540-58987-2

Vol. 200: **Kroemer, Guido; Martinez-A., Carlos (Eds.):** Apoptosis in Immunology. 1995. 14 figs. XI, 242 pp. ISBN 3-540-58756-X

Vol. 201: **Kosco-Vilbois, Marie H. (Ed.):** An Antigen Depository of the Immune System: Follicular Dendritic Cells. 1995. 39 figs. IX, 209 pp. ISBN 3-540-59013-7

Vol. 202: **Oldstone, Michael B. A.; Vitković, Ljubiša (Eds.):** HIV and Dementia. 1995. 40 figs. XIII, 279 pp. ISBN 3-540-59117-6

Vol. 203: **Sarnow, Peter (Ed.):** Cap-Independent Translation. 1995. 31 figs. XI, 183 pp. ISBN 3-540-59121-4

Vol. 204: **Saedler, Heinz; Gierl, Alfons (Eds.):** Transposable Elements. 1995. 42 figs. IX, 234 pp. ISBN 3-540-59342-X

Vol. 205: **Littman, Dan R. (Ed.):** The CD4 Molecule. 1995. 29 figs. XIII, 182 pp. ISBN 3-540-59344-6

Vol. 206: **Chisari, Francis V.; Oldstone, Michael B. A. (Eds.):** Transgenic Models of Human Viral and Immunological Disease. 1995. 53 figs. XI, 345 pp. ISBN 3-540-59341-1

Vol. 207: **Prusiner, Stanley B. (Ed.):** Prions Prions Prions. 1995. 42 figs. VII, 163 pp. ISBN 3-540-59343-8

Vol. 208: **Farnham, Peggy J. (Ed.):** Transcriptional Control of Cell Growth. 1995. 17 figs. IX, 141 pp. ISBN 3-540-60113-9

Vol. 209: **Miller, Virginia L. (Ed.):** Bacterial Invasiveness. 1996. 16 figs. IX, 115 pp. ISBN 3-540-60065-5

Vol. 210: **Potter, Michael; Rose, Noel R. (Eds.):** Immunology of Silicones. 1996. 136 figs. XX, 430 pp. ISBN 3-540-60272-0

Vol. 211: **Wolff, Linda; Perkins, Archibald S. (Eds.):** Molecular Aspects of Myeloid Stem Cell Development. 1996. 98 figs. XIV, 298 pp. ISBN 3-540-60414-6

Vol. 212: **Vainio, Olli; Imhof, Beat A. (Eds.):** Immunology and Developmental Biology of the Chicken. 1996. 43 figs. IX, 281 pp. ISBN 3-540-60585-1

Vol. 213/I: **Günthert, Ursula; Birchmeier, Walter (Eds.):** Attempts to Understand Metastasis Formation I. 1996. 35 figs. XV, 293 pp. ISBN 3-540-60680-7

Vol. 213/II: **Günthert, Ursula; Birchmeier, Walter (Eds.):** Attempts to Understand Metastasis Formation II. 1996. 33 figs. XV, 288 pp. ISBN 3-540-60681-5

Vol. 213/III: **Günthert, Ursula; Schlag, Peter M.; Birchmeier, Walter (Eds.):** Attempts to Understand Metastasis Formation III. 1996. 14 figs. XV, 262 pp. ISBN 3-540-60682-3

Vol. 214: **Kräusslich, Hans-Georg (Ed.):** Morphogenesis and Maturation of Retroviruses. 1996. 34 figs. XI, 344 pp. ISBN 3-540-60928-8

Vol. 215: **Shinnick, Thomas M. (Ed.):** Tuberculosis. 1996. 46 figs. XI, 307 pp. ISBN 3-540-60985-7

Vol. 216: **Rietschel, Ernst Th.; Wagner, Hermann (Eds.):** Pathology of Septic Shock. 1996. 34 figs. X, 321 pp. ISBN 3-540-61026-X

Vol. 217: **Jessberger, Rolf; Lieber, Michael R. (Eds.):** Molecular Analysis of DNA Rearrangements in the Immune System. 1996. 43 figs. IX, 224 pp. ISBN 3-540-61037-5

Vol. 218: **Berns, Kenneth I.; Giraud, Catherine (Eds.):** Adeno-Associated Virus (AAV) Vectors in Gene Therapy. 1996. 38 figs. IX,173 pp. ISBN 3-540-61076-6

Vol. 219: **Gross, Uwe (Ed.):** Toxoplasma gondii. 1996. 31 figs. XI, 274 pp. ISBN 3-540-61300-5

Vol. 220: **Rauscher, Frank J. III; Vogt, Peter K. (Eds.):** Chromosomal Translocations and Oncogenic Transcription Factors. 1997. 28 figs. XI, 166 pp. ISBN 3-540-61402-8

Vol. 221: **Kastan, Michael B. (Ed.):** Genetic Instability and Tumorigenesis. 1997. 12 figs.VII, 180 pp. ISBN 3-540-61518-0

Vol. 222: **Olding, Lars B. (Ed.):** Reproductive Immunology. 1997. 17 figs. XII, 219 pp. ISBN 3-540-61888-0

Vol. 223: **Tracy, S.; Chapman, N. M.; Mahy, B. W. J. (Eds.):** The Coxsackie B Viruses. 1997. 37 figs. VIII, 336 pp. ISBN 3-540-62390-6

Vol. 224: **Potter, Michael; Melchers, Fritz (Eds.):** C-Myc in B-Cell Neoplasia. 1997. 94 figs. XII, 291 pp. ISBN 3-540-62892-4

Vol. 225: **Vogt, Peter K.; Mahan, Michael J. (Eds.):** Bacterial Infection: Close Encounters at the Host Pathogen Interface. 1998. 15 figs. IX, 169 pp. ISBN 3-540-63260-3

Vol. 226: **Koprowski, Hilary; Weiner, David B. (Eds.):** DNA Vaccination/Genetic Vaccination. 1998. 31 figs. XVIII, 198 pp. ISBN 3-540-63392-8

Vol. 227: **Vogt, Peter K.; Reed, Steven I. (Eds.):** Cyclin Dependent Kinase (CDK) Inhibitors. 1998. 15 figs. XII, 169 pp. ISBN 3-540-63429-0

Vol. 228: **Pawson, Anthony I. (Ed.):** Protein Modules in Signal Transduction. 1998. 42 figs. IX, 368 pp. ISBN 3-540-63396-0

Vol. 229: **Kelsoe, Garnett; Flajnik, Martin (Eds.):** Somatic Diversification of Immune Responses. 1998. 38 figs. IX, 221 pp. ISBN 3-540-63608-0

Vol. 230: **Kärre, Klas; Colonna, Marco (Eds.):** Specificity, Function, and Development of NK Cells. 1998. 22 figs. IX, 248 pp. ISBN 3-540-63941-1

Vol. 231: **Holzmann, Bernhard; Wagner, Hermann (Eds.):** Leukocyte Integrins in the Immune System and Malignant Disease. 1998. 40 figs. XIII, 189 pp. ISBN 3-540-63609-9